ANCIENT
SIEGE
WARFARE

Paul Bentley Kern

ANCIENT
SIEGE
WARFARE

Souvenir

Press

First published in the USA by
Indiana University Press,
Bloomington, Indiana

First British edition published 1999 by
Souvenir Press Ltd.,
43 Great Russell Street, London WC1B 3PA

ISBN 0 285 63524 7

Printed in the United States of America

FOR MY WIFE, JULIE,
AND OUR TWO SONS,
CHRISTOPHER AND COLIN

CONTENTS

FIGURES

MAPS

ACKNOWLEDGMENTS

I would like to thank Cynthia Szymanski, Anne Koehler, and Jacqueline Cheairs of the Indiana University Northwest Library System Services. Without their help in obtaining many books through interlibrary loan I could not have written this book. I would also like to thank my colleagues John Gruenenfelder, James Newman, William Neil, James Lane, Ronald Cohen, Rhiman Rotz, Frederick Chary, Roberta Wollons, Xiaoqing Diana Chen Lin, Mark Sheldon, Steve McShane, and Jean Poulard, who gave me encouragement and advice. Gayle Gullett read an early draft on the Biblical image of women in siege warfare and made many helpful suggestions. Rosalie Zak and Ginger Wirtes provided excellent secretarial support. Tim Sutherland helped me with bibliographical searches. About halfway through the writing of this book, Indiana University Northwest changed from a mainframe computer to personal computers, requiring the first several chapters to be retyped into the PC format. Lois Libauskas minimized the damage by doing this with remarkable accuracy. Irma Arroyo Sher and Janet Taylor helped with the reproduction of the maps and drawings. Indiana University Northwest granted me two sabbaticals, just five years apart, without which this book would have taken even longer to write. I owe a great debt of gratitude to Indiana University Press. Joan Catapano offered encouragement early and late in the project and helped me in many ways. Jane Lyle smoothed the process of editing. Kenneth Goodall improved the book immeasurably with his skillful copy editing. Grace Profatilov helped sort out maps, illustrations, and permissions.

My wife, Julie, offered constant support. She and my sons, Christopher and Colin, were a source of love and encouragement.

ANCIENT
SIEGE
WARFARE

INTRODUCTION

> So the people shouted and the trumpets were blown. As soon as the people heard the sound of the trumpet, the people raised a great shout, and the wall fell down flat, so that the people went up into the city, every man straight before him, and they took the city. Then they utterly destroyed all in the city, both men and women, young and old, oxen, sheep, and asses, with the edge of the sword.
>
> —Book of Joshua[1]

For all its familiarity, the story of Jericho still shocks. All war is horrifying, but the idea that there are limits somehow reassures us. The massacre of women and children reminds us that the limits we try to place on war are fragile. Uncontrolled violence in which the distinction between warriors and the rest of society breaks down strikes at our deepest fears.

Even the restrained Thucydides was moved to a rare outburst of anger and pity when he described the fate of the small town of Mycalessus, whose entire population—again including even the animals—was killed by Thracian soldiers. Thucydides called the Thracians the "most bloodthirsty" barbarians and the massacre at Mycalessus "a calamity inferior to none that has ever fallen upon a whole city, and beyond any other unexpected and terrible."[2]

Siege warfare has often provided the moment for such uncontrolled violence, and the terror of a city violated and helpless before its enemies has long fascinated us. King Priam of Troy voiced the nightmare of a fallen city:

Oh, take
pity on me, the unfortunate still alive, still
  sentient
but ill-starred, whom the father, Kronos' son, on
  the threshold of old age
will blast with hard fate, after I have looked
  upon evils
and seen my sons destroyed and my daughters
  dragged away captive
and the chambers of marriage wrecked and the
  innocent children taken
and dashed to the ground in the hatefulness of
  war, and the wives
of my sons dragged off by the accursed hands of
  the Achaians.
And myself last of all, my dogs in front of my
  doorway
will rip me raw, after some man with stroke of the
  sharp bronze
spear, or with spearcast, has torn the life out of
  my body;
those dogs I raised in my halls to be at my table,
  to guard my
gates, who will lap my blood in the savagery of
  their anger
and then lie down in my courts.[3]

Priam goes on to imagine his dogs mutilating his genitals. This terrible scene captures in full the terror of siege warfare. His own dogs,
those most domesticated of animals, licking Priam's blood contrasts
the domesticity of his former life with its complete destruction. That he
is naked and the dogs are tearing his genitals reveal Priam's frightful
vulnerability. The contrast between the structured and creative sexual
world of the "chambers of marriage" and the raping of women and killing of babies shows a social and moral collapse that is complete.

Robert O'Connell likens the slaughter that can follow sieges, in
which "the victor casts aside any feeling of common humanity and in-
dulges in killing for its own sake," to the primitive urges of the hunt,
urges that may be rooted in the predatory nature of man.[4] Cities are the
product of human civilization, and when war bursts into them, plung-
ing their population into such chaos that that most human institution,
the funeral, breaks down and people die like animals, the distinction
between savagery and human culture disappears and our darkest fears
about human nature are realized. An ancient Sumerian lamentation
over the destruction of Ur expressed the horror of a breached wall by
brilliantly juxtaposing scenes of urban harmony with images of unburied corpses:

Its walls were breached; the people groan.
In its lofty gates, where they were wont to
    promenade,
dead bodies were lying about;
In its boulevards, where the feasts were
    celebrated,
scattered they lay.
In all its streets, where they were wont to
    promenade,
dead bodies were lying about;
In its places where the festivities of the land
    took place,
the people lay in heaps.[5]

As Carl von Clausewitz pointed out, the violence of war has no logical limit, and at Jericho, Troy, and Mycalessus violence reached its ultimate level, the complete destruction of the opposing society. Instinc-
tively people attempt to protect themselves from such horrifying
vulnerability.[6] For the warrior, armor serves this purpose. For a city, fortifications provide protection against sudden assault. Fortifications attempt to keep violence at bay, to interpose a barrier between the construction of a city and the destruction of war. Even if a fortified city
cannot hold out indefinitely, it can hope that the walls will discourage
the enemy sufficiently to bring about a negotiated settlement. In this
way fortifications try to control the violence of war.

But the failure of fortifications could lead to war without limits.
Walls placed a barrier between warriors that prevented them from displaying their prowess in hand-to-hand combat. Breaching a fortified
wall was largely a technical problem in which conventional warriors
often had to take a back seat to the engineers and the sappers. Once the
technical problem was solved, the assaulting troops found themselves
not on a conventional field of battle opposed by an army but in a maze
of streets and buildings opposed by an entire population. Often they
were under orders to sack the city, one of the few circumstances in
which military commanders countenanced indiscriminate violence.[7]

The stakes in ancient siege warfare were higher than in field battles.
Writing in the mid–fourth century B.C., Aineias the Tactician began his
manual on siege warfare by stating the difference between field battles
and siege battles:

When men leave their own territory to meet combat and danger beyond
its borders, the survivors of any disaster which strikes them, on land or at
sea, still have their native soil and state and fatherland between them and
utter extinction. But when it is in defence of the fundamentals—shrines
and fatherland and parents and children and so on—that the risks are to
be run, the struggle is not the same, or even similar. A successful repulse
of the enemy means safety, intimidated opponents, and the unlikelihood

of attack in the future, whereas a poor showing in the face of the danger leaves no hope of salvation.[8]

The finality of defeat in a siege was such that the convention of ransom often failed to obtain. Warriors captured in field battles could look to relatives, friends, or fellow citizens to pay ransom for their freedom. No such hope offered itself to a captive whose city had fallen because everyone else had also been enslaved.[9]

Siege warfare, then, was technical, unconventional, and total. It diminished the role of the traditional warrior with his conventional methods of fighting, methods designed to make possible the display of values such as honor and prowess. Conventional standards have always imposed the most effective controls on the violence of the battlefield. Siege warfare, with its technical nature and its unconventional circumstances, threatened conventional standards.

The most unconventional aspect of siege warfare was the involvement of women and children. In a criticism of the tear-jerking style of the historian Phylarchus, Polybius wrote:

> In his eagerness to arouse the pity and attention of his readers he treats us to a picture of clinging women with their hair disheveled and their breasts bare, or again of crowds of both sexes together with their children and aged parents weeping and lamenting as they are led away into slavery.[10]

Polybius went on to condemn the "ignoble and womanish character" of Phylarchus's style of history and accused Phylarchus of creating exaggerated and imaginary scenes to arouse the emotions of his readers. Polybius favored a pragmatic history that stuck to useful facts and explained the causes of events to the imaginary verisimilitude of the tragic poets, who sought to engage the emotions of their audience rather than instruct them in useful knowledge of the past.[11]

Although I have tried to avoid exaggerations, I confess that I have endeavored to arouse the pity of my readers for the women and children caught in siege warfare and that I have found literary sources, most especially the tragic poets, useful for that purpose. Women and children were an essential part of siege warfare. Their presence threatened the notion of war as a contest between warriors, undermined the conventional standards of honor and prowess that governed ancient warfare, and paradoxically made war less restrained by creating a morally chaotic cityscape in which not only the walls collapsed but deeply rooted social and moral distinctions as well. We cannot understand siege warfare without understanding the plight of women and children and the effect of their presence on war.

Controls on the violence of war come from two sources: the transcendent values of society at large and the practical consideration that the means should not exceed the ends.[12] Siege warfare needs to be placed in the context of both the larger values of the society in which it takes place and the problem of the means necessary to bring a siege to a successful conclusion.

This book attempts to study the conduct of siege warfare both on a moral and a technical level. What methods have been used and what values have guided the conduct of sieges? And what is the relation between the technical imperatives of siege warfare and its standards of conduct? Approached in this way, an understanding of sieges can perhaps help us understand something about the nature of war itself. That is because ancient siege warfare adumbrated the nature of modern war. Sieges were a form of total war, pitting entire societies against one another. All the characteristics of modern war—the blurring of the line between battlefield and society, the engulfing of women and children in the violence of war, the destruction of a society's infrastructure, the uprooting of entire populations—were anticipated in ancient siege warfare.

Even though the age of fortification and formal siege are long over, we can still recognize the face of siege warfare in the bombed-out cities, the stream of refugees, and the indiscriminate death of modern war. The fate of Jericho chills us because it reminds us of our own terrible vulnerability in the age of modern war.

PART ONE

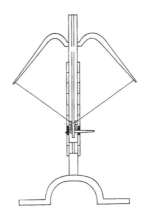

THE
ORIGINS

Chapter I

FORTIFICATIONS AND
SIEGE MACHINERY

Siege warfare is older than civilization itself. Long before the people of Jericho constructed the walls that fell down flat before Joshua, Jericho was a fortified city. Archeological excavations in the 1950s uncovered part of a massive wall ten feet thick and thirteen feet high, dating back to the early Neolithic times, some seven thousand years before Christ. The wall probably enclosed a city whose area was about ten acres and whose population was about 2,500. If so, the length of the wall was about 765 yards; and if we estimate the number of fighting men to be five to six hundred, it would have been possible to station one defender to nearly every yard of wall.[1]

This wall was part of an entire defensive system that included a moat almost ten feet deep and thirty feet wide dug into the rock at the bottom of the wall and a large circular tower twenty-eight feet high and thirty-three feet in diameter. Given the primitive flint tools available to the people of neolithic Jericho, the effort required to build such a system is truly astonishing.[2]

Two technological revolutions combined to spur the neolithic inhabitants of Jericho to build such fortifications. The first was the introduction of projectile weapons, such as the bow and the sling. Mesolithic cave paintings at Morela La Vella in Spain show that bows were in use as early as 20,000 B.C.[3] Every offensive weapon produces a defensive reaction, and the obvious defense against a projectile weapon was to impose a barrier between the projectile and its intended victim. In the beginning, warriors undoubtedly sought natural cover, hiding behind whatever protection they could find. Frequently it would have been necessary to improve natural defensive positions by constructing primitive ramparts from earth and stone. But nomadic mesolithic peoples did not remain in one place long enough to build fortifications on the scale of neolithic Jericho. The second technological revolution, the development of agriculture, provided the steady food supply necessary to support a permanent settlement and made possible the construction of permanent fortifications.

Arther Ferrill argues that the urge to fortify was so great after the invention of projectile weapons that the people of Jericho learned to farm for the express purpose of supporting permanent fortifications.[4] Jacob Bronowski reverses the sequence, arguing that agriculture came to Jericho first, bringing a prosperity that subjected the people to predatory raids from nomads, which in turn prompted the inhabitants of Jericho to build their famous walls.[5] Given the long gap between the invention of projectile weapons and the development of agriculture and the slow and serendipitous process of developing agriculture,[6] Bronowski's sequence seems more likely, but there can be no doubt about the close relationship between permanent fortifications and agriculture.

Jericho was not an isolated example of fortifications in neolithic times. Some towns, such as Çatal Hüyük in Turkey, built their houses contiguously so that the outside walls presented a continuous barrier to intruders. The houses had no doors; the exits were through the roofs and people moved about the town on the rooftops. If an enemy broke through the wall, he found himself in a single room with the defenders waiting for him on the roof. The inhabitants of the Mesopotamian town of Tell es-Sawwan built a buttressed fortification wall around 5000 B.C. By the middle of the fifth millennium, the people of Yalangach in the Transcaspian Lowlands had built round towers facing outward from their walls, enabling them to direct flanking fire against enemy besiegers.[7]

From the beginning the necessity of fortifications powerfully influenced the history of cities.[8] By the third millennium, when the archeological evidence becomes fuller, the art of building fortifications was well developed. The rise of wealthy cities and the development of centralized political authority created both the motive and the means for a proliferation of fortified cities.

Mesopotamian cities girded themselves with walls around 3000 B.C., about the same time that they placed themselves under monarchical authority.[9] The Mesopotamians prized their walls so highly that they gave them names and placed them under the protection of the gods. The city of Ur was famous for its walls, which were twenty-five to thirty-four meters thick at the base. The walls were so thick because southern Mesopotamia lacked either building stones or sufficient fuel to bake large bricks from which to construct freestanding walls. The thickness of earthen walls at the base had to be at least one-third the height of the wall and sometimes was up to two-thirds the height. Such walls were actually ramparts made of adobe brick with a vertical outward face of thin baked bricks.[10]

Mesopotamian engineers mastered the technique of building balconies and towers from which defenders could shoot straight down at the base of the wall and along the curtain. By placing the towers no farther apart than the range of such projectile weapons as bow, sling, and spear—about thirty meters—the architects not only ensured that there would be no dead space between the towers but also made the full length of the wall subject to flanking fire from two directions.[11]

Mesopotamian architects were also adept at designing walls on a zigzag pattern or breaking a curve into a series of right angles in order to increase flanking fire and make the walls more defensible.[12] In Palestine, walls were built in unbonded sections so that if one part of the wall collapsed from earthquakes or enemy action, it would not pull down the rest of the wall.[13]

Egyptian evidence, such as the late predynastic (early third millennium) palette of fortresses and the palette of King Narmer (c. 3100 B.C.), who may have been the first ruler to unite Egypt under a single king, shows that even though fortified cities were not yet common in Egypt, sieges had become a critical part of warfare by the end of the fourth millennium. The besieged cities depicted on these palettes have walls studded with bastions revealing the high level of sophistication that had been achieved.[14] King Narmer may have had a far-reaching strategy of expansion based on siege warfare.[15]

A particularly crucial part of any fortification was the gate, obviously the most vulnerable point in a fortified wall. The earliest pictorial evidence of siege warfare, a twenty-seventh-century wall painting from Dehashe in Egypt, shows soldiers trying to breach the gate by prying it open with poles.

Architects spared no effort trying to design secure gates. Entrances to the earliest fortified cities were usually winding and restricted, so if the attackers broke through a gate they would find themselves caught in a narrow, winding passageway that forced them to turn in a way that exposed their unshielded right sides. A good example of this sort of gate dates from the twenty-ninth century at Abydos in Egypt.[16]

After the invention of the chariot in the second millennium, gates had to be wide enough and the entrance accessible enough to allow chariots to move in and out of the city. This meant that there had to be double doors, which were weak at the center. A heavy wooden beam inserted horizontally behind double doors greatly strengthened them. Frequently, metal plating covered wooden doors to make them resistant to fire. Entranceways bristled with pilasters, bastions, towers, and balconies to provide maximum protection at the weakest point.

We find a good example of this sort of gate at Troy II (early third millennium), where one had to approach the gate between walls that extended from the main city walls. A roof supported by pilasters protected the gate itself. If an attacker managed to breach the gate, he found himself trapped in inner chambers from which it was difficult to exit against armed opposition.[17]

Archeology reveals a great deal about the art of fortification in the third millennium, but it is more difficult to learn how attackers overcame such formidable barriers. The Dehashe wall painting shows that one of the earliest means of overcoming a fortified position was escalade: scaling teams are assaulting the wall while archers provide covering fire. The success of escalade depended on the availability of scaling ladders built to the proper height. Obviously they had to be tall enough to reach the top of the wall. But if they were too long and projected above the wall, it was much easier for the defenders to push the ladders away from the wall.

Attackers liked to place the ladders at a rather flat angle to make it more difficult to push the ladder backward from the wall.[18] Placing the ladders at a flat angle, however, increased the possibility that it would break under the weight of the soldiers climbing it. The proper ratio of the length of the ladder to the height of the wall was 6:5, and the proper distance from the base of the wall to the base of the ladder was half the length of the ladder.[19]

Another remarkable Egyptian wall painting from the twenty-third century reveals that the Egyptians wrestled with these problems. It shows a scaling ladder equipped with wheels to make it easier to position properly. Warriors are standing on the ladder pounding on the wall with axes, a method that was effective against mud or brick.[20]

Defense against escalade required high walls. Scaling was ineffective against walls higher than ten meters; ladders longer than ten meters were too unwieldy.[21] That explains the huge scale of ancient walls. Their remarkable thickness was necessary to provide a broad base to walls high enough to discourage escalade. Because of the lack of perspective in ancient paintings, we cannot estimate the height of walls depicted in them. Archeological ruins preserve only the foundations. However, old Babylonian mathematical problems from the first half of the second millennium indicate that walls were usually around twenty meters high.[22] The Dehashe painting shows semicircular bastions, which not only strengthened the wall but also provided a line of flanking fire along the curtain for the archers defending the wall.

Mesopotamian sources are less revealing, but it is certain that siege warfare was important. Inscriptions from the third millennium refer to booty from captured cities. Victorious armies pursued the defeated enemy back into the city in the hope of breaking in on the heels of the fleeing troops. The high level of technical expertise demonstrated later in the construction of earthen ramps up to the tops of walls suggests that this method began in the third millennium.[23]

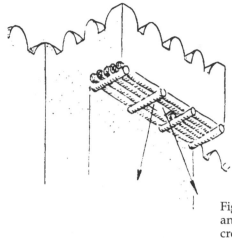

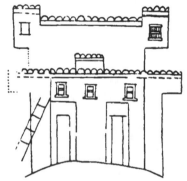

Figure 1. Egyptian representation of an early fortified city showing crenelation and a machicolated balcony. *From Yigael Yadin,* The Art of War in Biblical Lands, *p. 20. Copyright International Publishing Company Ltd., 1963.*

If the amount of evidence available is the measure, siege dominated warfare in the first half of the second millennium and the art of fortification reached an even higher level. An Egyptian wall painting at Beni Hasan shows a complete defensive system featuring high walls with battlements and machicolated balconies from which archers could direct flanking fire as well as shoot directly down on attackers below. The painting also clearly shows the placement of a sloped bank or glacis at the foot of the wall to make the approach to the wall even more difficult.[24] We also have Egyptian representations of defenses (figures 1 and 2).

The upper Nile in Nubia bristled with fortresses in the early second millennium.[25] Archeologists have discovered some of the most elaborate such fortifications at the Nubian town of Buhen in the Sudan (XIIth Dynasty, twentieth and nineteenth centuries B.C.). The main wall was five meters thick and about ten meters high. It had square bastions every five meters, and large towers protruded from each corner of the square-shaped fort. A lower wall protected the main wall. The lower wall had

Figure 2. Defenders of a besieged city, Beni Hasan, c. 1900 B.C. *From Yigael Yadin*, The Art of War in Biblical Lands, *p. 21. Copyright International Publishing Company Ltd., 1963.*

bastions every 10 meters. Both the wall and the bastions had embrasures with loopholes skillfully situated to provide a wide angle of fire. A dry moat 8.5 meters wide and 6 meters deep ran around the foot of the lower wall. Yet another low wall circled the outside edge of the moat, and its face was strengthened by an earthen glacis. The openings in the outer walls were in line with the gate in the main wall. Two large towers stretching from the main gate to beyond the outermost wall protected the entranceway. There may have been a drawbridge across the moat.[26]

Another impressive Egyptian fortification from the first half of the second millennium was the island fortress of Askut in the upper Nile in Nubia. Egyptian engineers skillfully adapted the walls to the rugged terrain of the island, taking full advantage of its natural defenses. Spur walls allowed the defenders a commanding position over the entire island. An open space along the inside base of the wall provided freedom of movement to the defenders. The walls themselves had towers, crenelations, and machicolated balconies to provide the full range of defensive fire. There were granaries and a covered way to the riverbank, providing the food and water necessary for a long siege. Askut was the anchor of a strategic system of fortresses that enabled the pharaohs to rule and defend Nubia.[27]

The scale of ancient fortifications is astonishing. At the Palestinian city of Hazor, the perimeter of the walls, built in the eighteenth century B.C., was three kilometers long. The huge moat was fifteen meters deep and forty meters wide at the top. The city was large enough to hold almost thirty thousand people.[28]

The great Hittite capital of Hattussas (modern Boghazköy) in Ana-

Figure 3. The King's Gate of Hattussas (modern Boghazköy). *From Yigael Yadin*, The Art of War in Biblical Lands, *p. 93. Copyright International Publishing Company Ltd., 1963.*

tolia had no less than six kilometers of wall. Hattussas represented the culmination of the art of fortification in the Late Bronze Age. It had a complicated system of walls, skillfully adapted to the contours of its hilly site. Unlike Mesopotamian architects, who preferred to avoid circular designs in favor of sharp angles, Hittite architects preferred making walls more circular than rectangular. Circular walls were more difficult to cover with flanking fire, but corners were more vulnerable to undermining.[29] The main walls were of the casemate type, i.e., double walls with a narrow space in between, perhaps filled with rubble. A lower outer wall, about twenty feet in front, protected the main wall. Underground posterns provided a means of staging surprise sorties against besiegers. All the posterns ran under the southern wall, the side opposite the most likely direction of attack. This suggests that they may have served as escape routes as well. Although posterns created a possible entrance into the city, enemy troops would hesitate to try to force an opening so narrow they would have to emerge one by one. Interior walls divided the city into sections, so an enemy breaking through gained access to only part of the city. Hattussas also provides some of the finest examples of fortified gates (figure 3). An inner citadel stood on a hill, offering a last line of defense.[30]

Clearly there was a powerful motive for expending so much labor, time, and money to build such fortifications. It is tempting to infer the threat of battering rams, and there is evidence to support such an inference. The most striking feature of the wall painting at Beni Hasan shows three men standing in a mobile structure that protects them

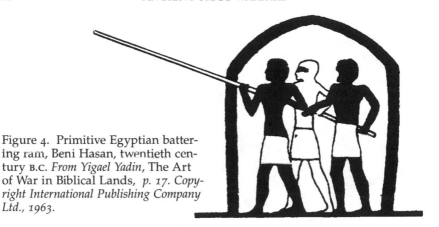

Figure 4. Primitive Egyptian battering ram, Beni Hasan, twentieth century B.C. *From Yigael Yadin,* The Art
of War in Biblical Lands, *p. 17. Copyright International Publishing Company
Ltd., 1963.*

while they batter at the wall with a long beam (figure 4). The beam
probably had a sharp metal point. Primitive as this battering ram was,
it clearly impressed the painter, who shows it as the main weapon in
the siege. The impressive fortifications shown in the Beni Hasan painting were surely a response to this new weapon.[31] The palette of the fortresses (c. 3000 B.C.) may show battering rams in use, but they may be
little more than picks with which the symbolic creatures are hacking at
the wall.[32]

We draw a blank on pictorial evidence for the battering ram elsewhere in the Middle East before the first millennium B.C. But there are
references to battering rams in the written records. The Old Babylonian
version of Tablet III of *Gilgamesh* compares the god Hamara to a "siege
engine."[33] Richard Humble interprets this to mean that battering rams
existed in Mesopotamia as early as 2500 B.C. Humble also infers that
Sargon the Great (c. 2325) possessed battering rams from his ability to
conquer the walled cities of Sumer.[34] Yigael Yadin, the Israeli historian
and perhaps the greatest expert on warfare in Biblical times, dates the
development of "a primitive type of battering ram" to the second millennium from the existence of walls with glacis and bastions such as
the ones at Troy in Anatolia.[35]

But our first clear evidence of battering rams comes from the eight-
eenth-century documents of Mari, a city on the upper Euphrates. Indeed, these records provide some of our best evidence for siege warfare
in the early second millennium, especially because they tell of the ex-
ploits of the Assyrian king Ishme-Dagan. Ishme-Dagan was the son of
Shamshi-Adad (1813–1781 B.C.), who had usurped the throne of Assur
and united Assyria for the first time. Shamshi-Adad had spent some
time in Babylonia before going to Assur and may have learned siege
methods there. In any case, the Mari documents show not only that battering rams had become powerful weapons in Assyria by the eighteenth

century but also that Ishme-Dagan had mastered a variety of siege tactics, including the use of siege towers, sapping, and the construction of siege ramps against city walls.

Two campaign reports show Ishme-Dagan's methods. The first tells of the use of battering rams and siege towers:

> Thus saith Ishme-Dagan, thy brother: "After I conquered [the names of three cities], I turned and laid siege to Hurara. I set against it the siege towers and battering-rams and in seven days I vanquished it. Be pleased!"[36]

In the second document, Ishme-Dagan reports the successful use of sapping:

> As soon as I had approached the town of Qirhadat I set up siege towers. By sapping I caused its walls to collapse. On the eighth day I seized the city of Qirhadat. Rejoice.[37]

Because we have no pictorial evidence of Mesopotamian siege warfare during this time, the way Ishme-Dagan used battering rams, siege towers, and sapping is not entirely clear. We do not know if his sappers tunneled under the walls, bored through them, or both. Tunneling protected the sappers from enemy fire but was a ticklish operation. If the sappers dug too deep, the tunnel would not collapse the wall. If they dug too shallow, they risked collapsing the walls on themselves. The method of tunneling at just the right depth so that the tunnel would collapse when the support beams were burned required considerable skill and experience. More likely, Ishme-Dagan's sappers dug through the walls, especially if the walls were made of adobe brick.

The purpose of the siege towers is also not certain. They may have provided a means of scaling the walls by dropping a boarding bridge from the tower to the wall.[38] But they also could have served in support of sappers. Battering and sapping walls exposed the men working at the base of the wall to deadly enemy fire from the wall. Without heavy covering fire, the task would be impossible. Egyptian paintings of sieges show support fire coming from siege towers, whose height gave archers and slingers a much better angle of fire at the defenders atop a wall than the angle from below on the ground. The effects of such tactics could be devastating: Ishme-Dagan reported taking a city in a single day.[39]

The same lack of pictorial evidence that causes uncertainty about the siege tactics of Ishme-Dagan prevents us from knowing what sort of battering rams he possessed. One scholar refers to both siege towers and the battering rams as "military dinosaurs," apparently in the conviction that they were huge.[40] Certainly the siege towers were higher than the walls, which means that they could have been more than twenty meters high. But we cannot know the size of the rams. Ishme-

Dagan's army transported its siege train by boat or wagon,[41] but we cannot infer great size from this fact. The "battering rams" may not have been much more than instruments to help the sappers pry or force bricks loose from the wall.

The Mari documents also provide us with our first close look at the techniques of building earthen ramps against a city wall. A report describing yet another siege by the indefatigable Ishme-Dagan makes clear the effectiveness of these ramps:

> The town of Nilimmar that Ishme-Dagan besieged, Ishme-Dagan has now taken. As long as the siege-ramps did not reach to the heights of the top of the city wall, he could not seize the town. As soon as the siege-ramps reached the top of the city wall, he gained mastery over this town.[42]

The engineers who supervised the construction of these ramps applied sophisticated mathematical calculations to the task, as an Old Babylonian mathematical problem reveals. The problem shows that Mesopotamian engineers were able to calculate the volume of earth necessary to construct a ramp if they knew the height of the wall and that they could calculate at any given time during the construction how much more earth remained to be hauled. This in turn allowed precise calculations of how many men and how much time would be necessary to complete the ramp. For example, one problem that we have allows us to calculate that if one man could haul two cubic meters of earth per day, it would take five days for 9,500 men, working twelve hours a day, to build a ramp to the top of a wall twenty-two meters high.[43] The ramp neutralized the wall, but assault teams had to fight their way uphill, a severe disadvantage in shock tactics. However, the limited space at the top of the wall prevented the defenders from drawing up in depth.[44] Finally, the time required to construct the ramp gave the defenders time for countermeasures, such as raising the height of the wall or building an interior wall to seal off the section of outer wall against which the ramp rose.

Such a large construction project required efficient organization of labor. The heart of the old Assyrian army was a corps of professional soldiers who did the fighting in war and carried out police duties in peacetime. For military campaigns, a muster swelled the army by a factor of perhaps four. These militia troops served less as a fighting force than as a labor force, providing logistical support to the army and hauling dirt for siege ramps.[45]

Despite the introduction of siege equipment, social organization remained more important than technology in siege warfare. Shalmaneser I (c. 1280 B.C.) was a great builder, restoring many temples, and also a devastating siege general, taking fifty-one cities in a single campaign.[46] There was a direct connection between these two abilities: both depended on the mobilization of labor. It is instructive to note, for example, that assault ramps were unknown in the Middle Ages, when armies were small and dominated by aristocrats. The plentiful labor supply necessary to construct ramps was simply absent.[47]

A Hittite document dated about 1500 B.C. provides a vivid description of siege warfare in Anatolia some three hundred years after the Mari documents:

> They broke the battering ram. The King waxed wroth and his face was grim: "They constantly bring me evil tidings. . . . Make a battering-ram in the Hurrian manner! and let it be brought into place. Make a 'mountain' and let it [also] be set in its place. Hew a great battering-ram from the mountains of Hazzu and let it be brought into place. Begin to heap up earth. . . . " The King was angered and said: "Watch the roads; observe who enters the city and who leaves the city. No one is to go out from the city to the enemy. . . . " They answered: "We watch. Eighty chariots and eight armies surround the city."[48]

The "mountain" refers to an earthen ramp, so we know that the Hittites had learned the old Mesopotamian skill. Apparently the Hittites had tried to avoid the arduous work of building a ramp and only resorted to it when their battering rams proved ineffective.

The close connection between the ram and the ramp may indicate that the Hittites constructed ramps to cross moats and low outer walls so they could deploy the battering ram against the main wall. The ramp would also bring the ram up toward the top of the wall, where it was thinner than at the base.

The king's orders show that his engineers could build a ram on the spot, but the reference to the "Hurrian manner" suggests a different type of ram from the one that broke. By this time the Hurrians had become the culturally dominant people of Assyria, so the Hittite king may have been demanding a ram such as those used by Shamshi-Adad and Ishme-Dagan, but the lack of pictorial evidence from this time prevents us from knowing exactly what sort of ram either the Assyrians and Hurrians or the Hittites used. It was most likely a levering device that could pry open a door or dislodge blocks from a wall.

In any case, the battering ram could not have been very powerful, because the casemate walls of the Hittites could not have withstood battering rams as powerful as those deployed by the Assyrians in the ninth century.[49] Although we know comparatively little about Hittite siege methods, they were effective, as the fall of the powerfully fortified city of Carchemish after an eight-day Hittite siege attests.[50]

It is interesting to note how the Hittites utilized chariots in a siege.

Chariots were useless against fortifications, but they were valuable in maintaining the blockade essential to wearing down the defenders.[51]

Oddly, the battering ram seems to have disappeared from the equipment of the Egyptian army in the New Kingdom (1570–1200 B.C.). Reliefs from the time of Rameses II in the thirteenth century do not show them: the soldiers are assaulting the wall with scaling ladders and the gate with battle-axes.[52] This suggests that stronger defensive systems succeeded in neutralizing the battering rams that were available in the second millennium.[53] For example, when Thutmose III's troops failed to maintain discipline after their great field victory before Megiddo (1458 B.C.), he had to resort to a passive siege complete with a wall of circumvallation to reduce the city. It took seven months.[54]

Siege warfare was the most dangerous form of war in the ancient world.[55] It was fought at close range, the defenders were well protected but the attackers terribly exposed, and the task of overcoming the barrier was agonizingly time-consuming, leaving the soldiers exposed for long periods of time. It took place before the eyes of the defenders, so surprise was difficult. Hittite annals indicate that unexpected timing of assaults, such as during the midday heat or even at night, could achieve some tactical surprise.[56]

Egyptian shields had a shoulder strap that enabled the soldier to carry a shield on his back, protecting him from the missiles raining down on him from the wall and leaving his hands free for climbing, hacking, and fighting, but the tasks of the assault troops cannot have been easy. A shield strapped to the back provided scant protection for such laborious tasks as scaling a wall or breaking through a heavy gate. The Egyptian soldier of Thutmose III had every right to brag:

> His majesty sent forth every valiant man of his army, to breach the new wall which Kadesh had made. I was the one who breached it, being the first of every valiant man.[57]

Although this boast suggests that the assault troops were elite units with a strong esprit de corps, we cannot know for sure by what methods commanders of this time motivated their men to undertake the hazardous task of assaulting a fortified city. Certainly the promise of booty, as the Mari documents make clear, was one inducement. The Mari documents also reveal that the Mari kings rewarded veterans with land,[58] and we can imagine that especially hazardous duty might gain a larger land bonus. Commanders could also use punishment as a means of forcing men to assault a fortified position. A Hittite document stated: "While they achieved nothing at all against the town, many servants of the king were beaten so that many of them died."[59]

Because of high casualties, assaults were practical only against inferior fortifications. But prolonged siege operations had their hazards as well. They were inordinately expensive. They required keeping the army in camp for a long time, presenting logistical problems in some ways more difficult than those of a well-stocked fortified town.[60] Disease could weaken the forces, and boredom could undermine discipline and vigilance. Troops besieging cities were vulnerable to sorties from the besieged troops or to attack by allies of the besieged. Sieges could continue almost indefinitely. Egyptian documents record a siege lasting for three years, and Hittite sources reveal that one siege continued for an incredible nine years.[61] Sieges of several months were common.[62]

Small wonder that commanders often relied on trickery to bring a siege to a successful conclusion. Legends of siege warfare almost always turn on some ruse. The story of the Trojan horse is the most famous, but an even earlier legend from Egypt tells a similar story. It seems that an Egyptian army of Thutmose III (1500 B.C.), under the command of Thot, was besieging Jaffa with little success. Thot feigned surrender, convincing the people of Jaffa to allow him to enter the city for this purpose. Thot hid two hundred men with a large supply of fetters in baskets. Another three hundred men carried the baskets into the city. Once inside, the men in the baskets emerged with their fetters and bound the people of Jaffa in them, thus delivering the city to the Pharaoh.[63]

By the end of the second millennium B.C., then, siege warfare, already at least five thousand years old, had reached a high degree of development. A variety of tactical methods, including the use of siege machinery, had prompted the development of astonishingly large and elaborate fortifications. Ancient engineers showed great ingenuity in constructing sophisticated, integrated defense systems that were effective against both escalade and early battering rams. More powerful rams would be developed, and the introduction of the catapult would add a powerful weapon to siege warfare; but in its essential elements, siege warfare would change little until the introduction of gunpowder in modern times revolutionized the ancient arts of fortification and siege.[64]

Chapter II

TREATMENT OF
CAPTURED CITIES

Although the technical dimensions of siege warfare before the first millennium B.C. are fairly clear, the social and moral context in which sieges were conducted is much more obscure. Certainly property, freedom, and life were at the disposal of the victors. Sacking was a standard practice. As one king wrote his commander;

> Let your troops seize booty and they will bless you. These three towns are not heavily fortified. In a day we shall be able to take them. Quickly come up and let us capture these cities and let your troops seize booty.[1]

Only strict regulations governing the distribution of the booty, mainly to ensure that the king received his due share, provided any control over the rampage that surely followed such orders.[2] Perhaps the only common modern American experience that faintly captures some aspect of a sack is the home invasion burglary. The fear of death, the possibility of rape, the feeling of violation, the loss of privacy and sanctuary, and the helplessness in the face of plunder represent in microcosm some of the experience of a sack.

The survivors of a sack were at the mercy of their conquerors. Primitive societies that could not produce an economic surplus most commonly killed all prisoners of war. Only in societies that produced a surplus were prisoners of war useful.[3] Early Mesopotamia was one of the first societies to produce an agricultural surplus, an achievement that made possible the construction of cities. It is likely, therefore, that patterns in the treatment of prisoners of war more complicated than routine slaughter first developed in the context of siege warfare. An early

Assyrian document from Mari perhaps reflected a transition from slaughter to economic exploitation. A king wrote in astonishment to his commander:

> You have written me that you have taken the city of Tillabnim and have not slain the citizens of that village but that you have pacified them and permitted them to go. This deed which you performed was very commendable, it was worth . . . talents of gold. Truly in the past when you seized a city one did not know this procedure.[4]

The earliest reference to the fate of captured cities comes from the Assyrian king Rimus (c. 2284–2275 B.C.), who conquered several towns in Babylonia and in the Zagros Mountains. He reported that he killed thousands of men in these conquests and took several thousand captives. The inscriptions are obscure, but they may suggest that the slain enemy were not all killed in battle but were executed after having been captured. We must doubt Rimus's figures, however. None of the early Mesopotamian administrative documents that listed numbers of prisoners reported more than around 200 prisoners and often the number was less than five.[5] The fact that Rimus himself made a gift of only six slaves to the temple of Enlil at Nippur belied his boast of "thousands" of captives.[6]

Two administrative texts from the reign of Bur-Sin (c. 2052–2043) of Ur provided more exact information, a precise list of prisoners of war. All were women and children. We do not know what happened to the men, but their absence suggests there was a distinction between male captives and women and children. Men would have been more difficult to control than women and children, so the early practice was to kill them and take the women and children captive.[7] The lot of the latter was not easy. Even though the rations listed in the administrative texts were adequate, the number of captives on the list dropped from 169 women and 28 children to 39 women and 10 children over a period of just a few months.[8] We do not know what caused such a high mortality rate, but disease is the most likely explanation. Whether it was simply an unfortunate epidemic and therefore not typical or whether mistreatment and exhaustion commonly left the captives so vulnerable to disease that a high death rate was inevitable we cannot say.

As the power of the state increased, the means of controlling male captives improved. Usually their captors immobilized them by tying them with ropes or yoking them with wood blocks. Although the evidence is problematical, blinding may have been another method of controlling male prisoners.[9] Kings often enrolled captured soldiers in royal bodyguards, thus creating greater means with which to control larger numbers of captives.[10]

All prisoners of war were at the disposal of the king, but their ultimate fate varied. The economy of early Mesopotamia was too weak to support masses of slave labor, so full slavery was rare.[11] Rather the exploitation of the human booty of sieges centered on some form of dependent labor. In order to preserve this source of labor, the king elevated the status of captives from unfree to semifree so they could support themselves. Indeed, war captives thus transformed into dependent laborers were second only to the native working classes as a source of labor in ancient Mesopotamia.[12]

King Su-Sin of Ur (c. 2043–2034) told of transporting the people of Simanum to a town near Nippur, where they labored in agriculture. This was probably the most common practice. Another common practice was to dedicate prisoners, especially women and children, to temple economies. Women captives also labored as artisans in the weaving trade and, at least in one case, in construction. Although we have no evidence of a specific instance, the king in all likelihood transferred some captives to his nobles.[13]

Frequently the conquerors destroyed captured cities. In the late second millennium, the Assyrian king Shalmaneser I (c. 1280 B.C.) regularly burned the cities he captured, but the inscriptions do not tell us the fate of the inhabitants.[14] A Babylonian chronicle said of another Assyrian king, Tukulta-Urta I (c. 1250 B.C.): "The wall of Babylon he destroyed, the Babylonians he put to the sword."[15] Tiglath-pileser I (c. 1100 B.C.) also burned many of the cities he captured. The annals of his first five campaigns recorded the destruction of cities no less than thirteen times.

The level of resistance offered by a city often determined its fate. To those who surrendered and embraced his feet, obviating the need for a siege, Tiglath-pileser offered mercy. For example, when Shadi-Teshub, the king of Urratinash, submitted on Tiglath-pileser's approach, he spared Shadi-Teshub's life. But Shadi-Teshub had to surrender his sons as hostages, pay a heavy indemnity, and submit to Assyrian rule.[16] In a similar situation, Tiglath-pileser, having taken hostages and exacted tribute, required the king of the Kumani to destroy the walls of his fortified city.[17] In all, Tiglath-pileser's inscriptions related seven cases when he spared the lives of those who surrendered. This revealed a pattern of destruction for those who resisted and mercy for those who did not.

We can learn something of Egyptian practices in siege warfare from the booty lists of Thutmose III (c. 1500). After the capture of Megiddo, he listed 340 living prisoners and 83 hands (the Egyptians cut off a hand from slain enemies as a trophy), 2,041 horses, 1,929 cows, 2,000 goats, 20,500 sheep, 1,796 male and female slaves and their children, and 103 pardoned persons who had escaped from the city during the siege because of hunger.[18] Another plunder inventory listed no fewer than 89,000 prisoners in addition to "their goods, without their limit; all small cattle belonging to them; all (kinds of) cattle, without their limit; chariots of silver and gold: 60; painted chariots of wood: 1,032; in addition to all their weapons of warfare, being 13,050."[19] From this very limited evidence we may infer that the Egyptian conduct of siege warfare aimed at booty more than massacre.

Fragmentary evidence suggests that the Hittites paid considerable attention to conventions of war. By the fourteenth century B.C., the Hittites did not embark on war without a declaration of war. The declaration always contained a justification for the war. But treatment of cities that resisted was harsh. The Hittites usually looted and destroyed conquered cities and carried the population to Hattussas to toil as serfs. One Hittite document recorded the transportation of 60,000 people to Hattussas. An early surrender, however, usually gained more lenient treatment, such as the reduction to vassalage and the imposition of tribute.[20]

Scanty evidence prevents us from drawing any fine distinctions about the conduct of siege warfare before the first millennium B.C. It is apparent, however, that all conquerors of cities considered both people and property at their disposal. Looting was universal, massacres or transportation common. Faced with such dire consequences, defenders sometimes chose to challenge the enemy in the open field rather than risk the entire city in a siege.[21] The decision between resistance or submission could divide a city. An obscure Hittite reference to saving a city by fire may mean that a faction opposed to resistance was so desperate it set fire to the city in order to deliver it to the Hittites.[22] Siege warfare was total war. When a city chose to resist a siege, it risked annihilation.

PART TWO

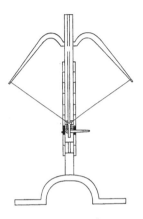

THE
MIDDLE EAST

Chapter III

ISRAEL, MESOPOTAMIA, AND THE PERSIANS

Israel

In the history of siege warfare, the ancient Hebrews do not loom large. Their ability to wage siege warfare was limited, and they contributed nothing to the art of siege. But if the Hebrews accomplished little of military significance, they left behind a historical source of unparalleled importance. Their religious writings constitute one of our longest early written records of warfare. Indeed, the Bible contains descriptions of a wider variety of battles than any other ancient Near Eastern literature.[1] Thus despite the Hebrews' modest role in the history of siege warfare, the Bible remains a source of prime importance.

Whatever the uncertainties, and there are many, about the Hebrew invasion of Canaan in the thirteenth century B.C., it is clear that the problem fortified places posed to the Hebrews determined the course of military events. Most of the cities of Canaan were fortified, and the Hebrews were inexperienced in siege warfare; they also lacked any sort of siege machinery.[2] Fortified positions on the main road from Egypt to Canaan had forced Moses to trek into the desert and follow a circuitous route to Canaan.[3] When the Hebrews finally approached Canaan, Moses sent scouts ahead to make a reconnaissance of the land. Their report that the cities were strongly fortified was so grim that the Hebrews wished they had never left Egypt, and it was all Moses could do to prevent a rebellion.[4]

The fortified cities in southern Palestine not only posed a formidable defense; they also posed an offensive threat by serving as bases for

chariot forces which dominated Palestinian warfare at the time.[5] The
Hebrews had no chariots and greatly feared them.[6]

Moses decided to seek a course of least resistance. He led the Hebrews on a long detour around the edge of the desert until he reached
the Amorite territory of King Sihon. This was newly conquered land,
and Sihon had not completely fortified his towns. The Hebrews were
thus able to defeat him and then push on into sparsely settled Gilead to
establish a base for the invasion of Canaan itself.[7]

Moses' successor, Joshua, faced the problem of fortified towns as
soon as he contemplated crossing the Jordan River into Canaan. The
strongly fortified city of Jericho blocked his invasion route, and he could
not bypass it because it controlled the water supply necessary for further
military operations in Canaan.[8]

Joshua compensated for the Hebrew deficiency in siege equipment
by exact reconnaissance.[9] He sent two spies to Jericho who sought out
an inn operated by a woman of dubious reputation named Rehab.
Somehow the Hebrew spies won the confidence of Rehab, who divulged
the important information that morale in Jericho was low. Rehab was an
especially important contact because her inn, like many inns in ancient
cities, was built into the city wall.[10] Thus it occupied a vital spot in
Jericho's defenses and offered a possible point of entry to the Hebrews.
In gratitude for her indispensable role, the Hebrews spared Rehab and
her family from the carnage that followed the fall of Jericho.[11]

Emboldened by the encouraging intelligence of his spies, Joshua
crossed the Jordan and besieged Jericho. According to tradition, the
siege lasted seven days, during which time the Hebrews, following their
Lord's commandment to Joshua, marched around the city one time each
day for six days. On the seventh day they marched around the city seven
times, ending with a great shout and a long blast from a ram's horn,
whereupon the walls of Jericho "fell down flat."[12]

Biblical scholars and military historians have expended much ingenious speculation in an effort to salvage this miracle for history. The
guesses range from the improbable notion that vibrations from the
rhythmic marching or sound waves from the shouting and the ram's
horn battered down the walls to the more plausible suggestion that
the marching routine lulled the Jericho defenders away from vigilance,
allowing a surprise assault to succeed.[13] Others have found the fall of
Jericho to a seminomadic band of invaders so fantastic they have
doubted that the battle took place. This view originated with the German scholars Albrecht Alt and Martin Noth, who suggested that the
Hebrews did not actually conquer Canaan but rather peacefully infil-
trated into the thinly populated hill country.[14] Gerhard von Rad, in an
influential book titled *Der heilige Krieg im alten Israel*, placed the story in
the context of a literary tradition that stylized the ancient cult of holy

war into a literary motif in which victories were won by Yahweh without human beings playing any role whatsoever.[15]

Archeological evidence suggests that although Jericho had been destroyed in the fifteenth and fourteenth centuries, it was not destroyed in Joshua's time, lending credence to the idea that the story is legendary.[16] Such an argument of silence cannot be decisive. Mordechai Gichon attributes the absence of archeological evidence for the destruction of Jericho in the thirteenth century to the washing away of the ruined walls.[17]

Similar problems confront us when we turn to the next battle in Joshua's Canaan campaign, the capture of Ai. The Book of Joshua provides an even more detailed and vivid description of this battle than it does of the more famous story of Jericho. Once again Joshua sent spies ahead to reconnoiter the city. When the spies returned, they complacently reported that Ai had so few defenders it would not be necessary to commit the entire army.[18]

This report, along with Joshua's decision to follow its advice, revealed the Hebrew inexperience with siege warfare. Very small numbers of defenders provided formidable opposition if they were protected by fortifications. For example, diplomatic correspondence between Canaanite kings and Pharaoh Akhenaton in the mid–fourteenth century, known as the Amarna letters after Akhenaton's capital city where the correspondence was found, contained frequent requests for reinforcements of as few as ten to fifty men, revealing that such small numbers were significant to the defense of a city.[19]

Apparently unaware of this danger, Joshua dispatched a small force (the Bible says three thousand, but we may assume a much smaller number) to Ai, where it was rudely repulsed. The men of Ai routed the Hebrews, killing thirty-six of them, a number that sounds authentic.[20]

Our author tells us that this setback thoroughly demoralized the Hebrews, who were clearly uneasy about taking on fortified towns. But the ever resourceful Joshua noticed that the eager men of Ai had pursued the Hebrews for some distance, and that suggested to him a new plan of attack.

Acting quickly to cut short the spreading demoralization, Joshua drew up a daring stratagem. Violating the principle of concentration, he decided to split his forces, thus risking defeat in detail. Joshua accepted this risk because he wanted to lure the defenders out of the city, leaving it open to surprise attack. He planned to send ahead a smaller force (again, Biblical numbers are unreliable, but they indicate that the ratio between the forces was about five to one) under cover of darkness to lie in ambush. When dawn came, Joshua would lead the larger part of the army up to Ai to draw the defenders out of the city.

Joshua's plan called for one of the most difficult of military maneuvers, a feigned retreat and then a sudden reversal to attack. He hoped the retreat would lure the defenders far enough away from Ai to leave the city open to the small ambush force hidden near the city. The larger force could then turn on its pursuers and, in a hammer-and-anvil maneuver, drive them back against the captured city.[21] Nothing could make plainer the difficulty of siege and the Hebrew inability to sustain a siege than Joshua's willingness to try such a risky and delicate maneuver.

Timing was of supreme importance, so Joshua himself led the main force. As he had hoped, the defenders of Ai came out of the city to attack him across a ravine behind which the Hebrews had encamped. Joshua led the army in retreat, and when he believed he had drawn Ai's soldiers far enough from the city, he signaled to the smaller force lying in ambush. It quickly occupied the undefended city and set it on fire. The Bible makes a special point that every one of the men of Ai had joined the pursuit of Joshua's force, leaving not a single defender in the city.

At just the right moment Joshua wheeled his army around and trapped the enemy between his main force and the ambushers, who were attacking from the rear.[22] Joshua signaled his army by raising his javelin. It has been suggested that the javelin may have had magical powers; if so, the battle of Ai turned on magic, just like the battle of Jericho.[23]

One of the most striking features of the battle of Ai is the vivid detail and accurate topography that the Book of Joshua provided. It is particularly disappointing, then, to learn that the entire story may be a legend. Once again, archeology is the killjoy, providing only negative evidence for the battle; indeed, the evidence tells us that there was no inhabited town at all on the site of Ai in the thirteenth century.[24]

This discrepancy between a vivid, detailed, and plausible account on one hand and the negative archeological evidence on the other has puzzled scholars. The American archeologist William F. Albright believed the battle actually took place at Bethel, where there is archeological evidence of a battle about that time, but that during the transmission of the story, the location of the battle shifted from the flourishing town of Bethel to the abandoned ruins of Ai.[25] Others have doubted that archaeologists have discovered the true location of Ai.

The most popular explanation was first put forth by Martin Noth. He argued that the story was a legend serving the etymological function of explaining how Ai came to be a heap of ruins.[26] Gichon, however, interprets the archeological evidence as indicating that Ai was in fact inhabited by Bethelites at the time of Joshua but that they had not completely rebuilt the defenses. Indeed, the Bible does not mention any walls at Ai, a fact that may help explain the Hebrews' initial complacency. Gichon envisions a hasty defense preparation behind the ruins of the town's old walls.[27]

In view of these uncertainties, what can we conclude about the role of siege warfare in the initial Hebrew invasion of Canaan? Clearly the fortified towns of Canaan determined Hebrew strategy, directing the invasion away from strong towns toward less densely populated and less strongly fortified areas in the highlands. This gave the invasion some of the characteristics of a migration, but Abraham Malamat is correct to insist that some military action accompanied this migration and that if the archeological evidence at Jericho and Ai is problematic, it also makes clear that the Hebrews destroyed "a significant number" of Canaanite cities in the thirteenth century.[28]

The Bible shows how that could have been done by good intelligence and clever stratagems without either siege equipment or the ability to sustain long sieges. There was a limit, however, to how much careful intelligence and guile could accomplish, and when the Book of Joshua tells us that Joshua was able to destroy only one fortified town, Hazor, we seem to be hearing the authentic voice of history.[29] The "cities" that Joshua destroyed would have been mostly small unfortified towns, barely more than villages, or towns whose fortifications had been ruined in recent wars with Egypt.[30] The strong, fortified cities remained unconquered; their presence delayed the Hebrew conquest of Canaan for three hundred years.

Hebrew backwardness in siege warfare continued through the period of the judges. Without a strong, centralized government it was impossible to organize the resources necessary to carry out a formal siege. Thus the fortified cities of the Canaanites remained a difficult obstacle for the Hebrews.[31] When Judah was unable to conquer the plain, it was not only because he lacked "chariots of iron"[32] but also because he was unable to reduce the fortified cities that served as the bases for the Canaanite chariots. Similarly, the Benjaminites failed to drive the Jebusites out of Jerusalem, and most of the other judges failed to drive the Canaanites out of their fortified towns.[33]

The inability of the Hebrews to take fortified cities had important consequences for the history of Israel. The continuing presence of the Canaanites in their midst exposed the Hebrews to Canaanite culture and threatened the purity of cult that the priests of Yahweh so highly valued.

The record was not one of unbroken failure, however. Betrayal, ruse, and surprise continued occasionally to compensate for lack of siege equipment. At Bethel, Joseph sent spies to reconnoiter the city. They found a man willing to betray his city in return for mercy for him and his family. This man showed the Hebrews how to get into the city.[34] Yadin interprets the Hebrew text to mean that he led them into Bethel through a postern. He finds the fact that the traitor took his family to the land of the Hittites, where we first find posterns, significant and thinks that the construction of posterns had spread to Palestine by this time.[35] However the Hebrews got into the city, it is clear that someone let them in and that this was brought about by the treachery of one of its inhabitants.

Another successful attack on a city described in the Book of Judges followed Joshua's methods at Ai. This was the Israelite defeat of Gibeah, the stronghold of the tribe of Benjamin, which had come into conflict with Israel. The Benjaminites, thanks to their strong base in Gibeah, had been able to defeat the men of Israel in an open field fight. But in the same way that Joshua took advantage of the initial defeat at Ai to draw the overconfident defenders away from the city, the men of Israel lured the Benjaminites out of Gibeah, leaving the city open to ambush. When smoke began to rise from the city, the retreating Israelites turned on the Benjaminites and defeated them.[36]

The most interesting siege in the Book of Judges was undertaken by the treacherous Abimelech. The account apparently combined two versions and is not clear in every detail. We do know that Abimelech made himself king of Shechem by exploiting Shechemite family connections and by murder. When a hostile faction drove him and his followers out of Shechem, he was able to retake the city by ruse and surprise. Having heard (perhaps through a traitor or spies) that the men of Shechem had left the town to work in their fields, Abimelech divided his men into three companies. He attacked the Shechemites in the fields with two of the companies while blocking the way back into the city with the third company. His force was able to kill all the Shechemites in the field, a loss that so weakened the city that Abimelech was able to take the city after a daylong battle.[37] The Bible does not tell us how he got into the city but implies that the fighting was around the gate, so it is a reasonable guess that he was finally able to force the gate.[38]

The outer wall of Shechem was not the only barrier Abimelech had to overcome. Many Canaanite cities had built a citadel or fortified tower within the outer walls.[39] Shechem had such a tower, and the defenders fled to it once Abimelech had broken into the city. And there they perished when Abimelech ordered his men to cut fagots with which to fire the tower. The Bible reported that about a thousand men and women died.[40]

After destroying Shechem, Abimelech attacked the town of Thebez, and again the population sought refuge in a fortified tower within the city. As Abimelech was attempting to repeat his success and set fire to the tower, a woman on the tower dropped a millstone on his head. Mortally wounded, Abimelech called out to his shieldbearer to kill him with his sword "lest men say of me, 'A woman killed him.' "[41] Abimelech's death shows the extreme hazards of operations at the foot of a fortification, and his fate became a proverbial warning against such hazards.

Two centuries later, when King David ordered his general Joab to see to it that the husband of the voluptuous Bathsheba, Uriah the Hittite, died in battle, Joab could think of no better way than to send Uriah

against the walls of Rabbah, an Ammonite city the Hebrews were besieging. Apparently there were standing orders not to go near the
walls because Joab feared David would be angry if he did not realize
that the only reason Joab had ordered men to assault the wall was to get
Uriah killed. It was Abimelech's death that Joab recalled when he gave
his careful instructions to his messenger to David:

> When you have finished telling all the news about the fighting to the
> king, then, if the king's anger rises, and if he says to you, "Why did you
> go so near the city to fight? Did you not know that they would shoot
> from the wall: Who killed Abimelech the son of Jerubbesheth? Did not a
> woman cast an upper millstone upon him from the wall, so that he died
> at Thebez: Why did you go so near the wall?" then you shall say, "Your
> servant Uriah the Hittite is dead also."[42]

Hebrew siege capabilities improved significantly under David. Although the early Hebrews were able to take some towns by a variety of
tricks, they lacked the resources to mount true sieges. It was only when
David centralized Israel under a monarchical government that this became possible. His ability to press sieges more relentlessly was an important reason why he was able to expand his power.

A key step in the creation of a united Israel was the establishment
of Jerusalem as the national capital. Jerusalem was ideally located for
such a purpose because the site was easily defensible and there was an
abundant supply of water from both a large spring and the plentiful
winter rainfall that could be gathered in cisterns.[43]

The Biblical account of David's capture of Jerusalem is obscure. Apparently he arrived before the city with such an overwhelming force
that the Jebusite inhabitants tried to cast a spell to stop him. They did
this by stationing blind and lame men on the walls.[44] Some believe this
was a gesture of contempt, reflecting the Jebusites' conviction that Jerusalem was impregnable, but Hittite sources reveal that the blind were
part of ceremonies sealing oaths that laid down a curse blinding any
violators of the oath. The Jebusites may have hoped this magical power
of the blind would cause the Hebrews to shy away.[45]

The magic apparently had some effect, because David had to offer
the special incentive of a major promotion to the man who would assault the wall first. It was Joab who dared to kill the blind and lame,
and this launched his career as David's most important general.[46] Second Samuel says that David's men got into the city through a water
shaft, but Yadin believes the archeological evidence shows the water
shaft to have been too narrow and steep for men to be able to get up. He
accepts an emendation of the text that allows the interpretation of the
Hebrew to refer to a three-pronged weapon rather than a water shaft.[47]
Friedrich Schwally has shown how much belief in magic, spirits, and
demons influenced the Hebrew conduct of war, and Yadin concludes

that David's greatest problem was not gaining entry to the city but his soldiers' fear of the spell cast by the blind and lame. That David immediately set out to improve Jerusalem's fortifications suggests that the Jebusite fortress left much to be desired.[48]

Whatever the difficulties surrounding our understanding of David's capture of Jerusalem, it is clear that he did not mount a full-scale siege but took the city either by assault or by a stealthy entry into the city. This contrasted sharply with the full-scale siege he brought against Rabbah, the first real siege that a Hebrew army conducted.

The Bible does not say how long the siege of Rabbah lasted, but we can infer that it was longer than any previous Hebrew action against a city. The longest siege reported in premonarchical times was the seven-day siege of Jericho, but the number seven is so pregnant with religious and magical significance that we may doubt that it actually served to number the days of the siege. More convincing is the report that the siege of Lachish lasted two days.[49] During the siege of Rabbah, Uriah traveled back to Jerusalem, spent at least three days in Jerusalem, then returned to die before the walls of Rabbah. An unspecified amount of time later, Joab sent notice to David in Jerusalem that the capture of the city was imminent. David took command of the militia, led it to Rabbah, and took the city. All this indicates that we should measure the length of the siege by weeks rather than days.

David was able to sustain a siege longer than the judges because he commanded enough resources to maintain a permanent army.[50] The Israelite army under the judges had been a militia. Short service and short campaigns characterize militia armies; the soldiers need to get back to their fields or their shops. Militia also tend to be poorly disciplined, and the desertion rate is high, a large disadvantage for a general trying to carry on a protracted siege. The logistics of militia armies are frequently inadequate to keep them in the field for long periods. In Israel, the families of the soldiers were responsible for feeding them on campaign, as we can see when Jesse sent David to carry food to his brothers when they were with Saul's army.[51]

For these reasons, the development of a professional army and the use of mercenary troops under the monarchy was of supreme importance for Israel's ability to wage siege warfare.[52] Significantly, Joab led only the regular army against Rabbah. The militia remained in Succoth in reserve.[53] Only when Joab believed the city was about to fall did he send word to David to bring the militia to Rabbah for the final assault.[54]

Although the siege of Rabbah was long, the Hebrews were still relatively backwards in siege warfare. There is no mention of siege equipment at Rabbah, and Joab avoided costly assaults. Instead he brought about the fall of the city by the indirect means of capturing its source of water.

Water supply was one of the most critical problems for a besieged city. Water is more difficult to store than food, and without it no city can withstand a siege. Because of this, water was always a primary concern in the design of fortified cities. Cisterns could gather rainwater, but only if it rained—and it frequently did not in the Middle East. Therefore either city walls encompassed a spring or a fortified way to a nearby spring protected access to it during a siege.[55] Rabbah apparently had the latter arrangement, and when Joab was able to discover the way to the spring and break through and capture it, he knew Rabbah's fall was imminent.

Joab's next siege showed further progress in the Hebrew ability to wage siege warfare. Not only did David become capable of sustaining longer sieges; he began to develop the art of siege also. When Joab besieged the rebellious Sheba in the city of Abel-beth-maacah, he built a ramp up against the city wall. This is the first time a Hebrew commander built siege works.

The Bible also reported that the Hebrews "were battering the wall, to throw it down." The Hebrew word translated as "battering" more literally meant "ruining," so we should not jump to the conclusion that the Hebrews used battering rams at Abel-beth-maacah.[56] If Joab did use a battering ram, it would represent a considerable jump from the time of Saul, when the Hebrews were dependent on the Philistines for weapons and other implements.[57] It certainly could not have been the powerful type of battering ram the Assyrians employed two hundred years later, because David's successor, Solomon, built casemate walls that could not have withstood such rams. If the Hebrews themselves had developed advanced battering rams during David's reign, Solomon's engineers would have designed stronger walls.

Although we cannot rule out the possibility that Joab used some sort of primitive ram, it seems more likely that the Hebrews "battered" the walls with pickaxes or hoes. Josephus interpreted the passage to mean that the Hebrews undermined the wall.[58] This had been a common method of collapsing walls at Mari, and the construction of the ramp shows that Joab may have had sappers capable of undermining. But undermining walls was a delicate operation, and Josephus's interpretation was probably anachronistic. The only thing that is certain is that this is the first instance we know where the Hebrews attempted to breach a wall.

Whatever means Joab employed against Abel-beth-maacah, he ended the siege by negotiation. When the city's inhabitants offered Sheba to him in return for sparing the city, Joab readily agreed. After the people of Abel-beth-maacah threw the severed head of Sheba over the wall to the Hebrews, Joab marched his army away.[59]

If the offensive-minded David used siege as an important means of expanding his power, it is not surprising that the defensive-minded Solomon built fortifications as an important part of his program of consolidation. Solomon was a great builder, most famous for the construction of Solomon's Temple in Jerusalem. Closely related to the construction of the temple were improvements in Jerusalem's fortifications, probably extending the walls to encompass the temple.

Solomon also built major fortifications at the strategic cities of Hazor, Megiddo, and Gezer.[60] Here, happily, modern archeology confirms the Biblical account. Solomon used a master plan for the walls of all three cities. The walls were of the casemate type, five-and-a-half meters thick. They featured impressive gates, probably patterned after the gate Solomon built for the court of the temple. Two square towers flanked each gate. The opening led into a twenty-meter-long entranceway. Three chambers on each side protected this entrance way. These impressive walls and gates reveal that Solomon's architects had mastered the most advanced techniques of fortification at that time. The casemate walls suggest the Hittites were the strongest influence.[61]

In addition to these impressive fortifications, Solomon developed an entire defensive system based on fortified towns. These towns served as bases for horse and chariot and as logistical bases for military operations.[62]

If a permanent army was the key to David's ability to wage siege warfare, a large, highly organized labor force was the key to Solomon's construction of fortifications. The only way that Solomon could mobilize enough labor to construct such works was by force.

The Bible makes it a special point that Solomon did not enslave Hebrews for this purpose but instead forced to the task the peoples subjected during the conquest.[63] However, Solomon did use Hebrew forced labor to build the temple, and this casts doubt on the apologetic account of fortification building.[64]

The social structure of Israel, still rooted in its nomadic past, was poorly suited for mobilizing vast labor-intensive building projects. But the defensive requirements of the state demanded fortification. This dilemma was an important reason for the short life of the united kingdom as the tribes soon rebelled against Solomon's labor exactions.

Siege warfare dominated Palestine in the years after Solomon. Solomon himself had set the policy. The era of expansion was over. Israel now occupied a strategic thoroughfare between such major powers as Syria, Egypt, and, most dangerous of all, Assyria. With only occasional exceptions, the military policy of both the Israelites and the Judean kings was to make their fortifications as strong as possible and to hope they could make conquest too expensive to tempt the stronger powers to an all-out campaign against them. Despite the split of David's kingdom, this policy preserved the northern kingdom of Israel for 150 years and Judah for more than 250 years.

King Omri (c. 882–870 B.C.) established Samaria as the capital of the northern kingdom of Israel and built strong fortifications there that formed a strategic center from which to defend the kingdom.[65] Archeological evidence reveals that there was a line of fortified towns in the north, which Mordechai Gichon calls the "Naphtali line," designed to prevent invasion from the north and provide a base of operations for campaigns in the north.[66] These preparations gave Omri's son Ahab enough confidence to stand up to the Syrian king Ben-hadad when he invaded Israel.[67]

In the southern kingdom of Judah, King Rehoboam (c. 928–911 B.C.) also based his military policy on fortification. He wisely decided to limit his defense to the Judean hills, where he built an elaborate system of fortifications to discourage an invasion of his heartland and protect the independence of Judah. Attending to every detail, Rehoboam made sure that the fortified cities were well stocked with food, oil, and wine, and he established armories in them.[68] That Judah outlasted Israel by 135 years shows that Rehoboam based his policy on sound political, military, and topographical principles.

Despite Rehoboam's efforts, a kingdom as small as Judah was necessarily vulnerable to invasion by powerful armies, as the plundering campaign of the Egyptian Pharaoh Shishak early in Rehoboam's reign showed. Shishak was able to capture a number of fortified towns in Judah and reached the walls of Jerusalem itself. Rehoboam bought him off with a huge tribute, but it had been a close call.[69]

No wonder that the Bible reported that the kings of Judah repeatedly strengthened their fortifications throughout the history of the kingdom. King Asa (c. 908–867) took advantage of a peaceful interlude to expand the fortifications of Judah, building cities with "walls and towers, gates and bars."[70] His son Jehoshaphat continued Asa's policy, building more fortified cities, stationing strong garrisons in them, and developing a system of store cities to provide a strong logistical base.[71]

After Rehoboam, King Uzziah (c. 786–758) was perhaps the greatest military organizer among the kings of Judah. He strengthened Jerusalem by building towers at key points along the walls. He also developed a system of towers in the wilderness to secure the many cisterns he had built and to provide a protective screen on the frontiers and along strategic routes.[72] Uzziah raised a large army through musters based on a census carried out jointly by his secretary and his military commander. He also organized an armaments industry to supply this army.[73]

Intriguingly, the Book of Chronicles tells us that Uzziah strengthened his defenses by placing "engines, invented by skillful men," on the towers and corners of Jerusalem's fortifications. These "engines" shot arrows and stones.[74] If true, this would be astonishing because catapults did not come into general use for another four hundred years. However, the translation of the passage is problematical, and most historians reject the idea that Uzziah had catapults.

Yadin provides the most ingenious explanation for the passage. He argues that the word translated "engines" refers not to catapults but to

wooden structures on the towers and walls. He believes that an Assyrian relief depicting the siege of Lachish about fifty years later shows
such structures. They were wooden frames that held the shields of archers and slingers, freeing their hands to fire their weapons from behind
their shields.[75]

Yadin's solution is far from certain, and it may be that we are deal-
ing with a simple anachronism resulting from a late authorship of the
Book of Chronicles.[76] But even if we cannot believe Uzziah used catapults, we can conclude that he was an extraordinarily resourceful military organizer who left the Kingdom of Judah stronger than he found it.

Uzziah's son Jotham also constantly expanded Judah's fortification
system.[77] Finally we find King Manasseh, during his long reign in the
first half of the seventh century, building a new, powerful wall at Jerusalem and strengthening his command over the fortified cities of
Judah.[78]

The Bible does not report such constant fortification building in Israel. But, given the greater strategic vulnerability of Israel, this silence
most likely reflects the Deuteronomic historian's lack of interest in the
kings of Israel rather than any lack of attention to fortifications on their
part. Indeed, archeological evidence makes clear that they assiduously
maintained and improved their fortifications.

The outstanding improvement in the art of fortification in the post-
Solomonic period was the abandonment of casemate walls in favor
of solid walls that were heavier and stronger than casemate walls. At
Megiddo, for example, a new stone wall 3.5 meters thick at the base replaced Solomon's casemate wall in the early ninth century. The upper
part of the wall was probably made of bricks. The wall had alternating
salients and recesses, creating a sort of bastion every six meters. These
bastions had balconies with crenelated parapets.

The ruins of the Judean town of Mizpah show that the Judean kings
also introduced this new, solid type of wall. At Mizpah the bastions
were especially strong, being ten meters wide and protruding three meters from the wall. A glacis provided further protection at the base of
the wall. Mizpah also featured a new type of gate, smaller and more
solid, that became characteristic during this time.[79]

The abandonment of casemate walls, which had offered the advantages of double walls, storage space within the wall, and overall thickness without unwieldy building blocks, revealed an overriding need
for greater strength in fortifications. The resurgence of the Assyrians
in the ninth century, a resurgence based primarily on their perfection
of siege war techniques and especially the use of powerful battering
rams, explains why the Israelite and Judean kings expended so much
effort strengthening their fortifications. This shift from casemate walls
to thick, solid walls was a classic example of a defensive reaction to a
new offensive threat.

Although military policy in both Judah and Israel focused on defensive fortifications in the years after Solomon, the Hebrews were capable of offensive action. Shortly after 850 B.C., for example, King Jehoram of Israel and King Jehoshaphat of Judah, with the support of the Kingdom of Edom, mounted an invasion of Moab because the Moabite king Mesha had ceased to pay the tribute that Israel had imposed upon him. After an arduous march through the desert, the combined forces of Israel, Judah, and Edom reached Moab and, encouraged by the prophecy of Elisha that they would "conquer every fortified city, and every choice city," they laid siege to the Moabite city of Kir-hareseth.[80]

This campaign tells us a good deal about Hebrew siege tactics in the ninth century B.C. Despite the prohibition against cutting down trees in Deuteronomy 20:19, the Hebrews concentrated on felling trees and stopping up springs. Their main weapon against the fortifications of Kir-hareseth was slingers, who can be very effective against defenders of walls.[81] These tactics were successful enough to force Mesha to lead a desperate effort by seven hundred swordsmen to break out of the town, but this sortie failed. Mesha then resorted to a breathtakingly cruel measure; he climbed to the top of the wall and, in full sight of the besieging Hebrews, sacrificed his eldest son and heir apparent as a burnt offering to the god Chemosh. Thereupon, the Bible reports, "there came a great wrath upon Israel; and they withdrew from him and returned to their own land."[82]

It is not easy to understand why the Hebrews withdrew at the moment when victory was within their grasp. The Bible may be correct in attributing their breaking off the siege to fear of divine wrath after so appalling a deed. Another possibility may be inferred from a Middle Eastern belief that child sacrifice warded off plague. If the Hebrews believed there was plague in Kir-hareseth, that could explain their abrupt departure.[83] Whatever his motives, Mesha's sacrifice of his son offered frightening testimony to the horrors of siege warfare.

Not long after the invasion of Moab—the chronology is exceedingly vague, but it was still during the time of Elisha, who lived until around 800 B.C.—Israel's capital city was placed under siege by the Syrians. That this siege failed is evidence of the strength of Samaria's fortifications; but the siege must have been long, because the Bible reports a famine in the city so severe that there was cannibalism.[84] This siege also ended with a sudden departure by the besiegers. When the king of Israel heard that the Syrians were gone, he suspected a ruse and did not relax his defenses until a thorough reconnaissance had confirmed that the Syrians had indeed returned home.[85] This shows that the Israelite king was experienced in siege warfare, because trickery remained one of the most common methods of capturing a fortified city.

Jerusalem was strongly fortified, but the kings of Judah were less belligerent than the kings of Israel. When the Syrians invaded Judah during the reign of Jehoash (or Joash, c. 812–810 B.C.), he preferred to pay a huge tribute to the Syrians rather than withstand a siege.[86] When one considers the horrors of the sieges at Kir-hareseth and Samaria, who can blame him?

Joash's son Amaziah was not so cautious. Buoyed by a bloody victory over the Edomites in which he murdered a large number of prisoners by ordering them thrown from a rock and dashed to pieces,[87] he challenged King Jehoash of Israel (not to be confused with Amaziah's father, King Jehoash of Judah). King Jehoash wanted no quarrel with Judah and counseled the young and hotheaded king of Judah to stay at home. But when Amaziah turned a deaf ear to this sensible advice, Jehoash marched into Judah and soundly defeated Amaziah in an open-field battle. This victory cleared the way to Jerusalem, which was apparently undefended, because Jehoash destroyed over six hundred feet of the wall and plundered the city.[88] Amaziah's blunder had enabled Jehoash to accomplish what every commander sought: to avoid a costly siege by defeating the enemy army in the field.

A little over seventy-five years later, another Israelite king, Pekah, joined by the king of Syria, did try to take Jerusalem by siege. Even with the help of the Syrians he failed, revealing how lucky Jehoash had been to have so feckless an opponent as Amaziah.[89]

If Pekah's siege of Jerusalem was a failure, it nevertheless had fateful consequences. King Ahaz of Judah was sufficiently alarmed that he offered generous treasure to the Assyrian king Tiglath-pileser III and placed himself under his protection.[90] This act brought the Assyrian wolf to the fold.[91] Tiglath-pileser responded by conquering Syria.

His successor, Shalmaneser V, invaded Israel and besieged Samaria. The Assyrians were carrying out an arduous siege of Tyre simultaneously, and that may have slowed operations at Samaria.[92] Nevertheless, it was a tribute to the strength of Samaria's fortifications that three years passed before the city fell to the Assyrians.[93]

We can only imagine the suffering in Samaria during the siege, but we know the fate of the Israelites when the city finally fell. Sargon II, who succeeded Shalmaneser the year Samaria fell, carried the Israelites away to Assyria, where he settled them in scattered and remote locations.[94] According to Assyrian records, Sargon transported 27,290 people from Samaria.[95] He completed the extinction of Israel by settling conquered peoples from other parts of the Assyrian Empire in the land that had been Israel.[96] King Ahaz may have been satisfied to see the destruction of his enemies in Syria and Israel, but his son Hezekiah paid the price for such short-sightedness.

In 701 B.C., the Assyrian king Sennacherib invaded Judah. King Hezekiah quickly repaired and strengthened the fortifications of Jerusalem. He also stopped up all the springs outside the city to deprive the Assyrians of water.[97] Despite these measures, Sennacherib was able to place Jerusalem under siege. The siege isolated Jerusalem from the elaborate fortification system of the Judean countryside, and Hezekiah desperately offered every bit of treasure he could scrape together to persuade Sennacherib to withdraw from Judah.[98] This offer failed to appease Sennacherib, and his army pressed on with the siege.

Now, however, the worth of Judah's fortifications became evident. Even though Hezekiah could not move from his base in Jerusalem to relieve the fortified points around the Judean countryside, Sennacherib found it necessary to reduce one fort after another. No less than forty-six fortified places required capture. Cut off from Jerusalem, most of these places undoubtedly fell easily enough to Sennacherib, but he was losing precious time.

One town, Lachish, managed to put up a strong defense; Sennacherib reduced it but at considerable cost.[99] Assyrian reliefs, Assyrian royal records, the Bible, and archeological excavations of the site provide a variety of evidence for Sennacherib's siege of Lachish that is unique in the history of ancient siege warfare.[100]

Lachish's fortifications were strong enough to force Sennacherib into full-scale siege operations. The town sat on top of a steeply sloped tel, so access to the city was extremely difficult. The high wall that surrounded Lachish had crenelated battlements. These, however, were apparently inadequate under combat conditions, because the Assyrian reliefs of the siege show that the Hebrews mounted wooden frames on the wall to hold shields that could provide additional protection. Square towers that also had crenelated battlements protected the curtain at regular intervals. A steep glacis supported by a revetment wall increased the difficulty of the approach.

A narrow roadway led to the main gate near the southwest corner of the city. An outer gate protected by two powerful towers led into a courtyard (or bastion, as the excavators dubbed it) surrounded by a low wall. If this gate were breached, the attackers would have to expose their unprotected right side as they entered the bastion and turned right toward the inner gate. Two towers also protected this gate. It led into a huge gatehouse, nearly twenty-five meters squared. Three chambers on each side of the gatehouse probably served as guardhouses for troops defending the gate. Within the city, a massive palace fort provided a last refuge for the defenders.[101]

Sennacherib himself supervised the siege of this powerfully fortified city. He established a camp on high ground some 350 meters from the southwest corner of Lachish, near where the gate was located and where access to the city was easiest. The Assyrians constructed a huge siege ramp up to the southwest corner of the main wall. A smaller ramp probably led up to the northwest corner of the outer bastion in front of the main gate. The Assyrians then attacked up the ramps with battering rams and almost certainly assaulted the gate with rams and fire.[102]

Sennacherib's lapidary account that he "conquered (them) by means of well-stamped (earth-) ramps, and battering-rams brought (thus) near (to the walls) (combined with) the attack by foot soldiers, (using) mines, breaches as well as sapper work" suggested the intensity of the fighting.[103] He was so proud of the capture of Lachish that he commissioned a series of reliefs in his royal palace at Nineveh to commemorate the deed.[104]

Despite the fate of Lachish, confidence in Jerusalem's fortifications must have been high, because when the Assyrians delivered an ultimatum in hearing range of the defenders and in their own language, the soldiers obeyed Hezekiah's order to remain silent.[105] Hezekiah lacked horses, chariots, and horsemen, as the Assyrian negotiator pointed out, so his refusal to surrender rested solely on the strength of his fortifications.[106]

Having failed to intimidate Jerusalem into surrendering, Sennacherib decided not to press the siege further. As the prophet Isaiah had foretold, Sennacherib did not personally come to Jerusalem and did not begin the construction of siege works.[107] (However, Assyrian sources report that Sennacherib's army did build siege works at Jerusalem.) Instead, the Assyrians abruptly broke off the siege of Jerusalem and returned home "with shame of face."[108]

The Bible reported that an angel of Yahweh slew 185,000 Assyrians in their camp before Jerusalem, leading some scholars to believe that a plague in the Assyrian army caused Sennacherib to lift the siege.[109] (Sennacherib's army was singularly unlucky. Herodotus reported that it suffered a defeat at the hands of the Egyptians after field mice overran the Assyrian camp during the night and ate all the Assyrian quivers, bowstrings and shield grips.)[110] Others point to a threat from Egypt to Sennacherib's supply lines as the reason for his withdrawal.[111]

Assyrian records hardly have Sennacherib withdrawing "with shame of face." In them Sennacherib reported that he deported over 200,000 Hebrews during the campaign, turned over some Judean territory to Philistine, and reduced Hezekiah to his vassal. He boasted that Hezekiah sent vast treasures, along with his own daughter and his concubines, to Sennacherib. "In order to deliver the tribute and to do obeisance as a slave he sent his (personal) messenger."[112]

Nevertheless, the facts remain that Sennacherib did not take Jerusalem and Hezekiah managed to preserve his throne. Whatever the reason Sennacherib withdrew, it is clear that the difficulties posed to him by the fortifications of Judah slowed his advance and left him vulnerable to angels of death, to plague, and to threats to his supply lines and amenable to breaking off the siege on terms. The kings of Judah had judged right when they based the defense of their kingdom on a strong system of fortifications.

Despite the strength of his fortifications, Hezekiah could be thankful that Sennacherib decided against a full-scale siege of Jerusalem, because the Assyrians had developed the art of siege to a very high level. In the end, the danger of Assyrian siege methods to even a strongly fortified city such as Jerusalem became apparent, although it was not the Assyrians themselves who conquered the city.

The Neo-Assyrian empire went into decline in the second half of the seventh century, and Babylon conquered Nineveh in 612 B.C. Judah took advantage of Nineveh's agony to break away, but in 598 B.C. the Babylonian king Nebuchadnezzar sent an army to Judah to bring the rebellious kingdom to heel. His army placed Jerusalem under siege.

When Nebuchadnezzar himself arrived before the city, King Jehoiachin realized that the Babylonians were committed to the fall of the city to a degree that Sennacherib had not been, and he surrendered himself and his family in hopes of saving the city. Nebuchadnezzar's terms were stiff. He carried away not only the king and his court but also the royal and temple treasures. In addition he took captive the Judean army and all the skilled craftsmen of Jerusalem, leaving behind only the lowest classes.

But Jehoiachin's surrender did prevent a bloodbath and allowed at least a shadow of the proud city to exist under the rule of Zedekiah, whom Nebuchadnezzar placed on the throne.[113] Jehoiachin survived thirty-seven years of imprisonment in Babylon, after which he finished out his days as a dependent in the court of the Babylonian king.[114]

Perhaps because Zedekiah was still in his twenties, he rashly raised the standard of rebellion against Babylon just seven years after Nebuchadnezzar had placed him on the throne, forcing Nebuchadnezzar to return to Judah and lay siege to Jerusalem again. This time there was no surrender. The Babylonians constructed siege works, but Jerusalem was able to hold out for two years. It was only after famine had weakened the defenders of Jerusalem that the Babylonians were able to breach the wall. Zedekiah tried to flee during the night with what was left of his army. They managed to get through the enemy lines, but the Babylonians pursued and soon routed Zedekiah's men and captured Zedekiah.

Failure after a determined resistance to a siege was always costly. The Babylonians slaughtered Zedekiah's sons before his eyes, then gouged his eyes out and carried him to Babylon. King Nebuchadnezzar sent the captain of his bodyguard, Nebuzaradan, back to Jerusalem to destroy the city. Nebuzaradan burned down the temple and all the houses and broke down the walls. The Babylonians thoroughly plundered the city. Most of the remaining population was transported to Babylon. Nebuzaradan sought out the priests in the temple and the military commanders and royal officials whom Zedekiah had left behind and took them to Nebuchadnezzar, who ordered them killed. Only a few of the poorest people were left behind in Jerusalem.[115]

Jeremiah's version is somewhat different, reporting that Nebuchadnezzar carried away a total of 4,600 people in three deportations over a sixteen-year period.[116] In any event the existence of an independent Jewish state had ended until modern times. The emergence of siege methods able to reduce strongly fortified cities had doomed the work of the great Hebrew kings who had built the fortifications of Israel and Judah.

Mesopotamia

The Neo-Assyrian Empire that conquered Israel was the greatest military power that ancient Mesopotamia produced. Although the Assyrians excelled in every area of warfare, skill in siege warfare was the key to their ability to conquer and rule an empire. Thanks to their preoccupation with war and the skill of their artists, the Assyrians present us with a vivid picture of their siege methods. These methods emphasized assault, and the most striking feature of Assyrian sieges was their diversity of assault methods. Like the Romans, the Assyrians excelled in war not so much because of technology as of organization.[117] The Assyrians utilized three assault methods simultaneously: battering rams, escalade, and sapping.

The battering ram was the Assyrians' most formidable assault weapon (figure 5). During the reign of Ashurnasirpal II (883–859 B.C.) this weapon, whose origins are so obscure, finally emerged into the clear light of day.

Ashurnasirpal's battering rams were extremely heavy machines, requiring six wheels for support (figure 6). The wheels gave the ram some maneuverability, but its weight sharply limited its mobility. A turret with a domed top protected the front. Embrasures in the turret provided visibility as well as openings through which to fire arrows. Wicker shields hung on the ram, providing further protection. The entire machine was about five meters long and two to three meters high. The tur-

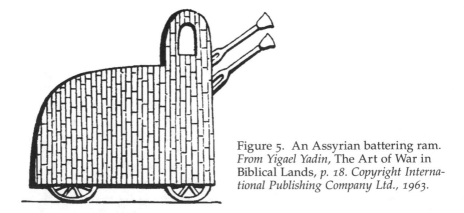

Figure 5. An Assyrian battering ram. *From Yigael Yadin,* The Art of War in Biblical Lands, *p. 18. Copyright International Publishing Company Ltd., 1963.*

Figure 6. Battering ram and mobile tower of Ashurnasirpal. *From Yigael Yadin*, The Art of War in Biblical Lands, *p. 314. Copyright International Publishing Company Ltd., 1963.*

ret added about three meters to the height of the front. The battering pole hung like a pendulum from a rope attached to the roof of the turret. It had a metal tip, flattened like a blade, which the crew could force in between the blocks of the wall. The pole then functioned as a lever, prying the blocks loose in an effort to collapse a section of the wall.[118]

With characteristic creativity, the Assyrians worked to improve their battering rams. Shalmaneser III (858–824 B.C.) employed a more stream-lined battering ram that may have been more mobile. The huge six-wheel model had poles extending from the rear that may have been tongues to which draft animals could be hitched. A lighter model had only four wheels (figure 7). Shalmaneser's artist's rendering of battering rams suggests that the battering poles, in sharp contrast to the pendulum type of Ashurnasirpal, may have been fixed.[119] If that is so, such a ram would have been so unwieldy that it is difficult to see how it could have been effective. In any case, such rams do not appear again.

Future Assyrian kings sacrificed weight for greater mobility. By the time of Tiglath-pileser III (745–727 B.C.), Assyrian battering rams were light enough to be supported by four wheels (figure 8). Also in the eighth century Sargon II (751–705 B.C.) maximized the impact of his battering rams by deploying them in groups against a single part of the wall (figure 9).[120]

Sennacherib (704–681 B.C.) deployed prefabricated battering rams that the crews could dismantle for easier transport and then assemble on the spot (figure 10). The relief depicting Sennacherib's siege of Lachish shows the Assyrians deploying no less than seven rams simultaneously.[121] Sennacherib also lengthened the battering pole to provide greater reach and leverage.[122]

As the battering rams attempted to breach the wall, Assyrian assault troops attempted to mount it with scaling ladders. Assyrian reliefs show

Figure 7. Four-wheeled Assyrian battering ram from the Gates of Shalmaneser III. *Copyright British Museum.*

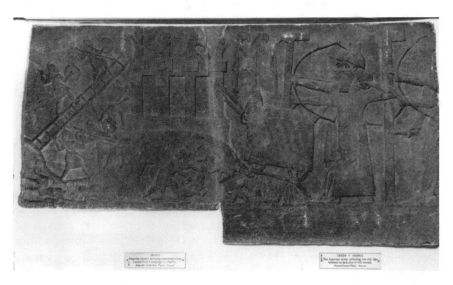

Figure 8. Assyrian relief from center palace, Nimrud, Tiglath-pileser III.
Copyright British Museum.

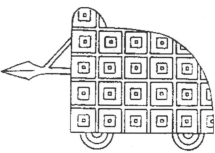

Figure 9. Battering rams of Sargon. *From Yigael Yadin,* The Art of War in Biblical Lands, *p. 315. Copyright International Publishing Company Ltd., 1963.*

Figure 10. Battering ram of Sennacherib. *From Yigael Yadin,* The Art of War in Biblical Lands, *p. 314. Copyright International Publishing Company Ltd., 1963.*

both lancers and archers climbing the ladders (figure 11). These troops could not wear the long Assyrian body armor that hung down to the feet because it would have restricted leg movement too much to allow rapid climbing. Instead they wore knee-length skirts that allowed maximum movement.

Figure 11. Ashurnasirpal II assaults a city. *Copyright British Museum.*

The reliefs show lancers protecting themselves with shields held in the left hand and the archers pausing on the ladder to fire their arrows. If these pictures are accurate, these soldiers were clearly highly trained elite troops, because climbing a ladder in the face of enemy opposition while wielding weapons and shields would be a difficult task indeed.[123] It is perhaps this difficulty that led to the construction of steps which troops could climb to the top of the wall. The relief from Sennacherib's palace that depicts this method shows spaces between sections of the steps, suggesting that the Assyrians placed prefabricated sections of steps against the wall.[124]

Overwhelming numbers were necessary for a successful escalade. The Book of Joel uses soldiers scaling a city wall as a simile for a swarm of locusts:

> Like warriors they charge,
>   like soldiers they scale the wall.
> They march each on his way,
>   they do not swerve from their paths.
> They do not jostle one another,
>   each marches in his path;
> they burst through the weapons
>   and are not halted.
> They leap upon the city,
>   they run upon the walls;
> they climb up into the houses,
>   they enter through the windows like a thief.[125]

While the rams battered at the walls and the assault troops attempted to cross over the walls, sappers worked at the base of the wall trying to set the gates on fire or to penetrate or undermine the wall. For this purpose they used a variety of sapping tools, such as bars, pickaxes, hoes, and drills.[126] As the sappers bored into the wall, they would prop it up with wooden supports until the hole had become large enough and deep enough to undermine the wall. They would then collapse the wall by setting fire to the supports.[127]

All of these methods of attacking the walls of a city had to be carried out in the teeth of opposition from the defenders. The outer range of projectile weapons, such as the bow, was about sixty meters. Effective range was about half that. Thus any soldier approaching the wall came into some danger within sixty meters and entered an intense fire zone at thirty meters. Protecting the troops in these zones, some of whom were engaged in the time-consuming tasks of battering and sapping, was essential to the success of an assault on the walls.

A primary means of protecting troops assaulting the wall was to mount an intense covering fire to drive the defenders from the top of the wall and minimize the fire they could direct at the assault troops. Archers played the main role in this task. Protected by shield holders, they approached within range of the wall and directed a withering fire at the defenders. The Assyrians developed a special shield higher than a man that rested on the ground and had a narrow roof to protect the archers (figure 12).[128] They also had towers that both protected the

Figure 12. Assyrian siege shields. *From Yigael Yadin*, The Art of War in Biblical Lands, *p. 295. Copyright International Publishing Company Ltd., 1963.*

archers and enabled them to shoot down at the wall, a much more effective angle of fire. The towers may have been mobile, but the reliefs do not show the bottoms clearly enough for us to know. They were certainly too large to be carried in a siege train, so they were either constructed on the spot or assembled from prefabricated parts.[129]

Covering fire became more effective in the eighth century with the addition of slingers to complement the fire of archers. The high-angled parabolic flight of the slingers' stones was especially effective against defenders crouched behind parapets.[130] The Assyrians also utilized chariots as a sort of mobile light artillery that could dart within range of the walls and loose a sudden volley of arrows before just as quickly ducking out of range.[131]

An especially difficult task was moving the ponderous battering rams up to the wall. The Assyrians deployed their rams in two ways. One was to push the ram up to a gate in the wall. This approach placed the ram at the weakest point in the wall. For this very reason, however, bastions usually flanked the gates, making them the most strongly defended point. A much more laborious method, but one that offered many advantages, was to construct an earthen ramp up which the ram could be pushed to the wall. An important advantage of the lighter rams was that crews could push them up these earthen ramps. The reliefs show the ramps as quite steep, but this is an illusion caused by the artists' effort to compress the ramp into a crowded space. In fact the ramps were inclined as little as possible; otherwise it would have been too difficult to push the rams up them.

Excavations at Lachish have discovered the main Assyrian ramps there. Although the entire ramp has not been uncovered, its gradient appears to have been about 30 degrees, becoming less steep toward the top.[132] The ramp was wide enough to accommodate five rams.[133] The Old Babylonian math problems always took sixty meters as the length of a ramp. Therefore the crews could safely assemble the rams and lift them onto the ramp just beyond bow and arrow range.

The ramps served several purposes. They could cross barriers such as low walls and glacis. They provided a smoother pathway to the wall for the rams, which, with their fixed wheels, were too immobile to be maneuvered over unpaved terrain. Finally, the ramps brought the ram to the upper part of the wall, which was not as thick and strong as the base and offered the best prospect of rapid success.[134]

As the rams neared the wall, the defenders attempted to pull the battering pole aside with chains, but special troops countered this tactic by hooking the chains with grappling irons and pulling them down by swinging on them with their entire weight. The defenders also attempted to set fire to the ram by dropping torches on it.

Fire was especially dangerous to besiegers, because siege engines and siege works contained so much wood. For example, when Esarhad-

don was besieging the city of Ubbume, the defenders managed to set
fire to his siege ramp. Luckily for Esarhaddon, a favorable wind blew
the fire away from the ramp onto the city walls, enabling the Assyrians
to take Ubbume, but it had been a close call.[135]

Not surprisingly, the Assyrians took special care to protect their
rams against fire.[136] They covered the rams with leather hides to ward
off flaming torches dropped from the wall.[137] One relief shows soldiers
in the mobile tower dousing the flames with water poured through long
spouts, and another relief shows a crew member in the battering ram
itself pouring water from a long-handled container onto the ram (see
figure 10).[138]

Most vulnerable perhaps were the sappers who had to labor at
the foot of the wall. Ashurnasirpal's sappers were heavily armored with
long padded or mailed covering reaching to the ankles and a pointed
helmet with mail protecting the face and neck. Sappers after Ashurnasirpal's reign were lightly armored with a helmet and a small round
shield that they held over themselves with one hand while they wielded
their tool with the other. This must have provided little protection
and made the sapping work difficult at best. Ashurbanipal (668–630
B.C.), whose reliefs curiously show no battering rams, provided better
protection for his sappers with a huge wicker shield curved at the top
so that the sappers could lean it against the wall, creating a protective
canopy under which they could work.[139] These huge shields may have
been strong enough to withstand stones dropped from the wall (see
figure 12).[140]

We cannot know how effective the Assyrians were in limiting the
casualties of the assault teams. Both Sargon and Esarhaddon list casualties from a siege as "1 charioteer, 2 cavalrymen, 3 sappers," apparently
a formulaic way of paying homage to the dead and not an actual count,
much like burying one "unknown soldier" to represent all unidentified
casualties.[141] Perhaps the proportions represent the order of danger, but
the absence of infantry from the list is inexplicable.

One indication of the danger for the assault teams is that the Assyrian kings did not lead them. If their own reports and the reliefs are
accurate, Assyrian kings led their troops into field battles, usually rid-
ing in a chariot.[142] One, Sargon II, may have died in battle.[143] But the
reliefs of the sieges show the king hanging back with the archers providing support fire.[144] The king always wore body-length armor, clear
evidence that he had no intention of climbing a ladder. Although considerations of social prestige clearly weighed heavily in the king's decision to lead the charge of aristocratic charioteers and not take his
place at the head of infantry assault troops, the role of the king suggests
that the Assyrians considered assaults on fortified positions considerably more hazardous than open-field fighting.

Assyrian assault methods could achieve sudden results. Esarhaddon reported that he took the Egyptian city of Memphis in half a day by "mines, breaches and assault ladders."[145] The small towns that were always clustered around Mesopotamian metropolises were especially vulnerable, and Assyrian kings regularly reported bagging large batches of these sort of towns.

Despite their impressive variety of assault tactics, however, the Assyrians were not always able to breach the walls of really strongly fortified cities. When assault failed, they resorted to passive siege, which they were able to sustain for remarkably long periods. Tiglath-pileser reduced the city of Arpad with a three-year siege.[146] Samaria fell to the Assyrians after a three-year siege. The Babylonian king Nebuchadnezzar sustained the siege of Jerusalem for two years before he managed to breach the wall.

Assyrian sources do not provide as vivid a portrait of passive siege as they do of assault methods. Adad-Nirari II (911–891 B.C.) dug moats around a city he was besieging, but it appears that the purpose of these moats may have been to undermine the wall rather than to isolate the city.[147]

The Syrian king Zakir of Hamat, in a description of a siege of one of his cities, mentioned that his enemies "made a wall higher than the wall of Hatarikka" and "made a moat deeper than its moat."[148] This may refer to a wall of circumvallation, but the emphasis on the fact that the wall was higher than the city wall suggests that the besiegers were attempting more than a passive siege by gaining an advantageous angle of fire. The "moat" may have been an effort to drain the moat around the city. Tiglath-pileser III reduced garrisons by constructing earthworks around them, and Sennacherib built earthworks around Jerusalem.[149]

Presumably chariots and cavalry played major roles in isolating cities from the outside world. This seems to have been the case when King Zedekiah fled Jerusalem and the Babylonians captured him, although the Biblical account unfortunately only refers vaguely to the "army of the Chaldeans" pursuing Zedekiah, so even here we must guess.[150]

Whatever the means, they were effective because we hear frequent mention of starvation within besieged towns. Famine was the main weapon of passive sieges, either forcing surrender or so weakening the defenders that an assault could succeed. Esarhaddon explicitly stated that he cut Tyre off from food and water when he besieged that city.[151] The horror of long sieges is evident from Ashurbanipal's laconic report that the people of Babylon "ate each other's flesh in their ravenous hunger" during his siege of that city.[152]

The Bible vividly portrays the horror of famine in besieged cities. Parents eating their children was a recurring motif. In 2 Kings' description of the siege of Samaria by Ben-hadad of Syria, we read:

Afterward Ben-hadad king of Syria mustered his entire army, and went up, and besieged Samaria. And there was a great famine in Samaria, as they besieged it, until an ass's head was sold for eighty shekels of silver. Now as the king of Israel was passing by upon the wall, a woman cried out to him, saying, "Help, my lord, O king!" And he said, "If the LORD will not help you, whence shall I help you? From the threshing floor, or from the wine press?" And the king asked her, "What is your trouble?" She answered, "This woman said to me, 'Give your son, that we may eat him today, and we will eat my son tomorrow.' So we boiled my son, and ate him. And on the next day I said to her, 'Give your son, that we may eat him'; but she has hidden her son." When the king heard the words of the woman he rent his clothes—now he was passing by upon the wall—and the people looked, and behold, he had sackcloth beneath upon his body. . . . [153]

This story shows the incredible inflation in the price of food during sieges, the demoralization of the population, and the way hunger drove people to violate the most deeply seated taboos.

During Sennacherib's siege of Jerusalem, the Assyrian demand for surrender made clear what he counted on to subdue the city when Sennacherib's negotiator pointed to the defenders as men "who are doomed . . . to eat their own dung and to drink their own urine."[154] The Assyrians tried to make good on this threat by forcing back into the city anyone trying to escape because in a passive siege large numbers are a liability to defenders trying to prolong their food reserves. When the Babylonians took Jerusalem, the Bible makes clear that it was only after "the famine was so severe in the city that there was no food for the people" that they were able to breach the walls and bring that siege to a successful conclusion.[155]

The prophets also associated famine with siege warfare. In his fourth lamentation, Jeremiah described the distress of a siege:

Happier were the victims of the sword
    than the victims of hunger,
who pined away, stricken
    by want of the fruits of the field.
The hands of compassionate women
    have boiled their own children;
they became their food
    in the destruction of the daughter of my people.[156]

Ezekiel summed up the probable fate of Jerusalem under siege:

A third part of you shall die of pestilence and be consumed with famine in the midst of you; a third part shall fall by the sword round about you; and a third part I will scatter to all the winds and will unsheathe the sword after them.[157]

Neo-Babylonian legal documents provide a less literary testimony to the horror of famine under siege. These documents reveal that it was a standard practice in Babylonia for parents to sell their children to the wealthy during sieges. The buyers obligated themselves to keep the children alive. After the siege, the parents could redeem the children for a price higher than the buyer had paid.[158] One such document reads:

> Take my small child (daughter) and keep (her) alive! She shall be your small child ([slave-] girl). Give me x shekels of silver so that I may (have something to) eat![159]

The documents imply that law or tradition defined precisely the level of inflation in the price of food during a famine at which it became permissible to sell children.[160]

When Babylon revolted from Persia during the reign of Darius I, the men sent away their mothers and chose one woman from their household to bake bread for them. The rest of the women they strangled in order to have fewer mouths to feed. Thus prepared, Babylon held out against Darius's siege for a year and seven months, and only a stratagem finally brought about the fall of the city.[161]

The frequent association in our sources between famine and siege demonstrates that passive methods were usually necessary against well-fortified cities. Because of the inherently long duration of passive sieges, they were never attractive to the Assyrians. After Shalmaneser drove King Hazael of Syria into Damascus, he decided against a siege, contenting himself with cutting down Hazael's gardens before continuing on.[162] Sennacherib's decision to withdraw from Jerusalem after accepting terms from King Hezekiah is another typical example of the way Assyrian kings often avoided long passive sieges.

Obviously Assyrian siege methods required a sophisticated logistical system and an elaborately organized army with a variety of highly specialized troops. Siege warfare alone deployed assault troops trained in escalade, archers, slingers, sappers capable of tunneling as well as cutting through walls, battering ram crews, and laborers to build ramps and other siege works. In addition to this, the Assyrians excelled in chariot warfare and developed the first cavalry troops. Moreover, they were able to deploy this complicated army in all kinds of terrain, from mountain to desert. The solution of the organizational problems inherent in sustaining such a large and diverse army was the key to Assyrian military power. Unfortunately, we know little about Assyrian logistics.[163] When Assyrian armies besieged a large metropolis, they captured the smaller towns in the vicinity, which often contained granaries, and, as Sargon put it, "let [the] army devour its abundant grain, in measureless quantities."[164]

At the beginning of the Neo-Assyrian Empire, the infantry con-

sisted of native Assyrians. The elite charioteers came from the aristocracy. Only the cavalry was drawn primarily from allies.[165] As the empire
expanded, this army raised primarily by a feudal levy proved inadequate and Sargon II created a standing army that was more royal and
less Assyrian.[166] A standing army made possible the long campaigns
demanded by an expanding empire.[167] It also allowed the greater spe-
cialization so important to such a technically demanding task as siege
warfare.

The standing army consisted of various branches based on arms-
charioteers, cavalry, archers (by far the largest contingent), shield-
bearers, lancers, and slingers.[168] Specialists, such as the engineers who
constructed siege works, were almost certainly auxiliaries permanently
attached to the central standing army.[169]

This standing army could strike very quickly. Sargon reports that
he was able to march against Ashdod quite suddenly, taking only his
cavalry, and besiege and conquer not only Ashdod but two other towns
as well.[170] However, it seems impossible that Sargon could have conducted real sieges with only his cavalry. Either he does not mention that
he was able to bring up his siege train or he was able to take the cities
by some means other than siege.

The relief sculptures do not have enough evidence to allow us to conclude that there was a special corps of engineers. But there very well
may have been.[171] Certainly the Assyrian army was capable of impressive engineering feats, such as constructing pontoon bridges and clearing roadways through forested mountainous terrain. Ashurnasirpal
said that he marched against Arzizu and Arsindu with his cavalry and
his pioneers, but it is not clear how the pioneers were organized.[172]

More clearly, there was a labor service that was indispensable to
such a labor-intensive operation as a siege. This labor service performed
a variety of tasks. It provided common laborers, drivers, craftsmen such
as armorers, and servants for officers. One group also served as couriers.

The sappers were drawn from this labor service. These men came
from the lower classes. The Assyrian army could tap this labor pool
through a sort of feudal levy that was the primary means of mobilizing labor in Assyria both for military duties and civilian construction
projects.[173] At least by the time of Ashurbanipal, and almost certainly
earlier, the practice of incorporating conquered manpower into the
Assyrian army helped fill the ranks of laborers who could serve as sappers.[174] These sappers also probably manned the battering rams.[175] Sapping and deploying battering rams required great skill, so we may assume that these men were highly trained.[176]

As the Assyrian Empire expanded, Assyrian kings filled the ranks
of their army more and more with troops drawn from the conquered
peoples of their empire. Although these troops may have been paid, they
were impressed soldiers rather than hired mercenaries. Promise of

booty was their main incentive to fight.[177] Sacking cities was the main source of booty. For this reason Assyrian military organization not only made siege warfare more effective; it also required siege to sustain itself.

Indeed, much of the organization of the Assyrian Empire came to depend on a military system that could provide the resources necessary to sustain the empire. Military campaigns not only served to protect and expand the empire, but they also augmented its material and human resources. Money to fill the royal treasury and support the temples came from the tribute the Assyrians imposed on the cities of its expanding empire. Under this system, military campaigns became a sort of tax collection backed by the threat of force.[178] Manpower for the army, horses for the cavalry, and scarce goods such as precious metals also flowed into Assyria through this system.[179] Because almost all Middle Eastern cities were fortified, skill in siege warfare was necessary to give the threat of force credibility.

The general prosperity of the Assyrian people also depended on this system. Soldiers carried the booty they had seized in conquered towns back to Assyria where it enriched the general economy of Assyria. There is some evidence that the Assyrian king himself took care to distribute the booty in a way that spread its benefits as widely as possible.[180] Ashurbanipal boasted that he brought so many camels back from one campaign that the price of a camel in Assyria dropped to less than a shekel, so that such people as tavernkeepers, brewers, and gardeners could easily afford camels.[181] Given these wide-ranging economic ramifications, it is no wonder that the Assyrian kings devoted themselves to the perfection of siege warfare.

The Persians

If the Assyrians built their power on siege warfare, they also lost it by siege. The Medes under Cyaxares dealt the Assyrians a fatal blow in 612 B.C. by conquering Nineveh in a three-month siege.

Herodotus promised to tell the story of Nineveh's fall, but unfortunately he never fulfilled the promise.[182] A Babylonian chronicle of this campaign does not say how the Medes took such a strongly fortified city as Nineveh in the relatively short time of three months. They may have been lucky. It was a Greek tradition that the tributary of the Tigris River on which Nineveh was located flooded and washed away some of the city's fortifications, but we find that story in Diodorus, whose account betrays an almost complete ignorance of the facts. However, the Hebrew prophet Nahum also alluded to a flood, so perhaps the story has some credibility.[183]

The fall of Nineveh cleared the way for the Persians to conquer an empire as large as the Assyrian Empire. After Cyrus brought the Persian

Achaemenid dynasty to power over the Medes and Persians, he pursued an expansionist policy that relied heavily on the ability to reduce cities.

The battle that brought Lydia into the Persian Empire was the siege of the Lydian capital of Sardis in 546 B.C. Croesus, having lost a battle to Cyrus, retreated into the city, where he was confident he could hold out long enough to summon help from his allies. But, according to Herodotus, the city fell in just two weeks. Cyrus promised a reward to the first man to scale the wall, and one of his soldiers discovered a way up a precipitous slope to a place the Lydians had left unguarded because they assumed it was inaccessible. Cyrus led the Persians up to this unguarded spot, and they captured the city.[184] Herodotus did not say how the Persians climbed over the wall. They may have used scaling ladders, if they were able to haul them up the cliff to the wall.[185]

The most eloquent testimony to Persian siege capabilities was the capture of Babylon, a feat they achieved twice. In 538 B.C., Cyrus drove the Babylonians into their city and besieged them. However, the Babylonians had anticipated a siege and were well provisioned. Herodotus tells us the siege "dragged on," but he does not say how long. Long enough, however, to discourage Cyrus.

But Cyrus then hit upon the idea of diverting the Euphrates, which flowed through the city of Babylon, making the river shallow enough to enable his troops to wade along it into the city. Walls ran along the river banks within the city, and the Babylonians could have closed the gates and cut the Persians to pieces if they had realized the enemy was within the city. But inexplicably the Persians took them unawares. Later the Babylonians gave Herodotus the excuse that they were celebrating a festival at the time and that the city was so huge the festivalgoers in the center of the city did not know the Persians had penetrated the outer defenses.[186] The explanation is unconvincing, but, in any case, the Persians gained entrance to the city by stealth.

The second Persian siege of Babylon took place during the reign of Darius when the Babylonians revolted against his rule. This siege went on for nineteen months without the Persians making any progress. Darius tried everything, including Cyrus's trick, but the Babylonians had learned their lesson and constantly maintained their vigilance, giving Darius no chances. It finally took one of the most elaborate ruses in the history of siege warfare to bring about the city's fall.

A Persian named Zopyrus, who was one of Darius's advisers, horribly mutilated himself, cutting off his ears and nose, a typical Persian punishment, and fled to Babylon posing as a much aggrieved Persian defector. The Babylonians thought such a man would be useful to them and placed him in a minor command. By a prearranged plan, Darius sent a thousand expendable troops into a trap set by Zopyrus, who slaughtered them all. Darius then sent two thousand men to the same fate. Finally he sacrificed four thousand men to Zopyrus, and by this time the Babylonians had such confidence in Zopyrus that they made him general in chief and guardian of the wall. When Darius launched a general assault against the walls, Zopyrus opened two of the gates and the city fell.[187]

In both the siege of Nineveh and the siege of Sardis, good fortune brought success, the flood in the case of Nineveh and a soldier observing a Lydian soldier climbing down the supposedly inaccessible precipice to fetch a helmet in the case of Sardis. Cyrus's stealth and the fanatical devotion of Zopyrus brought the two sieges of Babylon to successful conclusions. But we should not infer from this that the Persians were behindhand in siege warfare. At least by the time of Cyrus, they had mastered Assyrian siege methods.

Xenophon said Cyrus set up siege engines and prepared scaling ladders at Sardis.[188] Xenophon also had Cyrus lingering in Sardis after its capture attending to the construction of siege engines and rams as the first step necessary for the subjection of the rest of the Lydian Empire.[189] Herodotus did not mention the use of siege engines by the Persians until the siege of Miletus fifty-two years later. He did report that Cyrus's general Harpagus led a campaign against the Greeks immediately after the capture of Sardis in which he forced an entry into the cities by constructing earthen ramps against the walls.[190] Herodotus, who was no military man, was vague about these ramps. He did not mention battering rams, so he leaves us with the impression that Harpagus used the ramps to traverse the walls. Although Xenophon's *Cyropaedia* is more a work of imagination than of history, this is a rare instance where Xenophon is more convincing than Herodotus and we should picture the Persians under Cyrus using battering rams and other siege equipment.

The Persians also undermined walls, as they attempted to do at their siege of the Greek colony of Barca in Libya. Herodotus tells us that an ingenious metalworker of Barca discovered a way of locating the Persian mines. By walking around the base of the wall banging a bronze shield on the ground, he was able to detect the location of the mines from the different sound of the shield's reverberations. Aineias Tacticus reported that this method of locating tunnels was still in use in Greece in the fourth century, especially at night.[191] The Greeks then dug countermines and killed the Persian sappers. The Persians also attempted to assault the walls of Barca, but the Greeks were able to repulse them. In the end the Persians were able to take the city only by pretending to come to terms and then treacherously seizing control after the Greeks opened the gates.[192]

The Greeks received another taste of Persian siege methods around 497 B.C. during a Persian campaign in Cyprus. Archeological evidence at Paphos reveals that the Persians built a siege ramp across a ditch that protected the city. Apparently the Paphians dug countermines under the ramp and tried to weaken it with fire. The excavations uncovered large numbers of missile weapons, mute testimony to the fierce fire under which the Persian sappers labored.[193] We do not know the outcome of this contest over the ramp, but in the end Paphos fell.[194] The nearby town of Soli proved a tougher nut to crack. After over four months of siege, the Persians finally captured it when they managed to undermine the wall.[195] The Persians also undermined the walls at Miletus. Although we must assume that the Persians used battering rams more often than Herodotus indicated, undermining seems to have been their most reliable means of siege.

Undermining and ramp construction indicate that the Persians were proficient in mobilizing labor. Herodotus provided the only account of their methods when he described the digging of a canal across the isthmus at Mount Athos by Xerxes' army. This was a tremendous project, the isthmus being about a mile and a half wide, and it apparently took about three years to complete. The Persians based the army in the Chersonese, where it was easier to supply, and rotated the troops back and forth to the canal. The local population supplied additional laborers. Persian quartermasters shipped large amounts of grain to feed the workers, but it was also necessary to establish a market nearby to supplement the grain supplies. According to Herodotus, the Persians kept the soldiers at their task with whips. As they dug out the canal, lines of workers passed the dirt hand to hand up from the bottom. Herodotus reported that the sides of the canal frequently caved in, doubling the amount of work necessary. Only the Phoenicians were skilled enough to terrace the sides to prevent cave-ins. This suggests that the Persians overcame obstacles more through sheer manpower and discipline than skill in technique.[196]

Such a gigantic project was similar to a siege. The Persian ability to organize large numbers of laborers and provide logistical support for them was the key to their success in siege warfare. As always in ancient siege warfare, the spade in the hands of a well-fed sapper was the most effective siege machine.

Chapter IV

TREATMENT OF
CAPTURED CITIES

The Hebrews

The Book of Joshua presented a picture of siege warfare that was unspeakably barbaric. A band of wild seminomadic Hebrews crossed the Jordan, performed a bloody ritual of mass circumcision,[1] and then, having conquered Jericho, killed every living thing in it. That apparently satiated some of the Hebrews' blood lust, because when they conquered the city of Ai, they at least spared the animals, taking them as booty. However, they hanged the king of Ai, leaving his body on public display until sundown. The people of Ai did not have the opportunity to gaze upon this cruel sight because the Hebrews killed them all.[2]

The Bible made clear that in waging this war that can only be called genocidal, Joshua was following the command of Yahweh, the god of the Hebrews. The Book of Deuteronomy, for example, stated:

> But in the cities of these people that the LORD your God gives you for an inheritance, you shall save alive nothing that breathes, but you shall utterly destroy them, the Hittites and the Amorites, the Canaanites and the Perizzites, the Hivites and the Jebusites, as the LORD your God has commanded; that they may not teach you to do according to all their abominable practices which they have done in the service of the gods, and so to sin against the LORD your god.[3]

This genocidal command is deeply disturbing to modern sensibilities. Israeli military historians such as Yigael Yadin and Mordechai Gichon, who accept the events in the Bible as true, carefully ignore the genocide. The English historian Richard Humble emphasizes it and concludes the Hebrews were more ruthless even than the Assyrians.[4] Not surprisingly, these grim moments have caused a great deal of heartache among Biblical scholars.[5]

What conclusions are we to draw about the Hebrew conduct of siege warfare from these massacre stories? In the first place the Deuteronomic laws of siege drew a striking distinction between "cities which are very far from you" and "the cities of these people that the LORD your God gives you for an inheritance," i.e., the cities blocking the Hebrew conquest of the promised land. The latter were under a ban of utter destruction, but for the former Yahweh prescribed the following:

> When you draw near to a city to fight against it, offer terms of peace to it. And if its answer to you is peace and it opens to you, then all the people who are found in it shall do forced labor for you and shall serve you. But if it makes no peace with you, but makes war against you, then you shall besiege it; and when the LORD your God gives it into your hand you shall put all its males to the sword, but the women and the little ones, the cattle, and everything else in the city, all its spoil, you shall take as booty for yourselves; and you shall enjoy the spoil of your enemies, which the LORD your God has given you. Thus you shall do to all the cities which are very far from you, which are not cities of the nations here.[6]

What are we to make of the stark contrast between the fate of the cities in the promised land and the fate of cities far away? The rules for the faraway cities, such as the offer of peace terms before commencing a siege, may belong to a later period, during the reign of Josiah in the seventh century B.C. Its mention of long sieges and siege works presupposes a level of siege warfare far beyond what Joshua's army could mount, and Gerhard von Rad believes that these rules "completely contradict the spirit of ancient [i.e., premonarchical] war."[7] But the practical necessities of war lend some credibility to the sensible advice to avoid siege warfare whenever possible. It is the ban that arouses suspicion.

As we have already learned, the historical sources for the battles of Jericho and Ai are simply too problematical to support such remarkable stories of massacre. They fit into a pattern of hyperbole characteristic of ancient Middle Eastern conquest literature.[8] The same must hold true for the impressive list of cities whose populations are reported by the Book of Joshua to have been massacred by Joshua.[9] Not even the text fully supports such claims. For example, in Joshua 15: 15–16 we find Caleb warring against Debir, whose population Joshua had supposedly annihilated shortly before.

Although the treatment of captured cities in ancient times was harsh enough, as Yahweh's laws governing the capture of cities "very far from you" show, annihilation of the entire population was not common, and we must conclude that such stories in the Book of Joshua are not true. It is not in the context of history that we can understand these stories; instead we should turn to the realms of theology, epic, and myth. The rules for the cities "very far from you"—forced labor for those who surrendered immediately, the killing of the men and enslavement of the women and children of those who resisted—were common standards of conduct for siege warfare throughout the ancient world and probably reflected Hebrew practice from earliest times.

The Book of Deuteronomy not only prescribed rules for the treatment of the population of captured cities but also dealt with the disposal of its material resources. For example, it prohibited the cutting down of fruit trees:

> When you besiege a city for a long time, making war against it in order to take it, you shall not destroy its trees by wielding an axe against them; for you may eat of them, but you shall not cut them down. Are the trees in the field men that they should be besieged by you? Only the trees which you know are not trees for food you may destroy and cut down that you may build siege-works against the city that makes war with you, until it falls.[10]

We cannot rule out the possibility that this law may be rooted in a fear of demons or spirits that inhabited fruit trees, but there were practical considerations behind this humane law.[11] Cutting down fruit trees was a common practice in ancient siege warfare, partly because the wood was necessary to construct siege works and partly because it added the threat of severe economic damage to the incentive for the city to surrender. But gratuitous damage to a city's economy lessened its value when it fell, and felling fruit trees, especially olive trees, was laborious work of doubtful long-range efficacy.[12] Nonetheless, Deuteronomy's prohibition of the practice was a remarkable effort to control wanton destruction, even under the pressures of a long siege.

There were also customs regarding the handling of booty. The consecration of a part of the booty to Yahweh prevented a wild looting rampage. Again there were practical considerations at work, because no commander welcomed undiscipline among his troops and loss of control at the moment of victory could prevent the proper exploitation of success or even jeopardize the victory itself. The Hebrews took these customs seriously. When Achan violated them during the sack of Jericho, they stoned to death him and his family.[13]

The congruence between the archeological and the textual evidence for the reign of Solomon provides assurance that the Biblical account of the monarchy is more historical than the account of the premonarchical period. This means we can see more clearly what treatment the Hebrews meted out to conquered peoples.

David was not so harsh as Deuteronomy prescribed. The capture of

Jerusalem can serve as an example. The Bible is silent about the immediate fate of the Jebusites after the fall of Jerusalem. The Jewish historian Josephus, writing in the first century, said that David expelled the Canaanites.[14] Josephus may only have assumed this as common practice. Later we find David negotiating with the Jebusite Araunah for a threshing floor owned by Araunah, which David wanted for the site of an altar to the Lord.[15] This indicates at the very least that the Hebrews had not killed or transported all the Jebusites nor confiscated all their property. Araunah bowed down to David, but when the haggling was over David had paid him fifty shekels of silver for the property. Despite the mild treatment the Jebusites received from David, their position was precarious after the loss of Jerusalem, and the Bible lists them among those people whom Solomon put to forced labor.[16]

The people of Rabbah felt the pain of defeat more quickly than the Jebusites when David conquered their city. Again David killed no one, but he plundered the city and reduced its people to forced labor.[17] As we have seen, the siege of Abel-beth-maacah ended with a negotiated settlement that left its people completely free, but we must remember that Abel-beth-maacah was an Israelite city in rebellion. Joab's shrewd policy of an uncompromising demand for the head of Sheba in return for lenient treatment for the besieged city aimed at reestablishing Abel-beth-maacah's loyalty to David.

David expressed regrets about his life of bloody warfare and looked for a more peaceful reign for his son Solomon:

> My son, I had it in my heart to build a house to the name of the LORD my God. But the word of the LORD came to me, saying, "You have shed much blood and have waged great wars; you shall not build a house to my name, because you have shed so much blood before me upon the earth. Behold, a son shall be born to you; he shall be a man of peace. I will give him peace from all his enemies round about; for his name shall be Solomon, and I will give peace and quiet to Israel in his days. He shall build a house for my name. He shall be my son, and I will be his father, and I will establish his royal throne in Israel for ever."[18]

These regrets of David may contain a hint about why he did not massacre the Jebusites and others whose cities he captured. Their labor proved invaluable to Solomon's building program, and their technical expertise may have been indispensable. David dreamed of building a temple: could he already have been thinking of how useful the Canaanites would be for this great task? Although the evidence is limited, a reasonable tentative conclusion is that the standard Hebrew treatment of defeated cities was to reduce the population to forced labor.

The siege warfare described in the Bible could be much harsher, as the ghastly terms the Ammonites offered the Israelite city of Jabesh-gilead showed: that the Ammonites gouge out the right eye of all the Israelites, a condition designed deliberately to humiliate.[19] Fortunately, Saul arrived at Jabesh-gilead in time to break the siege and rescue the city.[20] Even worse were the actions of Doeg, an Edomite king in service to Saul. He carried out Saul's order to massacre the priests of Nob after Saul's bodyguard had refused. He then killed all the men, women, children, and animals in the city.[21] But the Bible clearly marked Saul's order as wrong, if not insane. In any case, Saul's application of the Deuteronomic ban to the priests of Yahweh was perhaps too ironic an inversion to be credible.

The killing of noncombatants was one of the most disturbing aspects of siege warfare. Abimelech's siege of Shechem, in which he burned to death the people after they had sought refuge in their keep, may serve as an example of the difficulty of maintaining distinctions between combatants and noncombatants, a difficulty that was especially acute in siege warfare.[22]

As we have seen, the Book of Deuteronomy drew a definite distinction between men and women in its rules concerning siege warfare,[23] but Abimelech's actions showed how weak such distinctions proved to be under the pressure of siege warfare. In the first place, Shechem fell under the ban Yahweh had placed on the cities of Canaan and therefore the distinction between men and women did not apply. But, as we have seen, the historical authenticity of the ban is doubtful and the moral stink surrounding Abimelech makes it clear that our author (or authors) did not consider him a man fulfilling the will of Yahweh. More to the point, the battle of Shechem was a civil war between political factions, always a ruthless and deadly business.

If we could reproach Abimelech for burning a thousand men and women to death, he would undoubtedly plead military necessity. He had no siege equipment with which to reduce the tower. There were still hostile forces in the area, making it dangerous and perhaps logistically impossible to starve the defenders into surrender. Abimelech needed to end the siege as quickly as possible, and firing the tower was the most effective means to that end. That a woman killed Abimelech showed that the women he burned to death had been dropping stones on the besiegers. When the Book of Judges pointed to the justice of Abimelech's death, it was because he had murdered his seventy brothers, not because he had burned to death the women and children of Shechem.[24]

Nevertheless, the slaughter of women and children challenged the Hebrew view of order. The Hebrews viewed Yahweh as the Creator who had triumphed over chaos and created shalom. *Shalom* is usually translated peace, but it meant much more than simply the absence of war. Paul Hanson defines it as "the cosmic harmony that exists where the world and all its inhabitants are reconciled with God . . . the qualities of the community living in harmony with God in covenant are variously described as prosperity, peace and righteousness, which taken together begin to describe *shalom*."[25] The sack of a city represented the exact antithesis to this communal harmony with Yahweh's cosmic order. When Yahweh warned the Hebrews of the consequences of disobeying his commands, siege warfare provided one of his most terrible threats:

> They shall besiege you in all your towns, until your high and fortified walls, in which you trusted, come down throughout all your land; and they shall besiege you in all your towns throughout all your land, which the LORD your God has given you. And you shall eat the offspring of your own body, the flesh of your sons and daughters, whom the LORD your God has given you, in the siege and in the distress with which your enemies shall distress you. The man who is the most tender and delicately bred among you will grudge food to his brother, to the wife of his bosom, and to the last of the children who remain to him; so that he will not give to any of them any of the flesh of his children whom he is eating, because he has nothing left him, in the siege and in the distress with which your enemy shall distress you in all your towns. The most tender and delicately bred woman among you, who would not venture to set the sole of her foot upon the ground because she is so delicate and tender, will grudge to the husband of her bosom, to her son and to her daughter, her afterbirth that comes out from between her feet and her children whom she bears, because she will eat them secretly, for want of all things, in the siege and in the distress with which your enemy shall distress you in your towns.[26]

Such a fundamental destruction of the community, in which walls that have shielded the people from violence and chaos have come tumbling down, expressed primal fears of utter chaos. A world in which starvation turned parents into predators who eat their little children was a world completely without security. The rules limiting the violence of siege warfare were an effort to preserve Yahweh's position as the conqueror of chaos, even under conditions that could hardly be further removed from shalom.

Conventions of war serve two important functions. First, they authorize the violence of war. By requiring that the Hebrews offer peace terms to a city, Yahweh legitimized the slaughter of the men if a city's refusal of terms forced a siege. Second, Yahweh's rules tried to limit the violence. By prohibiting the killing of women and children in certain cases, he reassured the Hebrews that they were not operating in a terrifyingly amoral world in which there were no distinctions and no limits. The point is not that these rules succeeded in controlling violence; conventions of war broke down all too frequently. Rather these conventions formed an important psychological precondition for the violence of war. The illusion that there were limits helped make the prospect of violence bearable.

The Assyrians

Assyrian sources provide a vivid picture of their conduct of siege warfare. It is a shockingly cruel picture. Listen to the words of Ashur-nasirpal:

> To the city of Suru of Bit-Halupe I drew near, and the terror of the splendor of Assur, my lord, overwhelmed them. The chief men and the leaders of the city, to save their lives, came forth into my presence and embraced my feet, saying: "If it is thy pleasure, slay! If it is thy pleasure, let live! that which thy heart desireth, do!" Ahiababa, the son of a nobody, whom they had brought from Bit-Adini, I took captive. In the valor of my heart and with the fury of my weapons I stormed the city. All the rebels they seized and delivered them up. . . . Azi-ilu I set over them as my own governor. I built a pillar over against his city gate, and I flayed all the chief men who had revolted, and I covered the pillar with their skins; some I walled up within the pillar, some I impaled upon the pillar on stakes, and others I bound to stakes round the pillar; many within the border of my own land I flayed, and I spread their skins upon the walls; and I cut off the limbs of the officers, of the royal officers who had rebelled. Ahiababa I took to Nineveh, I flayed him, I spread his skin upon the wall of Nineveh. . . . [27]

The nauseating cruelty of this document hardly needs emphasis, and it would be difficult to deny Arther Ferrill's judgment that the fate of Suru is "nearly as revolting as photographs of Nazi concentration camps."[28] But the Assyrians have found their defenders. Writing in the shadow of World War I, the great Assyriologist A. T. Olmstead concluded about the Assyrians:

> All in all, they seem pretty decent folk, not so very different from the men of our block in spite of different clothes, different speech, and a religion which never reached the Christian ideal. There were no saints among them, their religion was not of that type, but few of them appear such terrible sinners, if we judge them by twentieth-century practice and not by first-century preaching.[29]

One is tempted to remark that twentieth-century practice is hardly a high standard of judgment, but in our context the question we must ask is: do these atrocities tell us something about the nature of siege warfare or only something about the nature of the Assyrians?

The evidence is too fragmentary to allow us to draw a comprehensive picture of Assyrian policy toward conquered cities. Another difficulty is that the royal inscriptions, which provide most of the evidence, do not always make clear exactly what happened. Some phrases recur so often that we suspect they are more formulaic constructions than accurate descriptions.

Despite the unsatisfactory nature of the evidence, it seems best to analyze carefully what we have rather than draw impressionistic conclusions from a general reading of the Assyrian documents. We need to keep constantly in mind, however, that we are studying the way Assyrian practice is reflected in documents and reliefs that were not designed to be historical or factual in the modern sense and that are exceedingly fragmentary.

Intimidation was the main function of these documents and reliefs, most of which were located in palace halls where the Assyrian emperors received diplomatic representatives. Such intimidation was an important part of the psychological warfare in sieges. H. W. F. Saggs argues that the atrocities were a form of psychological warfare whose purpose was to avoid time-consuming and costly sieges.[30] A report from Shalmaneser III that when he marched against King Giammu, his people became terrified and killed him and opened their town to Shalmaneser showed that such psychological warfare sometimes succeeded.[31] Intimidation would not work, however, if those who viewed these terrible records did not know that such things could and did happen.

We do not have extensive records of Assyrian military policy until the time of Ashurnasirpal II (883–859 B.C.), and so we must begin with his reign.[32] Virtually every mention of captured cities in Ashurnasirpal's records related that he carried away heavy spoil. This was a universal practice in siege warfare, and the Assyrians were no exceptions. The plunder of captured cities was routine for all Assyrian kings.

Much more striking was the widespread destruction of captured cities, turning them into "mounds and ruins," in a favorite phrase of Ashurnasirpal's scribes. The total destruction of cities was recorded no less than twenty-one times in Ashurnasirpal's campaign records, which covered only six campaigns. Usually these reports spoke in the plural, making an exact count impossible, but it would certainly be in the hundreds. Many of these "cities" will have been quite small. Ashurnasirpal routinely reported that he destroyed all the towns within the vicinity of a large city.

The capture of a city was almost always bloody. Nine times Ashurnasirpal said that he slew the inhabitants of the town and six other times he said that he slew many of them. Probably there was no distinction between the two formulas. Only once did Ashurnasirpal explicitly state that he left no survivors.[33] In eleven cases, Ashurnasirpal reported the number of defenders whom Assyrian soldiers killed in the assault on the city. The high number is 3,000, the low is 50, the average is 1,079, and the median is 800. These numbers were probably arrived at by counting the number of heads severed from enemy bodies. A relief from the palace of Sennacherib shows a pile of severed heads at the feet of two scribes who appear to be recording the number of heads.[34]

Some defenders and inhabitants of a besieged city usually survived its fall, but their fate was not a happy one. They might be burned alive, impaled on stakes, flayed, mutilated, or buried alive. The number of reports by Ashurnasirpal of each of these was as follows: burned alive, seven times (a death especially reserved for the young; six of these times Ashurnasirpal said he burned the young men and maidens and once that he burned the young men); impaling, five times (usually captured soldiers); mutilation, five times (cutting off hands and fingers, twice; cutting off limbs, once; cutting off noses and ears, once); burying alive, once; flaying, twice (in both cases rebel leaders).

If the survivors did not fall victim to any of these atrocities, they faced transportation to some faraway place (twice mentioned in Ashurnasirpal's reports), forced labor (twice mentioned), or, the luckiest, imposition of an overseer with increased taxes and tributes for the city (twice mentioned).

The descriptions of the sieges in these records are so terse it is difficult to perceive a definite pattern in the treatment of conquered cities. In both cases when Ashurnasirpal imposed forced labor, the people had surrendered. In one case, a city surrendered as he approached, obviating the need for a siege. In the other case, men who had fled the city came down from the mountains to surrender. However, the only time that Ashurnasirpal made a point of having been merciful and sparing the lives of the inhabitants of a city, they had not surrendered until siege operations had begun.[35] It appears, though, that the surrender had come quickly. From this we can tentatively conclude that there was some connection between the level of resistance and the treatment handed out afterward. But what is most striking is the general brutality and cruelty of Ashurnasirpal's siege methods.

A somewhat less ferocious picture emerges from the records of Ashurnasirpal's successor, Shalmaneser III (858–824 B.C.) Shalmaneser made no mention of mutilation or flaying. There is one report of burning young men and maidens. It comes from the campaign in Shalmaneser's year of accession, and one suspects that the same scribe who recorded such bonfires for Ashurnasirpal was still at work here.

The reports of thirty-three cities Shalmaneser captured do not record their destruction. This does not mean, however, that he did not destroy them. Parallel campaign records show that in one place the scribe contented himself with recording the capture while in another account of the same campaign we are told that Shalmaneser destroyed the city. The records are too fragmentary to make certain that the absence of any mention of the destruction of a city anywhere in the records means the city definitely survived.

In any case, Shalmaneser records the destruction of cities twenty-six times. Many of these cases concerned large numbers of "cities." Shalmaneser, who was perhaps more interested in gathering tribute than destruction, regularly left central cities standing while destroying all the neighboring towns.

Shalmaneser's assaults were bloody. Eleven times he said that he slew the enemy warriors. Five times he more generally said that he slew the inhabitants of a city. Only once did he speak of carrying away prisoners.

Shalmaneser liked to build pyramids of heads severed from the bodies of slain enemies. This practice appears four times in the record. In perhaps his most gruesome atrocity, Shalmaneser once bound some captives and buried them in the mounds of severed heads. The records reported two times that he impaled his prisoners.

A relief depicting Shalmaneser's siege of Kulisi, a city near the source of the Tigris River, shows severed heads hanging from the walls of the burning city. A pole on which a prisoner is impaled rises next to the city. The impaled man's hands and feet are missing. The artist helpfully shows us how this came about by showing an Assyrian cutting off the hand of another prisoner whose other hand and both feet have already been severed. The same relief shows a different siege in which the Assyrians have built three stacks of heads by skewering them on stakes driven into the ground.[36]

Shalmaneser did not always press his sieges to the end. He reported twice having settled for gifts from the besieged city. Three times he said he merely took hostages and imposed tribute on cities that either had not resisted at all or offered only token resistance. Twice Shalmaneser was entirely frustrated by a city's fortifications (once by the Babylonian city of Gannanate and once by Damascus) and had to settle for cutting down orchards and harvesting grain.

Assyrian policy toward captured cities took a significant turn under Tiglath-pileser II (745–727 B.C.). Although he continued to destroy cities as his predecessors had done (his annals explicitly state that he destroyed cities fifteen times, frequently in large batches), Tiglath-pileser began carrying off conquered peoples and resettling them elsewhere in his empire. Although there had been scattered cases of mass deportations all during history of the Assyrian Empire, it was not until the time of Tiglath-pileser that this practice became a consistent policy.[37] This policy undoubtedly reduced the number of massacres following a siege. This does not mean that Tiglath-pileser was gentle. He continued to impale and mutilate prisoners, although such actions were mentioned rarely in his annals (impaling twice and mutilation once).

It is more difficult to determine the fate of the deportees. They were not destined for the slave markets; slavery was not a large source of labor in Mesopotamia. Resettlement was the policy.[38] The Assyrians believed that their resettlement policy would be more successful if they kept communities together to preserve their social cohesion rather than scattering peoples throughout the empire.[39] This policy will have been no small consolation to the uprooted.

Although the Assyrians attempted to keep small communities and families intact, they did not move the entire population of large cities to one place. Usually new deportees found themselves thrown in with deportees from other locations. Assyrian cities became quite cosmopolitan as a result of this policy. Even if families and small communities stayed together, then, there was cultural dislocation.

Saggs argues that the Assyrians tried to relocate people in similar environments so they could continue to pursue agriculture with their traditional methods. He cites the promise to the Jews trapped in Jerusalem by Sennacherib's army that if they surrendered they would be taken to "a land like your own land."[40] Such a policy would have made sense from an economic point of view, but Bustenay Oded's detailed study of Assyrian deportation policy does not entirely bear out Saggs's sanguine view. The Assyrians were proud of their ability to move whole peoples from one corner of the earth to another, and so some deportees found themselves in a very different environment. A mountain people might end up on the Mediterranean coast, for example. Most of the deportees were taken to Assyria proper, whether or not the environment was similar to their home.[41]

A long journey in the ancient world was difficult under the best of circumstances. The victims of deportation will have lost everything in the sack of the city, and some may have been malnourished after a long, exhausting siege. The violence of the siege, the slaughter of many inhabitants, the sight of impaled or mutilated prisoners will have traumatized many. Many of the women will have been recently raped. It is difficult to imagine worse conditions under which to embark on a long and hazardous journey into the unknown.

Counterbalancing these appalling hardships was the Assyrian king's desire to see his captives reach their destination as intact as possible. The Assyrians made a careful list of all captives, and the king demanded periodic reports on their condition during the march. The Assyrian records make clear that they did not intend the deportations to be a death march for the captives; rather they were to serve the economic and political interests of the king.[42] A special official had the responsibility of getting the deportees safely to their destination. The governors of the provinces through which the deportees traveled were responsible for providing food and supplies to them.[43]

The Assyrian kings considered the deportees their personal property, but more in the sense of villeinage than slavery. Most were settled on agricultural land to till the soil. Their legal and social status is obscure, but it apparently ranged from freeholding to serfdom.[44] As the Assyrian Empire expanded and the need for mass labor grew, many of the deportees ended up in labor gangs that the king could utilize in increasingly ambitious building projects.[45]

Sennacherib's palace at Nineveh contains reliefs showing laborers dragging heavy stones with ropes. They are dressed in various styles, indicating they are prisoners from different parts of the Assyrian empire. Assyrian overseers keep the laborers at their task with truncheons.[46]

Craftsmen often continued to ply their trade. Deportees of high social standing sometimes received training in Assyria and became part of the imperial bureaucracy. As we have seen, the Assyrians frequently incorporated soldiers into the Assyrian army.

Despite efforts to provide for the deportees, the Assyrian practice of systematically transporting entire populations on such a large scale was a drastic and cruel policy. Although the purposes of this practice must be placed in the broader context of Assyrian imperial policy, it was siege warfare that brought entire populations into the hands of the Assyrians, revealing that the stakes in siege warfare were often the very survival of an entire society. In ancient Mesopotamia, siege war meant total war.

The massive resettlement policy continued under Sargon II (721–705 B.C.) His inscriptions recorded eighteen cases of the resettlement of a captured city's population. In six of these cases, Sargon repopulated the city with captured people from somewhere else. Sargon pursued an expansionist policy and in six cases placed captured cities under an Assyrian governor. In all, his inscriptions mentioned the capture of fifteen cities without recording their destruction and explicitly reported the destruction of cities twenty-one times. In three cases, Sargon said he rebuilt the city. Twice Sargon recorded the flaying of rebel leaders, but there is no mention of mutilations. Once he killed a large number of rebel soldiers. Sargon boasted of extensive agricultural destruction four times.

Sennacherib's records followed a pattern similar to Sargon's. Sennacherib reported the destruction of cities thirteen times. Twelve times he recorded the capture of cities without saying that he destroyed them. Although Sennacherib reported one time that he killed the entire population of a city (Hirrimme), deportation remained the most common policy. Sennacherib listed ten deportations; in three of these cases, the towns were resettled with new population. Twice Sennacherib said that he hung the corpses of rebels on stakes. One of his reliefs at Nineveh depicted Assyrian soldiers hoisting naked impaled captives from Lachish on stakes.[47] Once Sennacherib told of flaying a rebel leader. There are two cases of agricultural damage in Sennacherib's records.

Two times Sennacherib gave reasons for his policies. In his third campaign, he said that he besieged, conquered, and despoiled those cities that "had not speedily bowed in submission at my feet."[48] This shows that no matter how ruthless Assyrian kings were, there was a policy guiding their path of destruction.

In the other case Sennacherib distinguished the punishment meted out after the conquest of a city that had rebelled against its pro-Assyrian king. He killed the leaders of the rebellion and hung their bodies on stakes around the city. The people who had collaborated with the leaders he deported. Those citizens who had not supported the rebellion were pardoned and allowed to continue living in their city under the rule of their previous king, whom Sennacherib restored.[49] We would like to know how Sennacherib distinguished the guilty from the innocent, but he did not say. Nevertheless, the passage is unusual in the annals of the Assyrian kings in the detailed accounting of the king's policy. It is reasonable to suspect that this provides a better insight into Assyrian policy than the constant formulaic reports of mass destruction.

Other evidence supports this conclusion. According to Olmstead, the Assyrians offered light tributes to cities before battle, but if a city's stubbornness forced a siege they exacted a huge indemnity in addition to the other horrors visited upon their victims.[50] Yadin believes that one of the palace reliefs shows an Assyrian official reading surrender terms from the turret of a battering ram.[51]

In any case, we have a lengthy account in 2 Kings of the negotiations between the Assyrians and the Judeans before Sennacherib's siege of Jerusalem.[52] Sennacherib's negotiator offered to spare the entire population and leave the people undisturbed in their possessions until the time came to transport them to some far-off but fertile land. But if the defenders rejected these terms, then the full power of Assyria would be applied and they would suffer the horrors of a siege with no hope for mercy. Significantly the Assyrian negotiator delivered this message in Hebrew so the common soldiers defending the walls could understand.

Total destruction, however, was not a figment of the scribes' imaginations. Sennacherib also left us an unusually detailed description of the destruction of a city, as well he might in view of the fame of this particular one: Babylon. After repeated troubles with Babylon, Sennacherib invested the city and conquered it. He then obliterated it.

> The city and its houses, from its foundations to its top, I destroyed, I devastated, I burned with fire. The wall and outer wall, temples and gods, temple towers of brick and earth, as many as there were, I razed and dumped them into the Arahtu Canal. Through the midst of that city I dug canals, I flooded its site with water, and the very foundations thereof I destroyed. I made its destruction more complete than that by a flood. That in days to come the site of that city, and (its) temples and gods, might not be remembered, I completely blotted it out with (floods) of water and made it like a meadow.[53]

The detail and vividness of this account show that it is not a mere formula and that when Sennacherib boasted that he could turn cities into "forgotten tels," it was not an empty boast.[54]

The inscriptions of Esarhaddon (681–668 B.C.) are too fragmentary and vague to be very useful. He mentioned the capture of cities only nine times, and some of these references are too obscure to be clear. In four of these cases he said he destroyed the cities. Five times he only mentioned plundering them. There is only one mention of transporting the people to Assyria, but the fact that Esarhaddon remarked that he rebuilt the royal palace at Nineveh with the labor of "people of the lands my arms had captured" shows that this practice remained common under his rule.[55] Esarhaddon mentioned building a pyramid of skulls once and, in an original stroke, he hung the severed heads of two rebellious kings around the necks of their nobles.

The last Assyrian king for whom we have extensive records is Ashurbanipal (668–630 B.C.). His reports told of destroying cities six times. He also said six times that he slaughtered people, although once the qualification that "not a man escaped" suggested the slaughter was of the men only. Ashurbanipal sometimes made distinctions, saying that only those who were not submissive were killed. Deportation remained common, occurring seven times in Ashurbanipal's records.

Ashurbanipal never specified that he killed women and children, as Ashurnasirpal did. But Ashurbanipal did boast of mutilations; ripping out tongues, dismembering, mass beheadings, and piercing (or cutting off) lips are all mentioned once. Impaling occurred twice in Ashurbanipal's records, as did flaying.

Ashurbanipal portrayed his tortures on a palace relief. The upper register shows Assyrian soldiers flaying two naked prisoners staked to the ground face down. To the right another Assyrian soldier carries away a severed head, suggesting that the Assyrians cut off the heads of prisoners after they had been flayed. In the bottom register, one Assyrian soldier firmly holds the head of a bound prisoner while another Assyrian reaches into the prisoner's mouth to tear out his tongue. In another scene from Ashurbanipal's reliefs, Assyrian soldiers carrying severed heads in their hands wade through the bodies of the slain enemy toward a tent where the heads are being piled.[56]

The ferocity of the Assyrians may leap out at us so strongly because the darkness of the ancient past has shielded us from knowing the worst. The palette of Narmer shows the Egyptian pharaoh standing over headless corpses, so we know that the Assyrians were not the first to sever the heads of defeated foes. The Mari documents from an earlier Assyrian period (c. 1800 B.C.) imply that the populations of captured cities were usually massacred, but there is no mention of torture or mutilation.

As we have seen, the Old Testament massacres were almost certainly legendary, but the hanging of the king of Ai and the public display of his body may be compared to the Assyrian display of impaled captives. However, even if the story of Ai is reliable enough to make the comparison valid, it would be to the favor of the Hebrews. The Ammonites proposed to gouge out the eyes of the people of Jabesh-gilead, but that grisly humiliation did not match the scale of Assyrian mutilations. Hebrews under Judah cut off the thumbs and big toes of King Adoni-bezek, the same treatment he had meted out to others, so mutilation was not unknown to the Hebrews.[57] After King Mesha of Moab captured the Israelite town of Nebo, he sacrificed all the seven thousand people, "men, boys, women, girls and maid servants," to his god, Ashtor-Chemosh.[58] Although the evidence of atrocity in the non-Assyrian world is fragmentary, there is enough to suggest that one does not have to turn to our unhappy century to find comparisons.

Nevertheless, we cannot escape the conclusion that Assyrian cruelty in warfare reflected what we would view as a general cruelty in Assyrian society. Assyrian law, for example, frequently prescribed mutilation as a punishment. Cutting off noses and ears was especially common.[59] Tearing out eyes, cutting off fingers, slicing off lips, castration, and pulling out hair all found their place in Assyrian law. Prisoners of war were not the only people the Assyrians impaled. They also impaled women who had aborted their pregnancies. Other Mesopotamian law codes, such as Hammurabi's famous laws, did not prescribe mutilation, except on a literal eye-for-an-eye basis. The same was true of Hittite and Hebrew laws. It is fair to conclude that although atrocity was no stranger to ancient Middle Eastern siege warfare and although it was not the universal Assyrian practice, the Assyrians were frequently unusually cruel and that this cruelty reflected the character of Assyrian culture.

The Persians

The Persians have a reputation as more enlightened imperialists than the Assyrians. Herodotus praised them for their generosity toward the sons of rebels,[60] and Isaiah's enthusiasm for Cyrus is well known.[61] But in treatment of captured cities the Persians were little different from the Assyrians. The Babylonian chronicle that recorded the Median conquest of Nineveh holds the same horror as the Assyrian inscriptions:

> A great havoc of people and nobles took place . . . they carried off the booty of the city, a quantity beyond reckoning, they turned the city into ruined mounds.[62]

Despite Herodotus's broad-minded appreciation of some Persian virtues, his History contains much evidence of Persian ruthlessness in siege warfare. After Darius captured Babylon, he impaled three thousand of the leading citizens, although he spared the rest of the population.[63] Cyrus ordered Sardis sacked after he had captured the city, and he prepared to burn Croesus and fourteen Lydian boys alive. Only Croesus's story of the Greek sage Solon's advice to him to avoid hubris persuaded Cyrus not to burn Croesus and the boys to death.[64]

While Herodotus's story about Solon and Croesus is improbable, his story of the sack of Sardis appears to reveal much about Persian policy. According to Herodotus, Croesus pointed out to Cyrus that the city the Persians were sacking now belonged to Cyrus and that it was his wealth his men were seizing. Croesus advised Cyrus to tell his men they must give up a tenth of their loot for Zeus, thus providing religious sanction to the requirement that they share the booty with Cyrus.[65] The tale may be a bit fanciful, but the problem of dividing loot was real enough and the figure of 10 percent for the king has the ring of authenticity.

Xenophon told a similar story of Cyrus's decision not to sack Sardis in the form of a dialogue between Croesus and Cyrus:

> "Listen, then, Croesus," said he [Cyrus]. "I observe that my soldiers have gone through many toils and dangers and now are thinking that they are in possession of the richest city in Asia, next to Babylon; and I think that they deserve some reward. For I know that if they do not reap some fruit of their labours, I shall not be able to keep them in obedience very long. Now, I do not wish to abandon the city to them to plunder; for I believe that then the city would be destroyed, and I am sure that in the pillaging the worst men would get the largest share."
>
> "Well," said Croesus on hearing these words, "permit me to say to any Lydians that I meet that I have secured from you the promise not to permit any pillaging nor to allow the women and children to be carried off, and that I, in return for that, have given you my solemn promise that you should get from the Lydians of their own free will everything there is of beauty or value in Sardis. For when they hear this, I am sure that whatever fair possession man or woman has will come to you; and next year you will again find the city just as full of wealth as it is now; whereas, if you pillage it completely, you will find even the industrial arts utterly ruined; and they say that these are the fountain of wealth. But when you have seen what is brought in, you will still have the privilege of deciding about plundering the city. And first of all," he went on, "send to my treasuries and let your guards obtain from my guards what is there."[66]

These tales suggest the dilemma of a siege commander, wishing to satisfy his troops and yet maximize his own profit. Cyrus undoubtedly had his own reasons for a moderate policy toward Sardis, but such moderation was not a general Persian policy, even under the great-souled Cyrus. The Greeks discovered this from the first Persian campaign against Ionia. Cyrus's general Mazares ruthlessly plundered the area around Magnesia all along the Maeander River and sold the inhabitants of Priene into slavery.[67]

Like the Assyrians at Jerusalem, the Persians often attempted to negotiate a surrender rather than resorting to a lengthy siege, especially if the city's fortifications were strong. Mazares' successor Harpagus demanded only a token submission from the Ionian city of Phocaea, telling the Phocaeans that he would lift the siege if they tore down a single tower and destroyed a single house. When the Phocaeans asked him to withdraw his forces for a day while they considered his offer, Harpagus agreed, even though he knew the Phocaeans intended to take advantage of the truce to make their getaway from the city. Not only was the entire population of Phocaea able to escape, but they also took all their movable property with them. Harpagus was left with an empty city, but he had avoided a siege of a city that was one of the most strongly fortified in Ionia.[68]

A more standard Persian policy was presented to the Ionian tyrants who had rebelled against Persian rule around 500 B.C. The Persians instructed the Ionian leaders to tell the Greek cities that if they submitted

> they shall suffer no hurt for their rebellion, and that neither their temples shall be burnt nor their houses, nor shall they in any regard be more violently used than aforetime. But if they will not be so guided, and nothing will serve them but fighting, then utter a threat that shall put constraint upon them, and tell them that if they are worsted in battle they shall be enslaved; we will make eunuchs of their boys, and carry their maidens captive to Bactra, and deliver their land to others.[69]

If angered, the Persians could massacre the inhabitants of a town. This happened at Samos after some Persian nobles, who believed the Samians had agreed to the Persian terms, were treacherously set upon and killed by soldiers led by a Samian whom Herodotus called crazy. The Persian general Otanes was so incensed that he disregarded Darius's instructions to preserve Samos and ordered his troops to kill all the Samian men and boys.[70] Another massacre occurred during Xerxes' Greek campaign in 480 B.C. After Xerxes' retreat to Asia following the Persian defeat in the naval battle of Salamis, his escort army, on its journey back to Greece to rejoin the Persian army Xerxes had left behind, besieged Olynthus, a Bottiaean town on the Chalcidic peninsula. The town quickly fell, and the Persian general Artabazus killed all the inhabitants. A nearby lake served as a mass grave for the corpses.[71]

Although the Persians did not practice resettlement on the same scale as the Assyrians, they were certainly capable of wholesale removal of populations. The fate of Miletus, which the Persians captured in 494 B.C., was especially hard. Most of the men died defending the city. The Persians enslaved the women and children. The surviving men were resettled near the mouth of the Tigris on the Persian Gulf. Apparently the Persians separated wives and children from these men. Miletus itself was resettled by Persians and Carians.[72] Archeological evidence suggests that the Persians destroyed the port of Miletus. It was never rebuilt, and Miletus never recovered its former glory.[73]

Herodotus tells us that the Athenians considered the fate of Miletus so horrible that they fined the playwright Phrynichus for reminding them of it in his play *The Capture of Miletus*.[74] However, later in his *History* Herodotus casts some doubt on his vivid description of the fate of Miletus by telling us that ten years afterward, a Milesian contingent in the Persian army played an important role in the Greek victory over the Persians at Mycale when it disobeyed orders and turned against the Persians.[75] This suggests that the Persians impressed some of the Milesians into the army, an old Assyrian practice. Another case of mass enslavement by the Persians took place on the island of Euboea when they carried off into slavery the inhabitants of Eretria after conquering it in 490 B.C. as a jumping-off place for Marathon.[76]

The campaign of the Persians through Phocis after they broke through at Thermopylae in 480 B.C. was a scorched-earth invasion of which an Assyrian king would have been proud. The Persians burned to the ground all the towns in their way. Herodotus named twelve towns that the Persians razed. Herodotus's account of this campaign was one of the few up to this time that mentioned the raping of women. He tells us that some women died from repeated rapings by Persian soldiers.[77]

Passing into Boeotia, the Persians burned Thespia and Plataea.[78] Continuing on to Athens, they found a deserted city, except for the acropolis, where a few die-hards made a last stand. The Persians tried to negotiate, but the Athenians refused. When, after considerable difficulty, the Persians finally reached the acropolis, some of the Athenians leaped to their death and others sought sanctuary in the temples. The Persians broke into the temples and massacred the suppliants. They then looted and burned the temples.

Xerxes may have had a bad conscience about the violation of the temples, because the next day he ordered some Athenian survivors to go up to the acropolis and offer sacrifices to the gods whose temples had been defiled. Perhaps his commanders had lost control of the troops during the rampage, a not uncommon circumstance during a sack. Herodotus thought his change of heart may have come from a dream during the night. Some modern scholars speculate that his motives were more political than religious.[79] In any case, the whole incident represented a particularly brutal example of the violation of temples, a frequent horror in siege warfare because temples often housed great riches.[80]

The Persians did not seem to have practiced mutilation as much as the Assyrians, but mutilation was not unknown to them. In the terrible Ionian campaign of 494–3 B.C., which had brought Miletus's agony, the Persians selected the best-looking boys and castrated them so they could serve as eunuchs in the Persian court. Pretty girls did not escape either; they were sent to Darius's harem. Here, too, the Persians razed the towns to the ground.[81]

When the Persians captured Barca to avenge the murder of their puppet tyrant, they turned the murderers over to the tyrant's mother, the notorious Pheretima. She had them impaled on stakes and also seized their wives, cut off their breasts, and displayed the severed breasts on stakes alongside those impaling the men. The Persians took most of the other Barcaeans as booty.[82]

Herodotus placed the blame for this atrocity squarely on Pheretima and expressed satisfaction at the justice of the horrible death she later met.[83] Herodotus's moral judgment was sound, but the punishments Pheretima inflicted on her political enemies were so much like Assyrian atrocities that one suspects Mesopotamian influence on her, mediated by the Persians.

Another piece of evidence that the Persians were capable of mutilation, although not one connected with a siege, is Herodotus's report that Cambyses, whom Herodotus presents as an unusually cruel Persian king, flayed a royal judge for taking a bribe.[84] We also have the gruesome account by Quintus Curtius of the four thousand mutilated Greek prisoners whom Alexander the Great found at Persepolis when he took that city in 331 B.C.:

> Some with their feet, others with their hands and ears cut off, and branded with the characters of barbarian letters. . . . They resembled strange images, not human beings, and there was nothing that could be recognized in them except their voices.[85]

Although these men had not been captured in a siege, their fate showed that the Persians mutilated prisoners.

From what evidence we have, then, Persian siege methods were not notably more humane than Assyrian. Persian practice included massacre, resettlement, enslavement, summary executions, mutilations, and razing of cities. There were cases of more generous treatment, but similar examples existed in the Assyrian records. Well might Aeschylus write:

> For by the will of the gods Fate hath held sway since ancient days, and hath enjoined upon the Persians the pursuit of war that levels ramparts low, the mellay of embattled steeds, and the storming of cities.[86]

Women and Children

One of Ashurnasirpal's siege reliefs shows an inner citadel on whose towers stand women raising their hands in terror and lamentation.[87]

This scene shows that the Assyrian artist was mindful of the presence of women in siege warfare, and he has brilliantly captured their fear. This fear was not unfounded. A mass grave uncovered at Lachish contained bones and skulls of at least 695 people. Some of the bones and skulls were burned, and some skulls were caved in. Almost certainly these people were casualties of the Assyrian siege of Lachish in 701 B.C. Two hundred and seventy-four were women and sixty-one were children.[88]

The Assyrian reliefs reveal how much the fate of women and children influences our image of siege warfare. Many of them show soldiers leading women and children off into captivity. Although it does not appear to have been the standard practice, the Assyrians sometimes stripped the clothes from the captives and led them away naked. The purpose was not only to humiliate them but also to leave them completely vulnerable. There was a logical progression from stripping away the protective walls of a city to stripping the clothes off its inhabitants.

The prophet Ezekiel saw the relation between nakedness and vulnerability as well as between sex and war in a strikingly sexual metaphor of harlots, symbols for Jerusalem and Samaria, who went whoring after the Assyrians only to be stripped naked and humiliated by them.[89] Nahum's celebration of the fall of Nineveh also spoke of nakedness:

> The river gates are opened,
>   the palace is in dismay;
> its mistress is stripped, she is carried off,
>   her maidens lamenting,
> moaning like doves,
>   and beating their breasts.[90]

Nahum implied that the women of Nineveh may have had to lift their skirts over their heads as they were marched away by their captors.[91]

For women, however, neither capture nor being stripped naked represented the final loss of control. Rape was the ultimate violation of women, marking the complete possession of them by the soldiers who had taken possession of their city. From the phallic shape of the battering ram trying to penetrate the walls of a city to priapic soldiers pillaging and raping in a violated city was a logical progression. All warfare has a strong sexual undercurrent, but siege warfare was an explicit battle for sexual rights. The defending soldiers were attempting to protect their sexual rights to their wives as well as to protect the sexual integrity of their tribe or ethnic group. The raping that frequently followed the fall of a city starkly symbolized total victory in a total war.

But if soldiers lusted for the sexual license provided by victory in siege warfare, they also feared the absence of sexual taboos that are such a strong part of our sense of order. Hebrew women did not follow the army, and Hebrew warriors apparently practiced sexual abstinence on campaign.[92] It is in this context that we should understand the laws of Yahweh providing guidelines for the treatment of captured women:

> When you go forth to war against your enemies, and the LORD your God gives them into your hands, and you take them captive, and see among the captives a beautiful woman, and you have desire for her and would take her for yourself as wife, then you shall bring her home to your house, and she shall shave her head and pare her nails. And she shall put off her captive's garb, and shall remain in your house and bewail her father and her mother a full month; after that you may go in to her, and be her husband, and she shall be your wife. Then, if you have no delight in her, you shall let her go where she will, but you shall not sell her for money, you shall not treat her as a slave, since you have humiliated her.[93]

The Hebrews regarded foreigners as unclean. Cutting the hair and paring the nails and putting off the clothes of captured foreign women were ritual cleansings necessary to remove the taboo against sexual relations with unclean women.[94] Such rules offered scant solace to captive women, but they nevertheless represented an effort to bring some moral order to the capture of women.

The Assyrians were also uncomfortable with rape. The same Assyrian sources that reveled in burning or burying people alive, impaling prisoners on stakes, slicing off heads, hands, feet, and lips, cutting out tongues, and building mounds of heads were silent about rape. The absence of women in the scenes of naked prisoners is another sign of inhibition. Indeed, we find in Assyrian law the same sort of provisions for Assyrian soldiers to marry captive women that we found in the Book of Deuteronomy.[95]

Rape loomed much larger in the imagery of the Hebrew prophets. The prophets did not doubt that rape was a certain prospect for fallen cities. Isaiah, envisioning the fate of the Assyrians at the hands of the Medes and Babylonians, prophesied:

> The infants will be dashed into pieces
> before their eyes;
> their houses will be plundered
> and their wives ravished.[96]

Zechariah had the same nightmare:

> For I will gather all the nations against Jerusalem to battle, and the city shall be taken and the houses plundered and the women ravished. . . .[97]

These nightmares, however, were prophecies rather than descriptions of events. The author of the fifth poem in Lamentations was the only person in the Bible to mention rape in the context of military defeat as past event, in this case the fall of Jerusalem to Babylonia:

> Women are ravished in Zion,
> virgins in the towns of Judah.[98]

Rape, then, represented a violation so horrible that it found a place in prophecies meant to be as terrible as possible. The image of rape in the Bible was a nightmare representing the collapse of human culture, a descent into savagery that was too terrifying to describe as actual event. It remained a prophecy, a mental construct, a horror of the mind. The prophecies of rape by the Hebrew prophets take us to an outer limit, but to cross this limit would reveal a world so violent and out of control that it would bring into question the existence of any divine structure at all. Such a world was too terrifying for either the Assyrians or the Hebrews to enter. And so rape did not find its place in the boasts of conquerors, but only in the Hebrew prophecies of doom.

Rape was undoubtedly the most common atrocity against noncombatants in siege warfare, but other atrocities occupied a similar position on a frontier that men were reluctant to cross in the representation of war. One was cutting open pregnant women.

An Assyrian poem, dating from around the turn of the eleventh century B.C., sang of a warrior king:

> He slits the wombs of pregnant women; he
>   blinds the infants
> He cuts the throats of their strong ones.[99]

E. Ebeling has identified the king as Tiglath-pileser I (1114–1076 B.C.) and associates the poem with a military campaign described on a royal cylinder inscription of Tiglath-pileser.[100] But the cylinder inscription did not mention cutting open the wombs of pregnant women; it contented itself with the boast, "I butchered their troops like sheep. . . . I burnt, razed and destroyed that city."[101]

We cannot rule out the possibility that Tiglath-pileser's soldiers really did rip open pregnant women, but the fact remains that the atrocity occurred in a product of the poetic imagination and not in the official report. There was no mention of ripping open pregnant women in any of the royal Assyrian reports that we have found.

Ripping open pregnant women was also a motif in Biblical representations of siege warfare. Again it was the prophets who saw the worst. The prophet Elisha explained to Hazael, soon to be king of Syria, that he wept because

I know the evil that you will do to the people of Israel; you will set fire
their fortresses, and you will slay their young men with the sword and
dash in pieces their little ones, and rip up their women with child.[102]

Hazael responded that only a dog would do such things, and we have
no report of the prophecy being fulfilled.

Hosea uttered an equally horrific prophecy about the fall of Samaria
to the Assyrians:

Samaria shall bear her guilt,
because she has rebelled against her God;
they shall fall by the sword,
their little ones shall be dashed in pieces,
and their pregnant women ripped open.[103]

The account in 2 Kings of the actual siege of Samaria was considerably
less lurid:

Then the king of Assyria invaded all the land and came to Samaria, and
for three years he besieged it. In the ninth year of Hoshea the king of
Assyria captured Samaria and he carried the Israelites away to Assyria,
and placed them in Halah, and on the Habor, the river of Gozan, and in
the cities of the Medes.[104]

In these two cases, the image of splitting open the wombs of pregnant women appeared in dire prophecies intended to shock people from
their complacency. Both Assyrian and Hebrew law placed a high value
on the unborn fetus. There was a preoccupation with miscarriage in the
Middle Assyrian laws, and the penalty for a man who caused one, "even
if her fetus was a girl," was death. The penalty for abortion was death
by impalement.[105] Hebrew society also had taboos about killing a fetus.
Although much less preoccupied with the possibility of men causing
a miscarriage in a brawl, Hebrew law also carried penalties for a man
who caused a miscarriage, the payment of a fine.[106] In both Assyrian and
Hebrew society, then, the killing of a fetus was considered peculiarly
abhorrent. The representation of such deeds in war served as a literary
motif about the moral chaos of war.

But two Biblical references to ripped-open pregnant women concerned events that had already happened. Amos said that the Lord
would punish the Ammonites because "they have ripped up women
with child in Gilead, that they might enlarge their border."[107] Amos was
a well-informed observer of international relations, so we should take
his report about the Ammonites seriously. In the second instance, the
Book of Kings flatly stated that Menahem, king of Israel, after the siege
of Tappuah, "because they did not open it to him, therefore he sacked
it, and he ripped up all the women in it who were with child."[108] Mordechai Cogan has suggested that the passage is an interpolation of a

later editor wishing to emphasize the evil of Menahem, drawing on stock prophetic formulas to do so.[109] He is not suggesting that such formulas had no relation to actual experience but that the formula cannot serve as proof for the specific event to which it is applied.

The passage does, however, carry conviction, because the author has given us a reason for Menahem's atrocity, one that will not be unfamiliar to students of siege warfare. Sieges were so difficult that commanders did everything possible to avoid them and often handed out savage retribution to cities that forced a siege upon them. Menahem had demanded that Tappuah open its gates to him. It did not, and Menahem's horrible deed fits into the practice of wreaking horrible vengeance on a city that refused terms.

Another horror that fits into the same pattern as rape and ripping open pregnant women is violence against babies. The usual formula in the Bible was "dashing out the brains of little ones." The image of a soldier seizing a baby by the foot and smashing its head open against a rock was a peculiarly terrible one.

Again it was the prophets who did not avert their eyes from such a horror. As we have seen, Elisha and Hosea coupled the dashing of babies with the slitting open of pregnant women. Isaiah once combined it with rape and mentioned it alone in another place.[110] In a slight variant, Hosea imagined mothers and children dashed in pieces together.[111] A psalmist blessed a hoped-for enemy of Babylon: "Happy shall he be who takes your little ones and dashes them against the rock."[112] All of these references to baby bashing occurred in prophecies or curses. Only Nahum, in describing the fate of Thebes, mentioned such an atrocity as past event.[113]

That these atrocities against women and children occurred mostly in prophecies and curses rather than in narratives of events, as well as their formulaic form, makes it difficult to conclude that such horrors were a common part of siege warfare. Rape was certainly common, and I think we must assume that smashing babies' heads and ripping open pregnant women were not unknown. But they seem to function in our sources as images reflecting a vision of a world without limits or structure or morality, in which men violated deep-seated taboos about sex, pregnancy, and survival. These primal fears almost always occurred in the context of siege warfare, and the Hebrew prophets were the first to reveal the full terror that siege warfare inspired. Only siege warfare engulfed women and children so commonly that we naturally associate their destruction with it. Because of this, ancient siege warfare represented the most complete form of war as social conflict, and in this way it offers us the truest image of what war really is.

PART THREE

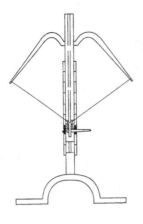

THE
GREEKS

Chapter V

EARLY SIEGES THROUGH THE PELOPONNESIAN WAR

The Early Greeks

The Greeks were a warlike people, but in methods of warfare they lagged far behind the Middle East. This was especially true of siege warfare.

Although Bronze Age Greeks may have conducted sieges (a Greek vase dated around 1500 B.C. shows a siege with both slingers and archers directing fire against a wall) and Homer's tale of the siege of Troy as well as Herodotus's story about King Minos meeting death during an unsuccessful five-year siege of the town of Camicus in Sicily may be distant echoes of such siege warfare, this capability was lost during the poverty of the Greek dark ages.[1] Moreover, Homer's account of a ten-year siege is a legendary story, and we can learn almost nothing from him about methods of siege warfare. However, when we consider that Troy only fell after ten years, it is clear that the Greeks did not consider siege warfare easy, even for legendary heroes.

Homer does not have the Greeks mount an actual siege of Troy. Instead they wage open-field battles against the Trojans, who conveniently make frequent sorties from the city. This suggests that siege warfare was hardly known to the Greeks of Homer's time (c. 750 B.C.), the very time when the Assyrians had perfected their methods of siege.

Even in classical times, siege war played a small role in Greece, at least until the rise of the Athenian Empire in the fifth century. Thucydides correctly identified the reason for the absence of siege warfare in early Greek history: lack of money.[2] The resources of early Greece were simply inadequate to sustain sieges. Greek society lacked the wealth and the elaborate organization of labor that were the foundations of siege warfare in the Middle East.

Thucydides placed great significance on the first fortifications in Greek civilization, seeing them as a significant reflection of economic progress. Before fortifications, he said,

> there was no mercantile traffic and the people did not mingle with one another without fear, either on land or by sea, and they each tilled their own land only enough to obtain a livelihood from it, having no surplus of wealth and not planting orchards, since it was uncertain, *especially as they were yet without walls* [my emphasis], when some invader might come and despoil them.[3]

Later, however,

> the cities which were founded in more recent times, when navigation had at length become safer, and were consequently beginning to have surplus resources, were built right on the seashore, and the isthmuses were occupied and walled off with a view to commerce and to the protection of the several peoples against their neighbors.[4]

But these early fortifications on which Thucydides placed such importance sufficed only to protect against piratical raiders. They could not protect Greek cities from enemies skilled in the art of siege.

The legends tell of Greeks besieging others, but the reality was that others besieged the Greeks. The first Greeks to suffer from Middle Eastern siege methods were the Ionians who had settled along the coast of Asia Minor. The rise of the Kingdom of Lydia in the interior of Asia Minor confronted these Greeks with a powerful neighbor. The first Lydian king to lead an expedition against the Greek city-states of Asia Minor was Gyges, who lived in the seventh century B.C. Although he succeeded in capturing Colophon, his power and siege capabilities were apparently not great, because that is all he accomplished.[5]

The Ionian cities enjoyed the advantage of a coastal location that made blockade difficult, as a later Lydian king, Alyattes (early sixth century B.C.) discovered in twelve consecutive annual campaigns against Miletus. He apparently lacked the means of breaching Miletus's fortifications, because he had to satisfy himself with the destruction of crops.[6] However, there is archeological evidence at Smyrna that Alyattes built a siege ramp there, and this may be a sign of improved siege methods, because Alyattes' son Croesus, beginning with a siege of Ephesus, succeeded in imposing tribute on all the Asiatic Greek cities.[7] Herodotus does not tell us how Croesus succeeded where his predecessors failed. We are left with the impression that his siege of Ephesus was so effective that the other cities quickly fell into line.

Croesus was a Hellenophile and proved to be an easy master for the Greeks. His fall to the Persians in 546 B.C. confronted the Greeks with a much more dangerous threat. The siege warfare that the Persians brought to Greece was new to Greek society.

Traditional Greek warfare did not aim at the conquest of cities and territory. Greek wars were mostly piratical raids in which invading armies sought to steal livestock and destroy crops. If the enemy's heavy infantry came out from the city to defend its fields, a set-piece battle between opposing hoplite phalanxes was fought. If no resistance was offered, the invading army simply ravaged and plundered the countryside and returned home. Heavy infantry so dominated Greek armies that they were short of the sort of troops necessary for sieges, such as archers and slingers for covering fire and sappers for siege operations. The farmers who made up the hoplite class could not sustain long campaigns because they could not be away from their farms for extended periods.[8]

Greek logistics were rudimentary. Soldiers took only a few days' ration with them on campaigns, and the large number of servants—one for every hoplite—that routinely accompanied Greek armies made their supply even more difficult.[9] Under these circumstances, sieges were rare and fortifications, if they existed at all, were rudimentary by Middle Eastern standards.[10]

The sites of early Iron Age Greek cities show that their inhabitants did not expect long sieges. Lacking the manpower and resources to build elaborate fortifications, early Greek settlements located on easily defensible sites, such as hilltops, peninsulas, or islands.[11] Many relied entirely on the natural strength of the position and, where there were walls, they protected only the most vulnerable approaches to the acropolis.

The walls were simple. Curtains lacked towers until the sixth century B.C. Gate defenses were only a little more elaborate, relying on bastions or overlapping walls for protection. Gate locations were often unfavorable for defensive purposes because the roads, which usually predated the fortifications, followed the most convenient route into the city and the Greeks seldom bothered to reroute roads to more defensible terrain.

Scanty evidence obscures the traces of early walls, but they may have featured some indentations or jogs for enfilading.[12] Stones and mud bricks were the most common materials the Greeks used for wall construction. Often a stone socle supported a mud brick wall. A layer of clay or lime plaster protected the face of mud brick walls from erosion, and wood and tiles protected the top. Later, stone walls became more common, but mud brick continued in use. Indeed, some Greeks thought mud bricks withstood battering rams better than stone walls.[13] Early Greek walls were not massive, ranging from 3.5 to 4.5 meters in height. Thickness varied considerably, ranging from 1.75 to 4.6 meters.[14]

The need to locate fortifications on naturally strong sites often left them without a supply of water, forcing them to rely entirely on cisterns.[15] The dangers of such a system were suggested by a tale Pausanias, the second-century travel writer, related about the famous Athenian law-giver Solon.

It seems that Solon attempted to force the surrender of Krisna by diverting a river that flowed through the town. The Krisnaians were indeed able to survive on cistern and well water, but not without considerable hardship. The ever-resourceful Solon then put a powerful purgative into the river and redirected it back into Krisna. When the thirsty Krisnaians greedily drank from the river, they got such acute diarrhea that they abandoned their walls and the city fell.[16]

The fortifications of early overseas settlements were on a somewhat larger scale. This was especially true of the colonies in Asia Minor that the Greeks settled in the seventh and sixth centuries B.C. There was a greater degree of social equality in such colonies, and none of the settlers wanted to leave their possessions unprotected. So the walls enclosed areas large enough to include the entire town.[17] The most successful colonies soon outgrew their walls and, as we have seen, few Greek cities were able to withstand the sieges of the Lydians and Persians.

Because Greek walls were so unsophisticated, their extension probably left them more accessible to Middle Eastern assault methods than the earlier fortified acropolis.[18] It appears that Croesus destroyed the outer walls of most Ionian cities, leaving them with only an inner citadel, much like the older cities on mainland Greece.[19] Even where walls existed, as late as the Peloponnesian War it was not unusual for Greeks to be careless about their maintenance.[20]

The shock of Lydian and Persian siege methods did lead the Ionian Greeks to strengthen the defense of their walls with two-story towers, a practice that soon spread to mainland Greece. But these towers usually only protected the gates. Often there was only a single tower, placed, if possible, on the left side of the gate (looking out) so the defenders could fire at the unprotected right side of assault troops.[21]

The most spectacular example of improved Greek defenses in Ionia during the sixth century was Samos under the rule of the tyrant Polycrates, who had the wealth and power necessary for impressive fortifications. He built walls protected by towers and a ditch.[22] Most famous was his aqueduct, built by the Megarian engineer Eupalinus to keep Samos well supplied with water during a siege. For protection against enemies, it cut through a hill in a tunnel that was almost a mile long and was an impressive eight feet wide and eight feet high.[23]

Another example of stronger fortifications was those of Buruncuk-Larisa. Its walls featured two-story towers at regular intervals. A particularly strong tower protected the gate. The gate itself was at the inner end of a gate corridor.[24]

Not only does the archeological evidence of early Greek fortifications suggest little siege warfare, but there are also few references in written sources to Greek sieges before the Persian Wars. Herodotus called a Spartan expedition to Samos around 525 B.C. a siege, but his account did not make it clear that any siege operations took place. The Spartans lingered for forty days, but it seems likely that they were merely hoping the Samians would come out and fight.[25]

The western Greeks may have been more experienced in siege warfare, which they may have learned from Carthage, a Phoenician city familiar with Middle Eastern siege methods. Herodotus said that Gelon proved to be a good siege general under the Geloan tyrant Hippocrates in various campaigns in Sicily during the 490s B.C.[26] We know nothing of their methods, but they employed mercenaries, which would have made possible the sustained campaigning necessary for sieges. However, a large city such as Syracuse was beyond the siege capabilities of Hippocrates and Gelon.[27] Archeological evidence suggests that the western Greeks were among the earliest to construct large-scale fortifications.[28]

Typical of Greek ineptitude in siege warfare was the experience of Miltiades, the genius of Marathon. After earning immortality by leading the Greek hoplites to victory at Marathon, Miltiades imagined that he could further enhance his glory by plundering Paros, an island city in the Cyclades. He easily persuaded the Athenians to equip a large expeditionary force and sailed to Paros. The result was a fiasco.

Paros was so weakly fortified that the Parians hastily had to double the height of their walls at their most vulnerable points. Even these make-shift walls sufficed to frustrate the Athenians, and Miltiades spent twenty-six ineffective days accomplishing nothing but the destruction of crops. He returned in disgrace to Athens, where his political enemies were able to place him on trial. The jury spared his life because of Marathon but fined him fifty talents. Shortly afterward, Miltiades died of a wound he had suffered at Paros.[29]

The Athenian Empire

Although Miltiades' ill-fated foray to Paros led to a siege, it fit into the traditional pattern of warfare in Greece before the fifth century B.C. It was a raid to settle an old grudge and to seize booty. After the second Persian invasion in 480–479 B.C., Greek warfare took a distinctly different turn. Hit-and-run raids would not suffice to drive the Persians from the Aegean; only the control of territory could accomplish that. Controlling territory required taking cities.

This new reality quickly became apparent with the siege of Sestos, a city on the strategically vital Hellespont then held by the Persians. Sestos was strongly fortified, and the Athenians had no means by which to breach the walls. Thus they resorted to blockade. The siege dragged on for months. Greek troops were not used to this sort of warfare, and the Athenians complained bitterly about the hardships as winter approached. It was only with great effort that the officers were able to sustain the siege. Fortunately for the Greeks the siege had caught Sestos by surprise, and it was not well provisioned. Famine reduced the inhabitants to boiling leather for food, and the city finally fell.[30]

A siege was also necessary to wrest from the Persians the town of Eion, which lay at the mouth of the Strymon River, a strategic point on the way to the Black Sea. Miltiades' son Cimon captured Eion in 475 B.C.[31] According to Pausanias, Cimon breached the wall of Eion, which was made of adobe, by washing it away with water diverted from the Strymon.[32]

Once Athens began to construct an Aegean empire as a buffer against Persia, siege warfare became indispensable to hold it together, and the empire provided the resources that made siege possible. Athens discovered this as early as 470 B.C. when Naxos revolted and a siege was required to keep it in the alliance.[33] It took another siege, this one lasting three years, to retain control of Thasos, which lay on the vital Thraceward route to the Black Sea.[34]

Even more threatening was the revolt in 440–439 B.C. of Samos, a much more powerful city than Thasos. The Athenians blockaded Samos on the landward side with three walls and, after a near-run war at sea, were able to blockade the city entirely. Samos fell after a nine-month siege.[35] By the standards of the day, it was not a long siege. Pericles, the Athenian commander at Samos, took great pride in having captured the city in such a short time, comparing the nine months with the ten years it had taken Agamemnon to take Troy.[36]

The free service of citizens could not have sustained such long sieges. Thucydides' information on the pay of Athenian hoplites at the siege of Potidaea in 433 B.C. is our first report of paid hoplites in Athens, but the practice must have begun earlier.[37] The imperatives of siege and empire made the Athenian army, as well as its navy, quasi-mercenary, with the tribute of empire providing the wages.[38]

Diodorus of Sicily, the first-century B.C. Greek historian, and Plutarch, who lived from around A.D. 50 to 120, both reported that the Athenians used battering rams and protective sheds, called tortoises, at Samos.[39] If true, this would be the first known use of battering rams by Greeks. Both Diodorus and Plutarch drew this information from Ephorus, the fourth-century B.C. historian whose work is almost entirely lost.[40]

Ephorus said that the Athenian siege machines were made by a

famous engineer from Clazomenae who was called Artemon Periphoretos (Carried About) because he was lame or rich and lazy. Plutarch
showed doubts, citing Heraclides Ponticus, a fourth-century B.C. product of Plato's academy whose dialogues influenced Plutarch. Heraclides
rejected the story because the sixth-century poet Anacreon had mentioned
Artemon Periphoretos in his poetry some one hundred years before the
siege of Samos, leading Heraclides to believe that Ephorus's chronology
was too confused to be credible. Polybius's dismissal of Ephorus as a
man who knew nothing about warfare on land also undermines Ephorus's credibility.[41]

Most modern scholars have doubted the truth of the story, both because Thucydides' account of the siege of Samos does not mention the
use of siege engines and because Diodorus's vocabulary for rams and
tortoises gives his account a suspiciously fourth-century ring. Although
we cannot flatly reject Ephorus's testimony, it is so problematic that we
are on safer ground to assume Thucydides would have mentioned such
a novelty in Greek warfare as battering rams if the Athenians had used
them against Samos.[42]

Even less convincing than Ephorus's account of the siege of Samos
is Pausanias's description of the siege of Oeniadae by the Messenians
who settled in Naupactus sometime between 454 and 438 B.C. Pausanias
has these Messenians deploying a full range of siege methods, from
escalade to undermining to the deployment of siege machines.[43] This
painfully anachronistic account is wholly unbelievable.[44]

If we cannot be certain that the Athenians were so far advanced
in siege methods that they used rams at Samos, it is clear that the imperatives of empire forced them into siege warfare, and they quickly
outstripped other Greeks in siege experience, a fact the Spartans reluctantly recognized around 460 B.C. when they called in the Athenians to
help in the siege of the helot fortress at Mount Ithome.[45]

The high tide of the Athenian Empire around the middle of the fifth
century B.C. was marked by sieges in far-off places, such as Cyprus and
Egypt. The great Athenian general Cimon dealt the Persian Empire a
stunning blow in Cyprus by reducing Citium and other Cypriot cities.
Salamis was proving more difficult for the Athenians when the Peace
of Callias cut short siege operations, but not before Cimon had died
during the siege.[46] The failure of the Athenian siege of the White Fortress in Memphis, which ended in the destruction of the Athenian expedition by a Persian relief force, marked the end of Athenian expan-
sionism.[47]

Another indication of the new direction in Greek warfare was the
expansion of fortifications. The experience of the Persian invasions revealed the danger of inadequate walls, and the rise of democracy made
it necessary to protect the entire urban demos. A fortified citadel on the
acropolis left democracies vulnerable to oligarchic coups d'état, which

usually began with the seizure of the acropolis. As the focus of defense shifted to the outer circuit, the acropolises became more of a political liability than a military asset; hence their rapid decline in the fifth and fourth centuries.[48] Walls not only became longer; they became stronger as well. Height and thickness increased, towers provided more protection, and gateways became more elaborate.[49]

The improvements of Athens' fortifications provides one example of this pattern. The old fortifications had not protected Athens' access to the sea and had not been strong enough to withstand Persian siege methods. When the Persian army approached in 480 B.C., the Athenians fled their city and the Persians destroyed its fortifications. After the war, the Athenians hurriedly rebuilt their walls, fortifying not only the city but the port of Piraeus as well.

Although the walls around Piraeus were not high, reaching only half the original plan, they were thick enough for two wagons to pass between the outer faces. The Athenians gave added weight to the walls by fitting massive stones, reinforced with metal clamps, into the space between the outer walls, instead of following the usual practice of filling the inner space with rubble.[50]

Later, when the Argives tried to build long walls to their port during the Peloponnesian War, they turned out the entire population, men, women, and slaves, to work on the walls.[51] We may assume the Athenians did the same thing. As their imperial policy unfolded, they extended the walls down to the port.[52] These "long walls" guarded the way to the sea. These extensions left Athens with a circuit of walls over twenty-one miles long.[53] But, although they were proof against any Greek army, Athenian fortifications were still too unsophisticated to withstand Middle Eastern siege methods.[54]

This was generally true of the long walls and extended city circuits of the early fifth century B.C. Gate systems remained quite simple.[55] Walls rambled around the countryside, following the contours of the terrain to seek the strongest defensive route instead of relying on the strength of the walls. This often left a city with more wall than it could defend. Posterns located in remote and difficult areas of the wall partly solved this problem by providing sally ports from which the defenders could drive away enemy troops before they got close to the wall.[56] The extension of walls to include the outer city or to reach the port provided a tempting target for siege engines in comparison to earlier Greek fortifications, whose inaccessible approaches made the deployment of rams unpromising.[57] Long walls often ran along level ground, so siege machinery could easily approach them.[58] Greek walls were rarely protected by ditches until after the Peloponnesian War.[59] However, the Greeks were slow to seize this opportunity and, with the possible exception of the Athenian siege of Samos, it is probably not until the Peloponnesian

War that we find the first halting efforts by Greeks to utilize battering rams.

The Peloponnesian War

The Peloponnesian War made clear that the Persian Wars and the rise of the Athenian Empire had changed warfare in Greece. The struggle for hegemony between the two major powers in Greece made war strategic in a way that Greek warfare simply had not been before the fifth century B.C. It was now necessary to control the strategic points that determined the balance of power. For this reason, the difficulty of taking a fortified position became one of the central problems of Greek warfare.[60]

THE SIEGE OF PLATAEA

The problem that a fortified position presented to Greek commanders emerged in the first action of the Peloponnesian War, the Theban attack on Plataea.[61] Thucydides' account of the struggle over Plataea provides us with our first detailed description of a siege in the ancient world. Thucydides devoted so much attention to the battle for Plataea because the innovative Spartan siege methods fascinated him and because the struggle reflected the deep divisions that tore apart Greek society during the Peloponnesian War. For these reasons the siege of Plataea commands our attention.

The very existence of an independent Plataea was a tribute to the inability of Greek armies to take fortified positions. Plataea lay on the southern frontier of Boeotia, just eight miles from Thebes, the most powerful city in Boeotia. Thebes had long sought to unite Boeotia under its own domination; but even though Thebes was capable of fielding an army ten thousand men strong, it had been unable to subdue tiny Plataea, whose entire adult male population was no more than a thousand and perhaps as few as five hundred.[62]

Plataea was fortified and allied to Athens. The old tactic of drawing the defending army out of a fortified city by threatening the crops had not worked against Plataea. The Athenian alliance and the terms of the Thirty Years Peace, which had ended the First Peloponnesian War in 445 B.C., had prevented the Thebans from ravaging the Plataean countryside. Plataea's fortifications had prevented sudden assaults, and the Athenian alliance had prevented long sieges. These hard political and military facts had continually frustrated Thebes and preserved Plataean independence.[63]

The threat of war between Athens and Sparta determined the Thebans to seize Plataea. Plataea flanked the road from Thebes to the Peloponnese. It thus threatened Thebes' line of communication with its

Spartan ally as well as affording Athens a jumping-off place for an invasion of Boeotia. If war came, it was essential for Thebes to possess Plataea.[64] For these reasons Thebes decided to force the pace and attack in early March, well before harvest time, the usual time of invasion in Greece. But the problem of the fortifications remained. By attacking while the Plataean crops were still green in the fields, the Thebans were denying themselves the opportunity to burn Plataea's crops, by far the most effective means of ravaging grain.[65] The limitations of Greek siege technology prevented a quick capture of Plataea by direct assault. These considerations turned the Thebans to an old strategy. They planned to take Plataea by treachery.

Thebes was an oligarchy and had united the towns of the Boeotian League in a broad-based oligarchic federal system.[66] Plataea was a democracy, and in Plataea democracy went hand in hand with the Athenian alliance. Some of Plataea's wealthy citizens resented popular government and despised Athens, under whose benevolent patronage the Plataean democracy existed.[67] They now saw an opportunity to throw off Athenian influence in Plataea and reclaim what they considered their natural right to rule under the protection of Thebes.

The leader of this aristocratic party, Naucleides, had secretly conspired with one of the most influential men in Thebes, Eurymachus. Eurymachus's family was well practiced in betrayal. His father, Leontiades, had led the Thebans over to the Persian side at Thermopylae.[68] His son, also named Leontiades, betrayed Thebes to the Spartans a generation later. Donald Kagan calls the family a "line of traitors and scoundrels"[69] and Herodotus's account of the betrayal at Thermopylae is indeed contemptuous and disapproving. But Herodotus was writing on a highly emotional subject, and his bias is evident.[70]

Xenophon was another notorious anti-Theban, but his account of the younger Leontiades' betrayal of his city made clear that factional politics were the overriding consideration.[71] When the Thebans defended their Medizing in the Persian War, they also portrayed a city riven by political factions.[72] Thucydides emphasized the frequency with which political factions in Greek polises called in outsiders for help.[73] The fact that Eurymachus's family was a political force in Thebes for three generations shows that the Thebans viewed their political methods as entirely normal. Little wonder that Greek commanders generally viewed betrayal as the quickest and most effective way to capture a fortified city.

Eurymachus and Naucleides worked out a treacherous plan. The Thebans would move against Plataea in two groups. An advance guard of a little more than three hundred men led by Eurymachus and two official magistrates of the Boeotian League called Boeotarchs would approach Plataea at night.[74] Naucleides and his supporters would open one of the city gates to this advance guard. Eurymachus and Naucleides counted on surprise and intimidation to stampede the Plataeans into abandoning Athens for an alliance with Thebes. The entire Theban army was to follow behind the advance guard to intervene if fighting broke out.

Completing the treachery, the conspirators chose to attack during a religious festival at Plataea.[75] They could count on surprise, because the Thirty Years Peace that prohibited a Theban attack on Plataea was still in effect, the Plataeans were preoccupied with the religious festival, and it was early March, well before harvest time and the normal campaigning season. As a result, the Plataeans had taken none of the usual precautionary measures, such as bringing the rural population into the city and posting guards.

At first all went well. The advance guard arrived at Plataea about ten in the evening, and the treachery of Naucleides brought it within the city walls. The Plataeans were asleep, and the noise of a rainstorm combined with the darkness of night enabled Naucleides to lead the Thebans to the marketplace undetected.

Naucleides wanted to make the treachery complete by proceeding without delay to the houses of the democratic leaders of Plataea and killing them. The Thebans, however, believed that a bloodbath would be a poor beginning for the new regime, which they most likely envisioned as a broad-based oligarchy incorporated into the Boeotian League rather than a narrow aristocracy. They preferred a negotiated settlement that at least had the appearance of a voluntary act by Plataea.

The Thebans sent heralds through the streets to wake up the startled Plataeans with shouts to join with Thebes in the name of Boeotian solidarity. Stunned by the discovery of the hated enemy within their walls and full of fear, the Plataeans hastened to negotiate. But as they parleyed with the Thebans and became fully awake, they began to realize that the Theban force was by no means too large to be resisted.[76]

Now the careful plans of Naucleides and Eurymachus began to unravel. The Thebans had underestimated the difficulty of persuading the Plataeans to join the Boeotian League. The Plataeans had allied themselves with Athens perhaps as early as 519 B.C.[77] They had honored the alliance at Marathon in 490 B.C. and, even though they had no experience at sea, at the naval battle of Artemisium in 480. In 479 they had fought alongside the Athenians at the Battle of Plataea.[78] Hatred of Thebes had probably been a stronger motive than hatred of the Persians. The years after the Persian Wars had been golden years for Plataea. After the Battle of Plataea the victorious Greeks had sworn an oath upholding the independence of Plataea. An age of prosperity and artistic flowering had followed.[79]

Most Plataeans were loath to submit to Thebes, and plans of resistance began as soon as they recovered their balance. Just before dawn, when they could still enjoy the cover of darkness, the Plataeans struck.

The Thebans, who had been up all night and standing in a pouring rain for hours, were hard pressed. Most were killed trying to escape the city. The Plataeans trapped one large contingent of Theban troops in a building abutting the city wall, which the Thebans had mistaken for a gatehouse. The enraged Plataeans almost burned down the house with the Thebans in it but settled for taking them captive. Out of over three hundred Thebans who had entered the city, 180 were now prisoners. Most of the others were dead.[80]

Where was the main force of the Theban army? Eurymachus had planned their arrival precisely to prevent the catastrophe that was befalling the advance guard, but the frictions of war had intervened. Rain and mud had slowed the march. The Theban army encountered considerable difficulty fording a flooded river, and by the time it had plodded through the thick quagmire that the road had become and reached Plataea the fighting in the city was over.[81]

When they learned about the Theban captives inside Plataea, the newly arrived Thebans determined to seize those Plataeans who lived in the countryside in order to have hostages to trade for their captured comrades. But just as the Thebans prepared to carry out their plan, a Plataean herald arrived to announce that if the Thebans committed any further hostile acts the Plataeans would kill all the Theban prisoners. Eurymachus and other leading Theban citizens were among the prisoners. The Thebans immediately withdrew their army from Plataean territory.[82]

Afterward the Thebans claimed that the Plataeans had committed themselves to release the prisoners if the Thebans withdrew and that they had sealed their promise with an oath. The Plataeans countered that they had only expressed a willingness to negotiate the release of the prisoners and denied making an oath. The point of the oath was important because oaths were sacred and, as we shall see, the Plataeans later found themselves appealing to their sanctity. In any case, as soon as the Theban army was gone the Plataeans hurriedly brought their rural people and their possessions to safety within the city. Then they killed the Theban prisoners.[83] We do not hear of Naucleides and his fellow aristocrats, but presumably they were killed as well.[84]

Thucydides expressed no preference on the two stories of the oath, and we can never know the truth. But even the Plataeans admitted that they had promised to negotiate the fate of the prisoners, a promise they immediately broke. F. E. Adcock's succinct judgment captures the essence of these terrible events: "Whether they [the Plataeans] added perjury to deceit may be left undecided. This affair, which crowded into twenty-four hours all the vices of war and civil strife, was a fitting prelude to the struggle that went far to undermine the spiritual greatness of the Greek people."[85]

Whether their act was justified or not, the Plataeans had acted foolishly. Hatred and a desire for revenge had blinded calculation. They had deprived themselves of Theban hostages, their surest defense against further Theban attacks. The Athenians recognized this instantly when they heard of the prisoners and rushed a messenger to Plataea to warn against hasty action against the prisoners. He arrived too late.[86] Now the Plataeans faced the certain prospect of attack by a much more powerful enemy whose hatred had been fanned by atrocity.[87]

The Plataeans counted on the Athenian alliance to protect them. They evacuated their women and children and the men unsuitable for military service to Athens. The Athenians sent food and a contingent of about eighty men. One hundred and ten women remained in the city to cook.[88] Plataea was now well prepared to defend itself. But events were not to justify the faith in Athens.

Thebes had suffered grievously, but there could be no immediate revenge. The attempt to overcome the limits of her siege technology by treachery had failed. Now the conquest of Plataea would require a full-scale and protracted military operation needing the resources not only of Thebes but of her Spartan ally as well. The first two summers of the war Sparta concentrated her forces against Athens. The Thebans had to wait.

In 429 B.C. Sparta, wishing to avoid the deadly plague that had broken out in Athens, finally responded to the importunities of its Theban allies. Around the middle of May a large Spartan force, under the command of the Spartan king Archidamus, arrived before the walls of Plataea. As soon as Archidamus encamped his army on Plataean territory and prepared to ravage the countryside, the Plataeans sent envoys to remind him of oaths the Spartans had made guaranteeing the independence of Plataea after the Greek victory over the Persians in the Battle of Plataea fifty years earlier.[89] Archidamus was deeply embarrassed. He was a cautious and pious man with a strong respect for tradition and custom. Before the war he had opposed a rash war policy against Athens, and in advising his fellow Spartans to move deliberately he had praised the Spartan tradition of wisdom, caution, and respect for law.[90] Now he stood accused by the Plataeans of violating Spartan oaths.

The question of the oaths was not the only embarrassment for Archidamus. There was also the question of Plataea's fortifications. He wished to avoid not only the violation of old oaths but also the arduous task of reducing a fortified town. For these two reasons he began with negotiations.

Appealing to the same spirit of liberty with which the Greeks had fought the Persians, Archidamus invited the Plataeans to join the fight to liberate Greece from the Athenian empire. Failing that, he urged the Plataeans at least to become neutral, allowing both sides free access to their territory and city. The Plataeans replied that they did not trust the Thebans and, in any case, they could not act without first consulting the Athenians because the Plataean women and children were in Athens. Archidamus responded with a remarkable offer. If the Plataeans would draw up an inventory of all their possessions, providing, for example, the exact number of their olive trees, and then turn their land over to the protection of the Spartans and migrate to wherever they wanted (Athens would be the obvious choice, and this would mean that the Plataeans could continue to fight with the Athenians in the war), the Spartans would guarantee that the Plataean land and possessions would be returned intact at the end of the war. For the duration of the war, the Spartans would pay the Plataeans a fair rent for their land.[91]

Archidamus's offer did not reassure the Plataeans and with good reason. Spartans were indeed famous for their rectitude among one another but, as the Athenians commented once, a whole book could be written about their conduct toward others.[92] Thucydides emphasized that the Spartans thought the oaths after the Persian War required Plataea's neutrality; therefore the continuing Plataean alliance with Athens released Sparta from all obligations, and the Spartans were moved by regard for Thebes in all their dealings with the Plataeans.[93]

The Plataeans attempted to buy time by requesting a truce so they could consult with Athens. Archidamus agreed, and Plataean messengers hastened to Athens. They returned with fighting advice from the Athenians, promises of help to the best of their power, reminders of the long history of Athenian-Plataean solidarity, and a clarion call in the name of the oaths that bound the two states together not to abandon the alliance.

That was all the encouragement the Plataeans needed. Although one historian has suggested that the Plataean women and children in Athens were virtual hostages binding the Plataeans to the Athenian alliance, tradition and hatred of the Thebans seem to have weighed more heavily in the minds of the Plataeans. In any case, they rejected Archidamus's offer.[94]

Athens' advice to Plataea, combined with what another historian has called the "pathetic loyalty" of the Plataeans, doomed the city.[95] At the beginning of the war Pericles had persuaded the Athenians to adopt a defensive strategy in Attica, avoiding land battles with the Spartans even at the cost of the countryside's devastation. If Athens had not protected its own countryside, how could it promise support to Plataea? The Athenian refusal to release Plataea from its oaths of allegiance was irresponsible and perhaps dishonorable, and modern historians have been harsh in their judgments. Alfred Zimmern, for example, found in the empty promise evidence of moral corruption after the leadership of Pericles ended: after Pericles the Athenians "care nothing for moral issues, and take no heed of saving thoughts."[96]

Sentiment played a large part in these decisions. The strong emotional bonds which had existed between the two cities since they fought together at Marathon produced brave words which proved empty in the deed. Moreover, Pericles lay ill with the plague that was to kill him a few months later, so his austere and selfless leadership was missing.[97] His critics, who advocated a more aggressive military policy, now asserted themselves and must have convinced the Platacans that they would abandon Pericles' defensive land strategy.[98]

Ill-fated as the Athenian advice was, it was not completely irrational. The 480 men garrisoning Plataea were ample to defend a wall that was only about 1,500 yards in circumference and in fact presented only about 750 to 850 yards of wall to terrain that was suitable for an assault.[99] Athenians were contemptuous of Spartan skills in siege warfare. They themselves had recently conducted a long siege of Potidaea at ruinous expense and undoubtedly believed Spartan and Theban resources were inadequate for a prolonged siege, an assumption that underestimated the rich agriculture of Boeotia.[100] Unwarranted optimism is an alluring temptress in war, and Athenians were more vulnerable to it than most.

The Plataeans found it wise not to send word of their decision to resist by herald. They shouted their defiance from the walls.[101] The immunity of a herald was one of the most sacred conventions of diplomacy and warfare in Greek society. The Plataeans' fear for the safety of their herald reveals just how deep the fear of treachery had become and is a sure sign of the collapse of conventional standards.[102]

Having failed to negotiate the capture of Plataea, Archidamus now began siege operations. The Spartans were inexperienced in siege warfare,[103] so their resourceful methods at Plataea, in which they built a siege ramp and employed battering rams, are surprising. From whom did the Spartans learn the innovative siege tactics they used at Plataea? Thucydides does not tell us. He simply attributed the tactics to the Peloponnesians. One explanation for this surprising silence may be that Thucydides' sources were Plataeans who could not tell him the inside story of Peloponnesian methods.[104]

The credit for Spartan innovation in the siege of Plataea must go to Archidamus.[105] Effective siege tactics offered a solution to the major strategic problem of the war for Sparta. Because the Athenian navy was able to guarantee the provisioning of Athens by sea, that city was invulnerable to siege. How, then, could Sparta bring the war to a victorious conclusion?

Archidamus had shown himself before the war to be one of the few Spartans who appreciated this strategic difficulty.[106] A buildup of Peloponnesian naval strength was one possible solution, and there is evidence that the Spartans gave more attention to naval strategy than most modern historians have allowed.[107] Aggressive siege tactics that did not depend on the passive method of blockade were another possibility, and there is evidence that Archidamus hoped to develop such tactics. In the first campaign of the war, he had laid siege to the Attic frontier town of Oenoe, but the delay caused by the siege brought sharp criticism from Spartans impatient to get on with the invasion of Attica and he had to cut short the siege before the place was taken.[108] Historians are unanimous in attributing Archidamus's siege at Oenoe to the political motive of delay which he hoped would buy time to avoid war. But the siege of Oenoe also showed that Archidamus realized from the beginning that siege tactics would be necessary for Sparta. The impatience with Archidamus revealed the lack of understanding among most Spartans for the importance of siege warfare and showed Archidamus to be a lonely pioneer in the art.[109]

Plataea's intransigence offered Archidamus another opportunity to improve Spartan siege methods. The leading expert on Greek siege warfare, Yvon Garlan, believes he succeeded, finding a Thucydidean contrast between word and deed in the Athenian reputation for siege warfare and the Spartan's actual accomplishments that revealed the Spartans as "at least their [the Athenians'] equal in deeds."[110]

The Spartans began with a conventional tactic, cutting down the olive orchards around Plataea.[111] Usually the main purpose of cutting down olive trees and other forms of agricultural ravaging was to draw the defending hoplites out of the city, but it is unlikely the Spartans expected the Plataeans to come out at this point. Their main aim was to get wood to build a palisade protecting them from unexpected sorties.[112] However, Archidamus wished to avoid a conventional siege involving the lengthy process of circumvallation. Rather than slowly sapping Plataean resistance with starvation, he hoped to break it with new tactics and new technology which would make possible active siege tactics instead of the passive method of circumvallation.[113]

First the Spartans set to work building a giant ramp up to the top of the Plataean wall.[114] This is the only time during the Peloponnesian War that a Greek army used this method, so common in the Middle East.[115] The Spartans chose the southern side of the city where the ground sloped upward, offering favorable terrain for the ramp.[116]

This enterprise required intensive labor, and Greek armies had no labor contingents such as Middle Eastern armies possessed. Thucydides made clear that the Spartans themselves did not do the sapping work, but only supervised it. Who served as the sappers?

There are three sources of labor on which the Spartans could have drawn. The Thebans may have recruited laborers from around the Theban countryside. Each Spartan brought several helots with him on campaign to serve as batmen; they could have provided some of the labor. And Theban and Lacedaemonian hoplites drawn from the perioikoi class probably pitched in to help build the mound; they were mostly farmers who would not have been unused to manual labor, and it is likely that they provided most of the labor.

Thucydides reported that the army worked in shifts around the clock for seventy days.[117] That is an incredible amount of time given that the longest Spartan invasion of Attica lasted only forty days, so modern historians have been skeptical. Evidence from the Old Babylonian math problems indicates that it should not have taken nearly so long to complete the ramp. The Babylonians could throw one up in five days. Either Thucydides or our text is probably wrong on the number of days, but that does not alter the fact that the Spartans showed remarkable determination at Plataea.[118]

A high mound of earth, wood, and stones, shored up on the sides with timber, soon began to rise up before the perplexed gaze of the Plataeans. But the Plataeans proved resourceful in defending against the Spartan methods. When they realized the purpose of the mound of earth they extended the height of their wall opposite the mound by constructing a makeshift wall on top of the permanent wall. They also slowed the construction of the ramp by opening a hole at the bottom of the city wall where the dirt of the ramp had piled up and hauling the dirt through the hole as fast as the Spartans and Thebans could bring it up the hill to the city. When the Spartans managed to plug the hole, the Plataeans undermined the ramp by digging a tunnel from the city to under the mound and hauling the dirt out from below. The Spartans and Thebans, both of whom the Athenians considered dull-witted, never realized that they were literally dumping the dirt into a hole.

As a final precaution the Plataeans built a crescent-shaped rampart bending into the city from the part of their wall where the Spartans were building the ramp so that even if the enemy raised the ramp high enough to surmount the wall, they would be confronted with yet another one that would virtually surround them, leaving their flanks terribly exposed to fire.[119] Sparta's aggressive methods had forced the Plataeans to abandon the traditional Greek line defense methods for walls in favor of a defense in depth.[120]

The Spartans now played their trump card. They brought siege engines forward to batter down the wall.[121] Here Thucydides gave the first reliable description of siege machinery in Greek warfare that is detailed enough to leave no doubt that the machines were battering rams, and this may be the first use of rams by Greeks.[122] Whatever the case, the unusual detail Thucydides provided concerning the Spartan siege of Plataea shows that he considered these methods highly significant, for both technical and strategic reasons.[123]

If Archidamus was indeed the author of these tactics, he must be acknowledged as one of the most creative military commanders of the Peloponnesian War. We can only speculate about where the Spartans learned to build and deploy siege machines. Perhaps Greeks from Sicily who had learned siegecraft from the Carthaginians taught them the art.[124]

The Spartans brought one of the battering rams to the top of the ramp, and it had soon knocked down a large part of the extended wall which the Plataeans had built. This unexpected weapon terrified the Plataeans. But terror can be the mother of invention, and they soon devised effective countermeasures.

The Spartan machines were too primitive to have the defensive covering required for rams to be effective. Thucydides did not mention any covering for the crews either, but they must have had some sort of protection because the Plataeans concentrated on disabling the rams rather than killing the crews.[125] In any case, the Spartans certainly did not have the tall mobile towers manned by archers and slingers that could secure the area before the wall from enemy fire and make easier the deployment of men and machines for breaching the wall.[126] Each time the Spartans pushed a ram against the wall, the Plataeans lassoed it with ropes and jerked it aside or smashed it with large timber beams which they dropped from the top of their wall. The battering rams failed to breach the Plataean wall.[127]

The Spartans had not quite exhausted their ideas. They now tried to fire the city by piling brushwood fagots on the top of the ramp and along its sloping sides so that they reached almost to the top of the wall. They then poured sulfur and pitch on the wood and set it on fire. The result was a conflagration so huge that those who witnessed it thought it must have been the largest man-made fire in history. The heat was so intense that the Plataeans could not get near it. The Spartans hoped for a favorable wind to blow the fire over the wall and into the city; but such a wind did not blow, and, to the immense relief of the Plataeans, the fire burned out without igniting the city.[128]

Despite imaginative and aggressive assault tactics, Archidamus had failed to capture the city. Failure to deploy the full range of methods, especially siege towers, archers, and undermining, was one reason.

The failure to use archers was especially critical. The Athenians had a contingent of 1,600 archers in their army, but Thucydides rarely mentioned archers in connection with other Greek forces.[129] The weapon seems not to have commanded much respect from the Greeks, perhaps because of the hoplite victories over Persian archers in the Persian Wars, perhaps because Homer viewed the bow contemptuously. The Cretans had the best archers, and they had the reputation of being liars.[130] When Mytilene was preparing to revolt against Athens, it had to recruit archers from the Black Sea area.[131] When the Spartans raised a force of archers after their stunning defeat on the island of Sphacteria, Thucydides called it "contrary to their custom."[132] Also, although the chorus in Aeschylus's *Seven against Thebes* expresses alarm at a "hail of stone" striking the battlements of Thebes, the Greeks seem not to have been skillful slingers.[133] Thucydides expressed surprise at how effective Acarnanian slingers were.[134] Another exception were the Rhodians; the

Athenians recruited seven hundred Rhodian slingers for the expedition to Sicily, as well as eighty Cretan archers.[135] Interestingly, the Boeotians sent for javelin throwers and slingers from the Malian Gulf during the siege of Delium, a curious move because they already had 10,000 light armed troops on the scene. The Boeotians were about to employ an ingenious siege machine against the wooden walls of Delium. Perhaps they remembered how the lack of cover fire had frustrated the siege machines at Plataea and were going to extraordinary lengths to find slingers.[136] These examples indicate that as siege warfare became more important during the Peloponnesian War, the role of slingers and archers also became more important.

Lack of experience and the innate Spartan suspicion of anything new also undoubtedly hindered Archidamus's efforts,[137] but lack of dash may have been the greatest reason for the Spartan failure. The citizen armies of the small Greek polises were extremely reluctant to incur heavy casualties. Assaulting a fortified position necessarily meant high casualties among the shock troops, and this sort of tactic was foreign to the Greek hoplite accustomed to the security of a closely drawn phalanx.[138] Heavy hoplite armor was unsuited to escalade, and hoplites may have been reluctant to undertake such a hazardous operation without their accustomed armor.[139]

The courage of the Spartans was beyond question, but standards of courage are conditioned by traditional expectations and traditional methods. Herodotus reported that the Spartans assaulted the fortifications of Samos in 525 B.C., but although the Spartans defeated a Samian sortie, they failed to take the city. Herodotus commented that

> had all the Lacedaemonians there fought as valiantly that day as Archias and Lycopas, Samos had been taken. These two alone entered the fortress along with the fleeing crowd of Samians, and their way back being barred were then slain in the city of Samos.[140]

Bursting into a fortification on the heels of a retreating enemy was one of the primary ways in ancient warfare of breaking through a wall. The Athenians demonstrated the method at the Battle of Mycale in 479 B.C. when they broke into the Persian stockade on the heels of fleeing Persian troops.[141] The Spartan failure to exploit this opportunity suggests both inexperience and a distaste for urban fighting. The Spartan mystique, as it came to be cultivated by such men as Xenophon in the fourth century B.C., disdained such fighting.[142] Plutarch's judgment on the death of the Spartan general Lysander at the siege of Haliartus in 395 B.C. caught this spirit:

> He received his death-wound, not as Cleombrotus, at Leuctra, resisting manfully the assault of an enemy in the field; not as Cyrus or Epaminondas, sustaining the declining battle, or making sure the victory; all these

died the death of kings and generals; but he, as it had been some common skirmisher or scout, cast away his life ingloriously, giving testimony
to the wisdom of the ancient Spartan maxim, to avoid attacks on walled
cities, in which the stoutest warrior may chance to fall by the hand, not
only of a man utterly his inferior, but by that of a boy or woman, as Achilles, they say, was slain by Paris in the gates.[143]

After the Battle of Plataea in the Persian War, the Spartans had assaulted the Persian fortified camp without success, a failure that Herodotus attributed to their lack of experience in attacking fortified positions.[144] But when the Athenians came up in support, the Greeks took
the Persian palisade by storm, one of the very few times we hear of
Greeks taking heavy casualties to overcome a wall, if we can believe that
Diodorus did not embellish his account when he said that "many were
wounded as they fought desperately, while not a few were also slain
by the multitude of missiles and met death with a stout heart."[145] It is
important to note, however, that these Athenian successes were against
palisades, not strongly fortified walls such as the Spartans faced at
Plataea.[146]

When the Spartan general Brasidas ordered an assault on a small
Athenian fortified position at Torone in the Chalcidice, he offered thirty
minae to the first man who scaled the wall.[147] Brasidas was commanding mercenaries, so it is not surprising that he offered money, but this
was a huge sum, the equivalent to three thousand days of hoplite pay.
Modern scholars have understandably questioned the figure,[148] but it is
paltry compared with the one hundred minae that Diodorus reported
Dionysius awarded in 397 B.C. to the first man who mounted the wall of
Motya.[149] Even if the figures are exaggerations, the two episodes give
some idea of the magnitude of the problem of motivating Greek soldiers
to assault a fortified position. Neither the technology nor the traditional
military values of the Greeks were suited to aggressive siege tactics. The
common practice of manning walls with superannuated or green
troops showed how little the Greeks expected sudden assaults on their
walls.[150]

By the time Archidamus's troops had carried out their unsuccessful assault on the Plataean wall, they had been away from home for over
a month and the Peloponnesian troops had already missed most of
the harvest. It was late June, and they were needed for threshing.[151] Although Boeotia was one of the richest agricultural areas in Greece, it
was necessary to reduce the size of the army to preserve provisions for
the siege. So Archidamus sent most of the army home. The remainder,
along with the Thebans, began the long, dreary labor of circumvallation
which Archidamus had so desperately wanted to avoid.

They worked at their task all during July and August and completed
the wall around the middle of September.[152] Thucydides' description of
this wall is the first we have of a wall of circumvallation.[153] It was no

mere barricade. The Spartans constructed two walls, both of clay bricks. These walls were about sixteen feet apart and provided protection both against sorties from Plataea and against a possible attack by a relief army from Athens. The digging of clay for the bricks left moats running along the outside edges of both walls. The Spartans roofed over the space between the walls and divided it into rooms to provide quarters for the troops. They constructed battlements on both walls, and at every tenth battlement a high tower spanned the walls.[154] Obviously the Spartans were preparing for a year-round blockade of considerable duration.[155]

Such was the wall that now entrapped the 480 men and 110 women defending Plataea. Having completed their work the rest of the Spartan army returned home, leaving a garrison only large enough to man half of the wall, which was about a mile in circumference.[156] The Thebans took responsibility for the other half, and the long wait began.

Winter and summer passed, and the garrison in Plataea waited in vain for the promised relief from Athens. In December 428 B.C., the second winter of the siege, hunger forced the Plataeans to attempt a daring breakout from the city. Thucydides' unusually detailed account provides a vivid description of this action, the first we have of a battle over a wall.[157]

The daring scheme was the plan of two of Plataea's boldest leaders, Theaenetus, the son of a soothsayer—soothsayers, perhaps reinforced by supreme confidence in their craft, made extremely brave soldiers— and Eupompidas, a Plataean general. Theaenetus and Eupompidas had studied the problem of a breakout thoroughly and developed a plan thought out to the smallest detail and based on excellent intelligence about the routine of the Peloponnesian and Theban besiegers. They counted on surprise and mobility for success.

Crucial to their plan was the construction of scaling ladders with which to surmount the siege wall. These ladders had to be made to exact specifications. If they were too short, they would not reach the top of the wall. If they were too long, they would extend over the top and be easily pushed away by the defenders or extend out from the wall at such a great angle that they would break under the weight of several men clambering up them at once. The Plataeans ascertained the exact height of the siege wall by counting the layers of bricks in the wall, which was close enough for them to see the individual bricks, and then calculating the height from the thickness of the bricks. To get as exact a count as possible, a large number of men counted the layers over and over again until they were satisfied that they had a true count. They were then able to construct ladders that were exactly the necessary length.

At first the plan called for a breakout by the entire garrison, but fear and doubt began to undermine confidence. Given the desperate situation in the besieged city, this reluctance to challenge a fortified wall is strong testimony to the hazards of such actions. The generals preferred not to lead doubters against the wall and called for volunteers. Their decision was undoubtedly influenced by the conviction that morale was more important than numbers and the knowledge that a smaller group would be more likely to achieve surprise. Those left behind could also create a diversion to cover the breakout. About 220 men volunteered for the breakout.

Ordinarily the besiegers manned the battlements of the siege wall continually, but after an uneventful year and a half they had fallen into the habit whenever it rained heavily of retreating into the covered towers located at every tenth battlement. The Plataean plan was to wait for a moonless, rainy night when the guards would be in the towers and, equipped with their carefully constructed scaling ladders, attempt to cross a section of the wall between the towers.

On a thoroughly miserable winter night when an easterly gale was howling and a mixture of snow and rain was falling, Theaenetus and Eupompidas led their volunteers out of the city and toward the siege wall. They had left their heavy armor behind because it was too clumsy and noisy. Lightly armed and keeping far enough apart to prevent their weapons from rattling against one another, they stealthily approached the wall. Despite the cold, each man had kept one foot bare in order to get a better grip in the slippery mud. Darkness enveloped them, and the howling wind drowned out any noise.

In the lead were the men carrying the scaling ladders. Thucydides does not tell us how they crossed the inner ditch. Perhaps they used the ladders as bridges. In any event, they reached the siege wall and managed to place the ladders undetected by the guards, who had, as expected, abandoned the battlements for the shelter of the towers.

The first man up was a doughty soldier named Ammeas, who led a dozen picked men armed only with daggers. Corselets were the only armor they wore. Their task was to seize the towers on both sides of the crossing point. They were supported by light infantry carrying short spears. Shieldbearers, who could hand the light infantry shields if they had to close with the enemy, followed closely behind.

One of these troops, in grabbing the top of the wall as he climbed up, knocked off some tiles, which fell with a clatter. The noise alerted the sentinels on top of the closest tower, who immediately raised a cry. A general alarm was sounded, and the Peloponnesians and Thebans rushed to their battle stations. When the Plataeans who remained in the city heard the alarm, they immediately made a diversionary attack against the siege wall on the opposite side of the city. In the darkness and noise, confusion reigned among the besiegers. Not certain what was happening, they stuck to their battle stations instead of concentrating at the point of the breakout.

The besiegers immediately sought to alert Thebes to the danger by lighting the signal fires that had kept them in regular communication with the nearby city. But here too the Plataeans were well prepared. They had built signal fires of their own in the city and now lit them to confuse the Theban signals.

Ammeas and his men had killed the guards in the towers on each side of them and the Plataean light infantry held the towers, thus protecting the section of wall in between. There was no longer any need for quiet, so the Plataeans tore down the battlements to make it easier to scramble over. The main body of Plataeans was rapidly crossing the wall.

Now, however, they encountered opposition. The Peloponnesians had maintained a 300-man reserve force with which to intercept just such a breakout attempt. After some groping about in the dark, this force now located the Plataeans and rushed at them. But Plataean archers held them at bay, shooting from both the top of the wall and the bottom. As each man reached the bottom of the ladder, he would momentarily halt to throw a javelin or loose a few arrows. This smoothly executed tactic kept up a steady fire of missiles. Moreover, the torches carried by the Peloponnesians illuminated them as targets, while they could scarcely see the Plataeans in the sleet and darkness. The result was a very one-sided fight in favor of the Plataeans.

Having crossed the wall, the Plataeans encountered a final obstacle. The ditch on the outside of the wall was quite deep, and the storm had filled it with water so cold that mushy ice had formed on top. Fording a ditch neck-deep in icy water while holding off the enemy was a harrowing task. The men who had guarded the wall while the main party crossed were especially hard pressed as they tried to cross the ditch. But so one-sided was the fighting that every man but one, who was captured, was able to struggle through the water and reach the far side of the ditch. A few of the 220 men in the breakout party had lost heart at the wall and sneaked back to the city, but 212 made it over the wall and ditch and escaped into the darkness.

The Peloponnesians set out in immediate pursuit, but in this too they were outwitted. Assuming the Plataeans would try to escape to Athens, the Peloponnesians started up the road to the mountain pass leading to Athens. But the Plataeans had planned to go in the direction least expected, down the road straight toward Thebes. As they set out they could look back and see the torches of the Peloponnesians receding up the mountain in the opposite direction. The Plataeans stayed on the road to Thebes for less than a mile and then made their way into the mountains and from there to Athens and safety.

Those who had turned back at the wall told the Plataeans in the city that the entire breakout party had been annihilated. At daybreak they sent a herald to request an armistice so they could recover the bodies according to Greek custom. Only then did they learn the truth.

That the Plataeans accomplished this feat with no casualties and only one captured was a tribute not only to their own daring and skill but also to the careful planning and resourcefulness of Theaenetus and Eupompidas. The key to their success was their reliance on surprise, mobility, and light infantry, especially archers. All of these elements were alien to hoplite warfare.

The breakout reduced the number of mouths to feed and prolonged the defense of the city through the winter, but by summer food was running out and the remaining defenders were starving. We can only imagine the horror within the city. Undoubtedly the women cooks suffered the most, as they would have been on the shortest rations and the first to be denied food altogether. By the middle of the summer food supplies had been completely exhausted. A reconnaissance in strength by the besiegers revealed that the Plataeans were too enfeebled to defend the city any longer. When the Spartans demanded that the Plataeans surrender, they reluctantly agreed.[158]

LIMITATIONS OF SIEGES

The Spartan siege of Plataea typified the Greek experience of siege warfare during the Peloponnesian War. For example, the failure of Spartan siege engines at Plataea was not unusual. The Athenians, the most experienced at siege among the Greeks, had accomplished nothing with whatever machines they had used at Potidaea.[159] Demosthenes utilized siege engines against a Syracusan counterwall in Sicily, but the Syracusans frustrated him by setting fire to the engines.[160] Toward the end of the war, the Athenians deployed siege engines against Eresus but had no more success with them than they had at Potidaea at the beginning of the war.[161] In fact, the only successful Athenian use of siege engines during the war was the capture of a fortified tower on the island of Minoa opposite the port of Nisaea by Nicias, who deployed these machines from the decks of ships.[162] These "machines" were probably scaling ladders.[163]

There is not a single successful use of battering rams by Greeks during the Peloponnesian War. The most impressive siege engine successfully deployed during the war was a flame-throwing device that the Boeotians turned against the fortified Athenian camp at Delium. By attaching a bellows to an iron-plated, hollowed-out wooden beam with a curved iron tube that reached from a cauldron of fire through the hollow beam, the Boeotians were able to blow an intense flame through the beam and against the wooden walls of the Athenian camp. The fire drove the defenders away from the wall, opening the way for the Boeotian assault.[164] The wall at Delium, however, was not a permanent fortification, but a hastily built wooden palisade.

Shortly afterward, Brasidas tried a similar device against the Athenians at Lecythus, a fortified place at Torone, whose weak, wooden fortifications promised success. The Athenians prepared countermeasures by building a wooden tower from which they intended to pour water on the fire. But the weight of the water collapsed the tower, and in the confusion the quick-witted Brasidas ordered an assault. Its success can only indirectly be credited to the flame thrower, which never reached the wall.[165]

Without effective siege engines and effective tactics, such as covering fire, it is not surprising that Greeks shied away from assaulting fortified positions. Although Thucydides does not always tell us how a town fell, there are no clear cases in his *History* of the capture of a fortified town by assault, except under very limited conditions.[166] He almost always explained assaults by reference to the absence of fortifications, the weakness of the fortifications, or their state of disrepair.[167] A strong force could take relatively weakly fortified places, as the Boeotians did at Delium and Brasidas at Lecythus.

On rare occasions, such as the Spartan capture of Iasus by sudden assault when the Iasians mistook the Spartan ships for Athenian, successful assaults were achieved by surprise.[168] Those less lucky than the Spartans at Iasus had to resort to desperate measures to catch a fortified position by surprise. Two of the most resourceful commanders of the Peloponnesian War, Brasidas and Demosthenes, tried night attacks, Brasidas at Potidaea in 422 B.C. and Demosthenes against Syracusan fortifications on Epipolae in Sicily in 413 B.C. Both failed.[169] Brasidas did capture Torone in a sudden assault, but Torone's walls were in a state of disrepair and traitors had opened the postern and market gates for him.[170]

Xenophon and, even more, Diodorus, described assaults more often than Thucydides. In the case of Xenophon this may reflect a necessity for assault imposed by the lack of financial resources for a siege in the later years of the Peloponnesian War.[171] But their accounts are not always credible. For example, Xenophon said that the Spartans took Methymna, a small city on Lesbos, by storm, a remarkable accomplishment, because Methymna had stood firm against a combined force of Spartans and Mytilenians in 428 B.C. and again in 411 B.C. against some Methymnian rebels supported by three hundred mercenaries.[172] But Diodorus told a different story. He reported that repeated assaults against Methymna failed and that the Spartans only gained entrance into the city by betrayal.[173]

Although Diodorus was more plausible in that case, in general he spoke of repeated assaults so often that it appears to be a formulaic phrase he employed whenever he was describing a siege. Diodorus's flawed account of the fall of Plataea shows that he was not reliable on these matters. He said that the escape of part of the garrison from Plataea so angered the Spartans that they mounted an assault the next day and forced the Plataeans to surrender.[174] Thucydides' detailed account, as we have seen, made it clear that Plataea held out for several more months until starvation had so weakened its defenders that they could no longer resist and that even then the Spartans preferred a negotiated surrender to an assault.

When the failure of assault methods at Plataea forced the Spartans to use a blockade, they were turning to the preferred siege method of Greeks before the fourth century B.C.[175] This was especially true of the Athenians, whose naval power and skill at circumvallation made blockade their most effective siege weapon. The Athenians were very good builders. When they and their allies began a wall of circumvallation around Epidaurus, only the Athenians managed to finish their section.[176]

The Athenian army was well organized for the construction of siege works. Each detachment was responsible for building a section of the wall.[177] Although the Athenians who landed at Pylos were caught without their sapping equipment, tools for building fortifications were normally a part of the Athenian army's equipment.[178] Often the Athenians utilized special craftsmen, such as masons and carpenters, to help build the walls, as they did at Nisaea and in Sicily.[179] In the latter case, Demosthenes even took his masons and carpenters on a nighttime assault against a Syracusan counterwall, such was his anxiety to throw up an Athenian wall as quickly as possible.[180] But hoplites did the bulk of the work.[181]

We have clear evidence that hoplites were not unwilling to labor as sappers. When the Athenians landed at Pylos in the Peloponnese, one of the generals, Demosthenes, wanted to fortify the place, but he was unable to persuade the other officers of the wisdom of his plan. As the Athenian force lingered at Pylos, however, the soldiers became bored and spontaneously decided to pass their time by constructing fortifications. The labor was backbreaking because they had no iron tools or hods, but the men worked so industriously that they completed some rudimentary walls in six days.[182]

The Athenians possessed large numbers of slaves, and they were present at both the siege of Potidaea and the siege of Syracuse, but there is no evidence that the slaves constituted a labor gang for sapping operations.[183] These slaves did not belong to Athens but were the personal servants of the hoplites. In a careful study of the role of Athenian slaves in warfare, Rachel Sargent identified the following tasks for slaves in the Athenian army: cooking, serving as guides, rescuing the wounded, carrying messages. Most of all, however, they served as batmen for their masters, carrying their baggage and armor and caring for the armor.[184] We cannot discount the possibility that these slaves may have assisted their masters in sapping work, but we have no direct evidence that they did. At most the Athenian slaves may have worked side by side with the hoplites, as we find both slaves and freemen working in the construction trade in Athens.

During the Mytilenian revolt, Athenian hoplites showed their toughness by first rowing the ships to Mytilene and then building a wall of circumvallation.[185] The dual role of hoplites as sappers and fighters was made clear by Thucydides when he said that when the Athenians arrived before Syracuse they built a fort at Labdalum so they would have a secure place to store their equipment "whether they advanced either to fight or to work on the wall"[186] and when he tells us that the Syracusans hoped their counterwall would force the Athenians to "cease from their work" to attack the Syracusan position.[187] If the hoplites were not essential to wall construction, it would have been possible to continue working on the wall while the hoplites attacked the Syracusan wall.

This readiness of the hoplite warrior to perform manual labor was the key to Athenian siege methods. Only a society that admired hard work could turn so easily to the labor of free men rather than the press gang. Hesiod's *Works and Days*, with its emphasis on the value and dignity of labor, provides an insight into the character of the hoplite class in Greece that produced these laboring warriors.

The greatest Athenian strength in fortification building was speed. At Nisaea, the Athenians had almost finished their blockade wall in a day and a half.[188] They had just about completed their fortifications at Delium in three days.[189] The fortifications at Pylos took six days to complete, even though the men had no tools (although much of the position was naturally fortified).[190] When the Athenians began to close in Syracuse with forts, the Syracusans were dismayed by the speed with which the Athenians constructed these forts.[191]

There then followed, however, the greatest failure in the history of Athenian siege warfare, the failure to complete the wall of circumvallation at Syracuse. Garlan believes the unequal progress on the Athenian walls at Syracuse reflected a general lack of enthusiasm by hoplites for this sort of labor, but there are no other such cases in Thucydides' *History*. The dispirited leadership of Nicias and the tenacity of the Syracusan defense are the most likely causes of the dilatory progress at Syracuse.[192]

The Spartans took two-and-a-half months to build their wall of circumvallation around Plataea.[193] The Spartans, however, were more thorough than the Athenians. Ever since they had hastily rebuilt the walls of Athens after the Persian Wars, the Athenians had shown themselves as masters of improvisation, able to construct walls quickly out of whatever materials were at hand.[194] At Pylos they were able to fit unshaped stones together.[195] At Delium they used wood from a nearby vineyard and tore down houses to get stones and bricks.[196] At Nisaea, the Athenians cut down orchards for wood, tore down houses for bricks and stones, and made other houses, which were located along the line of the wall, a part of the wall so that they only had to build battlements on the roofs.[197] The Athenians blockaded Calchedon with a simple wooden palisade.[198] There is no indication, with the exception of Delium, that these makeshift fortifications were inadequate for their purpose.

In short, as late as the Peloponnesian War Greeks rarely assaulted fortified positions, and when they did they usually failed. If betrayal or ruse did not work, the preferred method was blockade. Passive siege was less violent than assault, but it caused no less suffering. Starvation was its main weapon. Near the end of the Athenian siege of Potidaea, the defenders were so hungry they resorted to cannibalism.[199] The Plataeans surrendered only when starvation had so weakened them that they were physically unable to resist any longer.[200] Xenophon reported that during the Athenian siege of Byzantium women and children in the city died because the Spartan commander of the garrison gave all of the scarce food to the soldiers defending the city.[201] Thucydides mentioned a desperate famine in Chios during the Athenian siege of that city.[202] The Peloponnesian War ended with Athenians dying of starvation during the Peloponnesian siege of Athens.[203]

That a prolonged total blockade of a city was an ordeal for its inhabitants is obvious enough, but it was also a grim business for the besiegers. The logistics of maintaining an army in the field was sometimes more difficult than the logistics of the defenders. Generally Greek siege armies had to maintain themselves by a variety of expedients, such as pillaging the countryside, launching raids into neighboring territories, and establishing markets to attract traders.[204]

If circumstances prevented a siege army from using such methods, its position could be precarious. At Pylos the Athenians guarding the island of Sphacteria suffered greatly from food and water shortages. They were under siege themselves, and only one small spring was available. Most of the men had to dig for water on the beach. The Athenians supplied the army by sea, but the lack of a port made landing supplies a difficult and agonizingly slow task. On the other side, the Spartans succeeded in keeping the men who were trapped on the island relatively well supplied by paying boaters and swimmers to break the blockade.[205] In this case, the besieged could easily have outlasted the besiegers. The Athenians had to turn to assault to bring the siege to a successful conclusion. The Athenians suffered from similar logistical problems in Sicily after they lost their forts at Cape Plemmyrium, which guarded the entrance to the harbor.[206] On the other hand, there is no evidence that Syracuse was ever short of supplies.

Plague was one of the greatest dangers of prolonged siege operations. Crowded and unsanitary conditions, often with an inadequate water supply, made besiegers vulnerable to infectious diseases. The Athenian general Hagnon lost 25 percent of his troops to the plague in just forty days at the siege of Potidaea.[207] The terrible suffering of the troops was a main reason why the Athenians there accepted a less than

unconditional surrender. In perhaps our first evidence of the gulf dividing the home front from frontline troops, the Athenians back home,
with no comprehension of what the soldiers had endured, were furious over a settlement they considered a sellout.[208] The Athenian army at
Syracuse, especially after it based itself at Plemmyrium, which lacked
adequate drinking water, suffered terribly from disease.[209]

A herald in Aeschylus's *Agamemnon* expressed the Greek soldier's
view of siege warfare when he complained about his duty at Troy:

> Ashore there was still worse to loathe; for we had to lay us down close
> to the foeman's walls, and the drizzling from the sky and the dews from
> the meadows distilled upon us, working constant destruction to our
> clothes and filling our hair with vermin. And if one were to tell of the
> wintry cold, past all enduring, when Ida's snow slew the birds; or the
> heat, what time upon his waveless noon-day couch, windless the sea
> sank to sleep . . . [210]

Cost added to the difficulty of sustaining a long siege.[211] Greek cities lacked the financial resources available to Middle Eastern imperial
powers, which could draw tribute from a wide area. Only Athens commanded the resources of empire, but sieges pressed even Athens to the
limit.

During the siege of Potidaea, the Athenians paid their hoplites two
drachmas a day, one for the hoplite himself and one for his batman.[212]
They maintained three thousand hoplites for the duration of the siege,
so the siege cost Athens six thousand drachma (one talent) a day just in
hoplite wages. The siege lasted over two years and cost two thousand
talents.[213]

Pericles estimated the annual revenues of Athens at around six hundred talents a year. In addition, there was a reserve of six thousand talents of silver, and five hundred more talents could be scraped up from
miscellaneous sources. Finally, in extraordinary emergencies, there was
the gold plating on the statue of Athena in the Parthenon, worth forty
talents.[214] Normal expenditures consumed most of the annual revenues.
The Athenians set aside one thousand talents from the reserve at the
beginning of the war for use only in the extreme emergency of a naval
attack on Athens itself.[215] In short, the siege cost two-fifths of Athens
available reserves.

The staggering cost of the siege forced the Athenians to impose an
unusual levy on the citizens to raise two hundred talents. They also
sent out a naval squadron to exact a special levy from the empire.[216] It
is no wonder that the cost of siege was very much on the minds of Athenian leaders such as Diodotus in urging a moderate policy toward Mytilene and Demosthenes in urging an Athenian withdrawal from Sicily.[217]
By the last years of the war, Alcibiades had to admit to his men that
Athens could no longer afford a siege and that they would have to

assault the walls of Cyzicus.[218] The ruinous cost of siege is a leitmotif running through Thucydides' *History*.[219]

Given these difficulties, Greek armies tried to avoid sieges. Carrot and stick was the preferred method. The Spartan general Brasidas, during his Chalcidic campaign in 424 B.C., offered complete freedom to those cities that quit the Athenian empire and surrendered to him, but he threatened to ravage the territory of those that resisted.[220] This policy was highly successful.

When sieges were unavoidable, the Greeks tried to cut them short. With adequate equipment and tactics for assault lacking, by far the most effective means was the encouragement of treachery within the besieged city. The Theban plan to take Plataea began with treachery, and if it failed in that case, such plans often succeeded. More cities fell through betrayal than any other means during the Peloponnesian War.

One of the more colorful examples of victory through ruse and treachery was the Athenian capture of the long walls that connected Megara with its port at Nisaea. For weeks before the Athenian attack, Megarians who wanted to betray their city to Athens had been carrying a boat through the gates every night, claiming they were making night raids on the Athenian ships blockading Megara. Toward daybreak they would bring the boat back within the walls, explaining they did not want the Athenians to find it. This nightly charade soon became a routine for the unsuspecting guards at the gate. One night, Athenian hoplites hid in a ditch near the walls and an advance company of light armed troops crept up closer to the gate. When the Megarian traitors carried their boat back through the gate, they killed the guards, and with perfect timing, the light armed troops burst through and secured the gate for the hoplites who charged through and captured the long walls.[221] Although this plan was more elaborate than most, it was not untypical of the preferred way to capture a fortified wall. Soon afterward, the Spartan Brasidas captured Torone in a similar way. Traitors admitted into the city peltasts, who then captured the gate, opening the way for Brasidas's main army to storm the town and take it.[222]

One of the most blatant cases of treachery occurred in 427 B.C. when the Athenian commander Paches captured a fortress at Notium that was defended by Arcadian and foreign mercenaries. Paches invited the Arcadian commander Hippias to negotiate, promising him that he could return to the fortress if they failed to agree on terms. When Hippias came out to parley, Paches seized him and launched a sudden attack on the fortress. This unexpected assault succeeded, and the Athenians captured the fort. They killed all the defenders. As for Hippias, Paches brought him back into the fortress "safe and sound" as promised and then ordered Athenian archers to shoot him down.[223] Thucydides' tone is disapproving in his description of this episode, but such treachery was not unusual in siege warfare.[224] Perhaps the hypocrisy of Paches' formal fulfillment of his promise to Hippias, rather than the treachery itself, annoyed Thucydides.

Treachery was commonplace among the Greeks. Odysseus, whose cunning and resourceful lying and trickery were much admired, was the archetype. A. W. H. Adkins has shown how the Greeks owed their loyalties primarily to family and friends rather than to the polis. The wise and moderate Solon prayed forthrightly that he might be able to bring suffering to his enemies and sweetness to his friends. In a political context, *enemies* meant one's political opponents.[225] Plato has Meno echo the conventional view that *areté*, the concept of excellence which was central to Greek ethics, meant demonstrating in the political arena the ability to help friends and harm enemies while avoiding harm to oneself.[226] Indeed, Socrates was the first to suggest that it is better to suffer than to do wrong. In pre-Socratic Greece anything was permissible in trying to strike a blow against an enemy.

Adkins tellingly describes the moral roots of Greek political behavior:

> In both foreign and domestic politics . . . it was essential for any individual or state that claimed to be *agathos* [noble or good] not to submit to any situation which entailed, or appeared to entail, defeat, reduction of *eleutheria*, freedom, or *autarkeia*, self-sufficiency; for any such situation reduced or abolished one's *areté*, the most important quality. One might have to bide one's time; but so soon as an opportunity offered, the most powerful values of the society demanded that one should worst one's enemies, within or without the city, whatever the rights and wrongs of the case reckoned in terms of *dikaiosune* [righteousness or justice].[227]

This weakness of the Greek polis rarely manifested itself in hoplite battles because the hoplites constituted a relatively homogeneous social and political group. The demands of phalanx warfare encouraged solidarity and cohesiveness.[228] Siege warfare, which encompassed the whole polis society and placed it under extreme stress, had the opposite effect, often fracturing the polis along its social and political fissures.

One method of bringing a polis under stress and undermining its morale without resorting to siege was to establish a fortified position in the countryside. This placed the burden of taking a fortified position on a city's defenders and amounted to an offensive use of fortifications. From such a position, the invaders could control the country surrounding the city and prevent agricultural production altogether. The devastating economic consequences of this tactic are obvious. But it also had an important strategic social and political goal.[229] This is apparent in the two most notable examples of this strategy during the Peloponnesian War.

When Demosthenes established a fortified position at Pylos in Messenia on the western coast of the Peloponnese, he hoped that it would become a refuge for escaped helots and even serve as a base for a general helot revolt. He thus aimed to strike against the most dangerous fault line in Spartan society.[230] When the Spartans fortified Decelea in Attica, they exploited the split in Athens between conservative agrarian interests, which wanted to protect the countryside, and the urban, democratic, imperialistic interests, which had followed the Periclean strategy of remaining behind the walls rather than defending Attica.[231] The walls of both Athens and Plataea protected democracy and symbolized the division between democrats and oligarchs. Indeed, an important part of the aristocratic ethos after the Peloponnesian War became the rejection of fighting from behind walls in favor of the heroic open-field battles between hoplites.[232] It is no accident that civil disturbance and revolution racked Athens in the years after the Spartans fortified Decelea. Moreover, just as Pylos became a refuge for helots, Decelea was a magnet for Athenian slaves. As many as twenty thousand escaped Athens to Decelea.[233]

Aineias Tacticus, writing around the middle of the fourth century, devoted the bulk of his *How to Survive under Siege* to the prevention of treachery, and he offered a rich variety of ways in which treachery could occur. Aineias was probably an Arkadian general from Stymphalos, and he probably wrote his manual of siege warfare in the 350s B.C.[234] His treatise makes evident the tremendous tensions in a city under siege by focusing for the most part on the danger of treachery. Treachery could come from a number of sources: social divisions within the polis, political exiles, families of hostages, mercenary troops. His book provides a portrait of polis society under siege.[235]

Defensive siege strategy had social and political implications. Aineias tried to mediate between the old hoplite strategy of marching out to meet an invading army in order to protect the crops and the urban democratic strategy of abandoning the fields to fight from the fortified town.[236] Because this strategic dilemma reflected the class divisions within a Greek polis, Aineias saw the central defensive problem in siege warfare as maintaining the unity of the polis.[237] Aineias thought plotters within the city were more dangerous than the enemy without.

The first task in preparing a city's defenses was to identify loyal troops and commanders. Aineias thought those most "satisfied with the status quo" and with "most to fear from a change of regime" would be the most reliable.[238]

Guards and gatekeepers occupied especially critical posts. The safety of the city lay in the hands of guards and Aineias feared their potential for betrayal. Their watches should be short, not only to keep them alert but also to allow less opportunity for them to make contact with the enemy. Many guards should stand watch to lessen the danger of a cabal among them. Commanders should rotate to avoid cozy relations between the guards and their officers. Guards should never know their schedule ahead of time.[239] The most sensitive spots should be guarded by "the wealthiest and most highly respected men, those with the largest stake in the community."[240] Access to the top of the wall should be blocked. This would make it difficult for guards to abandon their positions, for plotters to seize a section of the wall, and for enemies to get from the top of the wall into the city.[241] Also no open spaces in the city should be left unoccupied, because they might serve as a collecting place for plotters. Aineias even advised the digging of trenches around open areas to limit access to them.[242]

Aineias gave a long list of examples of how gatekeepers had betrayed their cities.[243] Not only should the gatekeepers be vigilant and hard to trick, they should also "be well-to-do individuals with something at stake in the community—children and wife, I mean—and not men whom poverty, or the presence of obligations, or desperation of some other kind might leave open to being persuaded to join a revolution."[244]

If his frequent advice to place the well-to-do in the most sensitive positions suggests that Aineias thought the land-owning classes were the most reliable during a siege, his strategy by no means was subordinated to their interests. To be sure, he assumed the rural population would remain near their crops during harvest time to protect them, and he advised only how to bring them into the city at night.[245] But if a superior force approached the city, then there was nothing to do but withdraw into the city. Aineias even advocated the destruction of standing crops and poisoning of wells to render the enemy's logistical position difficult.[246] If possible, the well-off could save their livestock and slaves by placing them in the custody of a friendly neighboring town, but under no circumstances should the citizens bring them into a besieged city.[247]

When members of the rural population came into the city, they should bring all their property with them so they would have no incentive to betray the city in order to save their estates.[248] Indeed, rural landowners might pose a threat to discipline, and commanders should take special care to prevent disorganized sallies from the city by landowners desperate to protect their possessions.[249]

Nevertheless, Aineias did not recommend a passive defense. Well-organized sallies properly timed to strike the enemy when he was most vulnerable—in the midst of looting, drunk with stolen wine, and laden with booty—were an important part of Aineias's defensive strategy.[250] If the enemy managed to escape with booty, a well-planned ambush might recover it.[251] Also Aineias advocated opposing the enemy's initial approach to the city in order to avoid a siege altogether.[252] His strategy was a nuanced one that tried to balance the interests of the citizens with practical military considerations.

All Greek polises had exiles, and Aineias considered them a dangerous threat during sieges. He advised the city to forbid any contact with exiled citizens and to set a bounty on the heads of powerful exiles. Officials should read all incoming and outgoing mail. No one should leave the town without an identification token.

Foreigners presented a special problem. They should be disarmed. Innkeepers should keep registers of all guests, and the authorities should lock up inns from the outside every night. Vagrants should be expelled from the city. Shops and markets should be closed at night and a blackout imposed. Anyone needing to leave home should be required to carry a lantern. Informants should be encouraged by monetary rewards for reports of any plots.[253]

Aineias thought it was imperative to rid the city of malcontents during a siege. Sending them on official missions abroad was a good way to accomplish this without precipitating violence.[254] If the enemy held any hostages, their relatives could be another source of subversion. It was best to get them out of the city, but if that was not possible they should not be given important duties and should be kept under constant surveillance.[255]

Aineias knew that measures of strict control alone could not maintain unity. He emphasized the importance of morale during a siege and favored positive steps to keep it high, such as providing relief to debtors by canceling interest payments or, in times of extreme crisis, canceling debts altogether. The city should also see to the needs of the poor so that their discontent should not become a source of friction during a siege.[256]

Commanders should avoid harassing their men and instead encourage them to maintain a high morale. Even sleeping guards should be handled with circumspection.[257] Commanders should never lose their temper:

> Give appropriate encouragement—a word of praise here, an appeal there—to each of the men fighting on the wall. But do not make the rank-and-file even more disheartened by [losing] your temper.[258]

Aineias was especially solicitous of combat troops, suggesting the use of wheeled transport to bring them to the battlefield fresh. The wagons could then carry the wounded back into the town.[259] The knowledge that provisions had been made for the wounded undoubtedly would help the morale of the fighting men.

Aineias Tacticus's aggressive defensive tactics were relatively new in Greek siege warfare. Before the Peloponnesian War the defensive tactics of Greek cities under siege tended to be passive.[260] The lack of posterns in Greek walls until at least the late fifth century B.C. reflected passive defense methods. Posterns served mainly as exits for sallies against besieging forces, and their absence shows that Greek defenders preferred to remain behind their walls.[261] In the sieges of Potidaea and Plataea at the beginning of the Peloponnesian War, the defenders relied on the strength of their walls and their staying power and defended their cities only from their walls.

As the war progressed, defensive tactics began to change away from a linear defense of the wall to a defense in depth. Thucydides' description of the defense of Syracuse shows this trend. When word of the Athenian expedition reached Syracuse, the assembly met to ponder this intelligence. Hermocrates, the most far-sighted and energetic Syracusan leader, favored an aggressive strategy of sending the fleet to Italy to challenge the Athenians as they crossed the Ionian Sea, but it is not clear what fleet he had in mind because the Syracusans had none. It would have taken two months to build one.[262] Even Athenagoras, the democratic leader who did not believe the report, envisioned meeting the Athenians at their point of landing in Sicily.[263]

The Syracusan generals adopted a vaguer but implicitly more passive policy of simply mobilizing for war, and in the event the Syracusans opposed neither the Athenian voyage to Sicily nor their landing on the island.[264] Even reconnaissance voyages by Athenian ships into the Great Harbor at Syracuse encountered no opposition because the Syracusans still had not begun to prepare a fleet.[265] Fortunately for Syracuse the Athenian invasion developed slowly, giving the Syracusans time to prepare their defenses.

One Athenian general, Lamachus, had wanted to move against Syracuse immediately, hoping to catch its rural people and their supplies still in the countryside, the most auspicious beginning to a siege.[266] Thucydides knew the importance of morale in siege warfare and keeps us consistently informed about the state of mind in Syracuse. He believed that Lamachus's plan offered the best opportunity for an Athenian victory, because the Syracusans were at first intimidated by such a large expedition against their city. Maximum pressure immediately might have resulted in a surrender. But the Athenians did not follow Lamachus's recommendation, allowing the Syracusans precious time.

Instead, Nicias led the Athenians on a cruise along the northern coast of Sicily, where their only accomplishment was the capture of the small town of Hyccara. This represented no small gain, because the Athenians enslaved the entire population and sold it into slavery for 120 talents. But after returning to their winter camp at Catana, the Athenians failed to take the Sicel town of Hybla Geleatis.

Greatly encouraged by this demonstration of the limits of Athenian siege capacity and the Athenian failure to move against Syracuse immediately, the Syracusan attitude toward the Athenians turned from fear to contempt. The people demanded that the generals take the offensive. This eagerness enabled the Athenians to lure the Syracusan army away from the city, allowing the Athenians to board their own army on ships, sail into the Great Harbor, land unopposed, and establish a fortified camp beneath the walls of Syracuse. When the outwitted

Syracusans hurried back, the Athenians defeated them in a hard-fought battle in which 260 Syracusans died. The Athenians lost only fifty men. However, the Athenians did not follow up on this victory. On the next day, they boarded their ships and sailed back to their camp at Catana.[267]

Lack of cavalry caused the Athenians to make this controversial decision, which surrendered any strategic advantage gained by the tactical victory. Although the Athenians were well-equipped for wall building and had brought 480 archers and 700 Rhodian slingers, they needed cavalry to protect their sappers against harassment from the Syracusan cavalry.[268] Therefore Nicias, the cautious Athenian commander who doubted Athens' capacity to defeat Syracuse, decided to withdraw to winter camps at Catana and Naxos and send to Athens for more cavalry.[269] During the winter, Nicias satisfied himself with gathering bricks and iron for a circumvallation.[270]

The Syracusans made good use of this respite. Hermocrates raised their spirits by pointing out that lack of experience rather than lack of courage had brought defeat. He convinced the Syracusans to expand their hoplite force by arming those who could not afford the panoply of the hoplite and to spend the winter in training. He also convinced them to streamline their command under the authority of three generals.

Most important, the Syracusans extended their walls to take in a larger area in order to make it more difficult for the Athenians to circumvallate Syracuse. In addition, the Syracusans built forts at Megara and Olympieum to guard against Athenian landings and erected palisades along all the possible landing places on the shore of the Great Harbor.

However, they did not confine themselves to passive defense. When the Athenian army sailed away from Catana to Messene in the hope that their allies in the city would betray it to them, the Syracusans immediately marched against Catana, ravaged the countryside, and destroyed the Athenian camp there.[271]

The Syracusans also began a diplomatic campaign to isolate the Athenians, sending envoys to the Peloponnese to seek help from Sparta and Corinth.[272] They sent Hermocrates to Camarina to oppose an Athenian envoy who was attempting to persuade the Camarinaeans to support Athens. Thucydides tells us the Camarinaeans were more inclined to Athens than Syracuse but out of fear tilted toward Syracuse, telling evidence that Athens had lost its aura of invincibility. Finally, the Syracusans sent out garrisons to their Sicel allies to thwart an Athenian campaign against them.[273] In the spring, the Syracusans received further encouragement when the Athenians failed to take a Syracusan fort at Megara.[274]

Soon afterward, however, the situation took a turn for the worse for the Syracusans. The Athenians received cavalry reinforcements, which eliminated the cause of their relative lack of action against Syracuse. Upon hearing this, the Syracusans moved to guard Epipolae, a strategic

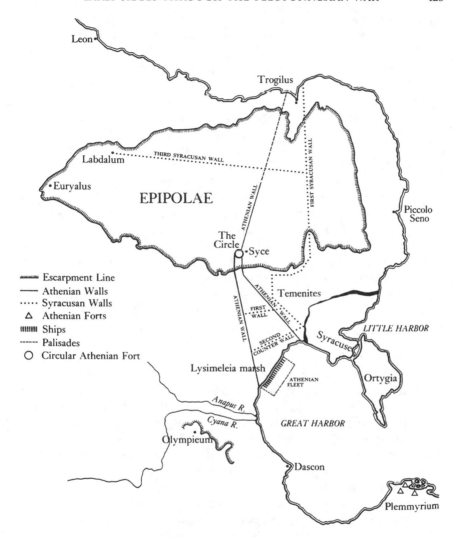

Map 1. Athenian siege of Syracuse. *Used by permission of the publisher, Cornell University Press, from Donald Kagan,* The Peace of Nicias and the Sicilian Expedition, *copyright 1981 by Cornell University Press.*

high plateau overlooking the city that the Athenians would have to control in order to build a wall of circumvallation (see map 1).[275] This move was not unusual in Greek tactics. The defenders of Scione in 423 B.C. had also gone out to defend a hill, the possession of which was necessary for the Athenians to invest the city.[276] The Athenian victory on the hill sounded the death knell for Scione. However, the Syracuse move against Epipolae came a moment too late. The Athenians beat them there, and the position was so naturally strong that the Syracusans were unable to drive them off despite a determined, if somewhat disorganized, effort. Three hundred Syracusans died in the battle.[277]

Even if we can credit excellent Athenian intelligence for this coup, the failure of the Syracusans to protect the strategic terrain above their city was unforgivable.[278] To be sure, as soon as Hermocrates became general he moved to secure Epipolae.[279] But Hermocrates' influence had been great ever since the Syracusan defeat the previous fall, and he had convinced the Syracusans to take a number of steps in defense of the city. He cannot be entirely exempted from the blame for this incredible blunder. The fortification of Megara and Olympieum instead of Epipolae suggests the Syracusans expected the Athenians to attempt another amphibious landing close to the city rather than to attack Epipolae. Only the arrival of the Athenian cavalry seems to have jarred them out of their illusion, but by then it was too late.

With Epipolae in firm grasp, the Athenians were able to begin the laborious task of circumvallation. They built a fort at Labdalum, a place on the northern edge of Epipolae, to serve as a base of operations and a circular fort in a more advanced position on Epipolae to serve as an anchor for a wall of circumvallation projecting north and south from the fort.[280]

Galvanized by the stunning loss of Epipolae and terrified at the speed with which the Athenians were constructing their fortifications, the Syracusans strained every nerve to prevent the circumvallation of their city. Rather than passively allowing the Athenians to build a wall of circumvallation as the Plataeans had allowed the Spartans and Boeotians to do, the Syracusans contested the terrain in front of their walls. However, the undiscipline of their hoplite ranks forced the Syracusan generals to abort a march in force against the Athenians in favor of a cavalry sortie against Athenian siege works. The newly raised Athenian cavalry beat it off.[281]

The lack of discipline in the ranks of Syracusan hoplites revealed that their winter training had been inadequate. Hermocrates therefore launched a strategy of counterwalls. This was partly the defensive analogue to the passive offensive strategy of circumvallation. But it also reflected new defense-in-depth methods that were replacing the traditional linear wall defense methods during the Peloponnesian War.[282]

The Syracusans built their counterwall along a line that intersected the line of the Athenian wall of circumvallation stretching southward to the Great Harbor. The Athenian fort at Labdalum and the circle fort would have made work in the northern sector hazardous. A southern counterwall enabled the Syracusans to work with no threat in their rear, and, tied in with their fort at Olympieum, it would keep open a route into the interior of Sicily.[283] The wall was not weak. The Syracusans built wooden towers along it and also constructed stockades to protect the counterwall.

This counterwall presented the Athenians with a dilemma. If they detached troops from wall building to attack the Syracusan counterwall, they risked losing the race to complete the walls. An attack would leave their left flank exposed to a counterattack from the city. Faced with these uncertainties, the Athenians chose to concentrate all their efforts on continuing their wall northward in the opposite direction from the Syracusan counterwall. This had the advantage of securing their supply route from Thapsus, where the fleet was stationed. The Athenians also cut the underground water pipes going into Syracuse, but the city had wells and springs, so this only caused inconvenience.[284]

Inexplicably, the Syracusans became complacent and left their stockades too lightly guarded. Perhaps they suffered from a manpower shortage in face of the need to defend their extended walls, man the palisade, and work on the counterwall at the same time. This tempting opportunity drew the Athenians away from their walls.

A surprise attack at midday caught most of the Syracusans taking a siesta in their tents (some had even returned to the city to eat with their families) and drove them away from the stockade into the city. The Athenians pursued the fleeing Syracusans so hotly they were able to burst through the gates of Syracuse's outer walls on the heels of the Syracusans. However, the Syracusans drove them out. The Athenians then tore down both the stockade and the counterwall and, leaving off work on the northward wall, commenced the construction of the southern wall in order to cut the road leading past the fort at Olympieum into the interior of Sicily.[285]

A second Syracusan effort to build a counterwall had to follow a more difficult route through a swamp farther to the south. In this terrain it was impossible to build an actual wall. The Syracusans dug a trench, which naturally filled with water, forming a sort of moat, and built a stockade in front of the trench. This line of defense also failed when the Athenians were again able to destroy the stockade.[286] In this action, however, the Syracusans managed to kill the Athenian general Lamachus, a success that emboldened them to make a sortie against the circle fort on Epipolae. The attack caught the Athenians dispersed on the plain below and only the quick thinking of Nicias prevented the Syracusans from taking the fort.[287] These Syracusan defeats cleared the way for the Athenians to complete two walls from their fort on Epipolae south to the Great Harbor, and Syracusan morale reached its nadir.[288]

Despite the Syracusan failure to stop the Athenian wall construction, the Athenians had not achieved a complete investment of Syracuse. The southern wall did not yet quite reach the sea, and the northern half of the wall was incomplete.[289] On the seaward side of the city, the Athenians again seemed to have the upper hand, but their naval blockade also remained incomplete.

As a result of these shortcomings, when a Peloponnesian relief force under the command of the Spartan Gylippus arrived at Syracuse two months later, it was able to reach the city by both land and sea. Gylippus proved to be a resourceful commander, and Thucydides believed his arrival was the turning point in the siege of Syracuse.

Modern historians blame Nicias, who had assumed the sole command upon the death of Lamachus, for allowing the initiative to slip away from the Athenians. Nicias was ill, and this led to some lassitude in the Athenian effort. More vigilance on both the landward and the seaward sides would have made the blockade more effective even before the walls were completed.[290] Admittedly, the circumvallation of Syracuse was no small task. The extension of the Syracusan walls over the winter forced the Athenians to traverse a peninsula five kilometers wide.[291]

There are no archeological remains of the Athenian walls and Thucydides' account of the topography raises many problems, so it is impossible to know their exact route, but the best estimate is that it ran from 6.4 to 7.5 kilometers, depending on whether the double wall began at the circle fort or did not start until the line dropped off Epipolae to the lower ground beneath.[292]

The Athenians captured Epipolae in mid-April.[293] Gylippus crossed over to Italy in June and probably arrived in Syracuse about two months later, in August.[294] The Athenians therefore had about four months to complete their walls from the time they arrived on Epipolae to the arrival of Gylippus. This was surely enough time, given favorable circumstances.

Georg Busolt estimates the Athenians were capable of building 1.5 kilometers of wall in two weeks with a force of nine thousand men.[295] With that rate of construction, the Athenians could have completed 7.5 kilometers, the maximum possible for the route of their walls, in ten weeks. Even considering the frictions of war, the sixteen weeks that they had should have sufficed.

For those reasons, the judgment of modern historians on Nicias has been harsh. Only Edward Meyer, who relished perverse judgments, dared to challenge Nicias's critics, calling them armchair generals whose only criterion for judgment was success.[296]

We should not excuse Nicias, but there were mitigating circumstances. Although the Athenians had consistently defeated the Syracusans, there was no time when the Syracusans were not contesting the terrain around the city, either by sorties or the construction of counterwalls. The Athenian logistical position was precarious. Indirect evidence suggests this.

The first indication is Nicias's decision to complete both southern walls before resuming construction on the northern wall. If he had built a single wall the entire width of the peninsula first, clearly he would have had a better chance of completing the investment.[297] The decision only makes sense if we assume that his supply route from Thapsus would have been inadequate if the fleet had been moved to the Great Harbor, as would have been necessary for a naval blockade of Syracuse. The transfer of the fleet to the Great Harbor at this time confirms that necessity, and the logistics of that move required a secure landing zone on the shore of the Great Harbor.

Another possibility is that fighting and building on the low ground close to the Great Harbor extended the line from Thapsus too far, forcing the establishment of the landing zone in the Great Harbor. Perhaps a palisade could have sufficed to protect the Athenian wall on the lower ground next to the Great Harbor, but with the Syracusan fort at Olympieum threatening his rear, Nicias apparently felt more secure with a double wall. From this we may conclude that the defensive tactics of the Syracusans were more effective than Thucydides allowed and that the arrival of Gylippus, although certainly a significant event, was not quite the dramatic turning point that Thucydides created.[298]

The influence of Gylippus in the defense of Syracuse was not so much on the military situation as on the internal politics of the city. In Greek siege warfare, psychology was all-important. Like all Greek commanders faced with the need to capture a fortified city, Nicias had hoped for political betrayal within the city to hand it over to him. Even when the Athenians were on the brink of catastrophe, he clung to this hope, and Thucydides confirmed there was a pro-Athenian faction in the city.[299]

After the Athenian victories that destroyed the first two Syracusan counterwalls, the Syracusans replaced Hermocrates and his two colleagues with new generals. The defeatist party was on the verge of taking control.[300] We even hear of a slave revolt at the prospect of a difficult siege, although the story may not be reliable.[301] Indeed, this prospect of the democratic faction taking control and betraying the city may have caused Nicias to hesitate rather than aggressively pushing circumvallation because the democrats were arguing that it would be best to come to terms before the circumvallation was complete.[302] The arrival of Gylippus, bearing the prestige of Sparta, strengthened the war party in Syracuse and steadied the people behind the defense of the city.

What followed was a continuation of the Syracusan tactics of sorties and counterwalls. The self-confidence and experience of Gylippus undoubtedly stiffened the Syracusan hoplites, but essentially he simply took advantage of the toll the Syracusan defense tactics had already exacted from the Athenians. His first success was to capture the Athenian fort at Labdalum on the northern part of Epipolae. The Syracusans could now contest not only the low ground before their city but the high plateau of Epipolae as well. The loss of Labdalum cut the Athenian supply route from Catana and forced Nicias to shift his main base of operations to Plemmyrium at the mouth of the Great Harbor, a position that strengthened the Athenian control of the Great Harbor but compounded the Athenian logistical problems because of the lack of drinking water and wood.[303] Labdalum provided a strong anchor for a third counterwall the Syracusans now began in that direction in order to intersect the unfinished northern Athenian wall. Thucydides believed that Nicias now abandoned all hope of investment and began thinking entirely in terms of a naval war.[304]

The Syracusans were able to use the stones the Athenians had gathered along the line of their northern wall, and this accelerated the progress of the Syracusan wall. Every day Gylippus drew his hoplites up in front of the wall and challenged the Athenians to attack, but Nicias declined. Then Gylippus blundered by attacking the Athenians in the cramped space between the two walls, where he was unable to deploy cavalry and light infantry. The Athenian hoplites demonstrated once more their superiority over the Syracusan hoplites, but the Athenian victory was strictly a tactical success. They did not capture the wall, whose inexorable progress continued.

Finally, as the Syracusans pushed their wall toward the point where it would intersect the line of the Athenian wall, thus depriving the Athenians of their last chance to invest Syracuse, they were able to force the Athenians to come out to fight in the open field away from the fortifications. Here the Syracusans were able to deploy their cavalry and javelin throwers on the Athenian flank and turn the Athenian left wing. Oddly, we hear nothing of the Athenian cavalry that Nicias had spent so much precious time to acquire. After the battle, the Syracusans worked all night on their counterwall and the next day it crossed the line of the Athenian wall.[305]

This Syracusan success reversed the military situation. As Nicias put it in a report to Athens, "so it has turned out that we, who are supposed to be besieging others, are rather ourselves under siege."[306] The Athenians now not only had to win a field battle to restore the opportunity to invest Syracuse; they would also have to storm a fortified wall. Without massive reinforcements, that would be an impossible task.

Although the Syracusans had forced Nicias into an embarrassing position, they were by no means complacent. The huge costs of the siege weighed heavily on the city. Not only were the Syracusans dependent on expensive mercenary troops; they had also raised and equipped a fleet. Numismatic evidence indicates severe financial strain in Syracuse at this time, and Thucydides said that Nicias's knowledge of financial straits in Syracuse was accurate.[307]

The problem of cost in a siege was different, however, for the besieged than for the besiegers. The latter were there by choice, so lifting the siege if the expense became too great was always a real consideration. But the defenders had no choice; defeat meant the loss of their city

and their wealth, either through sack or indemnity, so no financial sacrifice could be too great.[308] But financial strain could affect morale and
exacerbate divisions among the defenders. Successful offensive operations not only kept the besiegers off balance; they silenced the defeatists
in the city.

Accordingly, the Syracusans continued an aggressive defense against
the Athenians, especially after they heard that Athens had agreed to
send Nicias massive reinforcements. Helped by reinforcements of their
own from Corinth, Ambracia, and Leucas, the Syracusans consolidated
their control of the northwestern part of Epipolae by completing the
counterwall and building a system of forts there.[309] Then they dealt Nicias a stunning blow by capturing Plemmyrium, turning his difficult
logistical situation into a desperate one.

The Syracusan cavalry roamed the countryside freely, preventing
the Athenian troops from foraging. Again Thucydides made no mention
of the Athenian cavalry, and one wonders if poor logistical support limited its activities. Gylippus organized attacks against the Athenian wall
from both sides while the Syracusan fleet, using a new tactic of head-on
ramming with reinforced prows that the Corinthians had successfully
tested against an Athenian fleet in the Corinthian Gulf, challenged the
Athenians in the Great Harbor.[310]

By the time considerable reinforcements arrived under Eurymedon
and Demosthenes, matters had deteriorated so far that Demosthenes realized that the situation was untenable so long as the Syracusan counterwall gave the Syracusans such a strategic advantage. He therefore
moved immediately to capture that wall and was of the opinion that if
the effort failed, the Athenians should evacuate Sicily.[311]

Demosthenes was the opposite to Nicias. He was resourceful and
bold to a fault. His first move against the Syracusan wall was to bring
battering rams against it. This was a rare Greek assault on a fortified
position, and it failed. The Syracusans had organized their defense of
the wall very efficiently and were not caught by surprise. They were
able to set Demosthenes' machines on fire.[312]

Demosthenes' next effort was also rare: he determined to outflank
the wall by marching under cover of darkness around behind Epipolae
and, ascending by the same route both Nicias and Gylippus had followed, to make a night attack on the wall. Night fighting was unusual
in Greek warfare. Under the cover of darkness, Demosthenes was able
to achieve surprise. This had been the key to the Plataean success in
the escape from Plataea, but the Plataeans were only trying to cross the
wall, not capture it. Surprise enabled the Athenians to capture the fort
guarding the way up to Epipolae, and they even seized a part of the
wall and began tearing it down. Edward Freeman envisions the masons
and carpenters whom Demosthenes had brought along dismantling the

wall.[313] If so, they had followed on the very heels of the assault troops and would have been much exposed to danger.

Although the night attack had caught the Syracusans and their allies by surprise, they were by no means unprepared. They had stationed an elite force of six hundred men trained by Hermocrates to protect their counterwall, and this was the first force to respond to the attack. The Athenian hoplites defeated the Six Hundred but Thucydides said their resistance was vigorous, so they must have slowed the Athenian advance.

That bought enough time for Gylippus to lead his troops from outworks near the wall against the Athenians. The presence of Gylippus on Epipolae that night shows the importance he attached to the defense of the Syracusan counterwall. Once the surprise was over, the night conditions did not favor the Athenians, especially the newly arrived reinforcements who were unfamiliar with the terrain. Their attack soon began to unravel under Gylippus's counterattack. Chaos and confusion, the nightmare of every military commander, afflicted the Athenians. The result was a rout in which at least two thousand Athenians lost their lives.[314] The battle provided bloody proof that the Athenian reinforcements had not shifted the strategic balance in their favor and that it was too late for boldness, wit, and vigor to suffice.

This battle was the last offensive effort by Athens against Syracuse, and in that sense it ended the defense of Syracuse. The defenders had not only prevented the investment of the city but had forced their attackers into an untenable position. The only question now was whether the Athenians could escape.

Demosthenes strongly urged an immediate evacuation, but Nicias, with curious fatalism, refused. His excuse was the hope that the economic strain of the siege on Syracuse would enable the pro-Athenian party in the city to betray Syracuse to him. Such was the history of siege warfare in Greece that this seemed plausible enough to blunt the determination of Demosthenes and Eurymedon, and they acquiesced to Nicias's refusal to evacuate.[315] When Nicias wavered in his decision, an eclipse of the moon convinced him and his soothsayers that the moment was not auspicious for a withdrawal.

The Syracusans now held the initiative. They were determined to destroy the Athenians and kept up vigorous offensive action against the Athenian walls and fleet. They closed the mouth of the Great Harbor with ships and chains, cutting off the Athenian force entirely. This forced the Athenians to attempt to break out. They abandoned their fortifications on Epipolae and gathered all their forces on the shore of the Great Harbor.

But the Syracusan fleet beat back an Athenian effort to escape by sea, and when the Athenians attempted to flee by land, the Syracusan army blocked the way and then hunted the Athenians down and destroyed them. Thucydides minced no words in describing the magnitude of the Syracusan victory:

> This event proved to be the greatest of all that had happened in the course of this war, and, as it seems to me, of all Hellenic events of which we have record—for the victors most splendid, for the vanquished most disastrous. For the vanquished, beaten utterly at every point and having suffered no slight ill in any respect—having met as the saying goes, with utter destruction—land force and fleet and everything perished, and few out of many came back home.[316]

Modern historians have followed Thucydides in giving Gylippus most of the credit for the Syracusan success and Nicias most of the blame for the Athenian failure.[317] Certainly they are correct that Gylippus made the most of his opportunities and Nicias failed to exploit all the Athenian opportunities. The ability, intelligence, and health of individual commanders can have a decisive impact on battles and undoubtedly influenced the outcome of this one. But from our perspective, we have to place the siege of Syracuse in the context of Greek siege warfare.

Before the end of the fifth century, Greek siege commanders were cautious, shunning assault methods in favor of blockade and looking first to treachery within the city as the best way to bring about its fall. Moreover, siege commanders were accustomed to a passive linear defense. Defenders sometimes ventured out at the approach of an enemy to defend some strategic place in front of the walls, but if they failed to keep possession, they generally retreated behind the walls to try to outlast the besiegers.

The evidence suggests that Nicias did not believe the Athenians could win a long passive siege and, given their poor logistical position and the swampy terrain so conducive to disease, he was probably right. Thus the assumption that the completion of the wall of circumvallation would have made the fall of Syracuse inevitable is doubtful. We can argue about whether increased pressure on Syracuse would have produced the desired betrayal, but Nicias's strategic conception that Syracuse could only fall through betrayal was sound. His cautious and self-serving personality and his painful illness did indeed lead to mistakes, but his real failure was not the failure to conquer Syracuse; it was his failure to extricate his army when the siege became untenable.

And more credit is due the Syracusans. Nicias failed to adjust to the defense in depth that they mounted from the very beginning. Even before the arrival of Gylippus, they waged an unusually aggressive defense with sorties and counterwalls. Greek siege armies were not accustomed to such harassment, and because hoplites did double duty as fighters and builders, the need to guard against sorties took soldiers away from wall building. The equipment of a hoplite soldier was too heavy and cumbersome for men to wear while working on the walls, so they could not turn from that duty to fighting at the spur of the moment. The initial lack of Athenian cavalry was critical in this regard. Moreover, the technique of cutting a wall of circumvallation with a counterwall may have been new to Greek commanders. The protection of the sappers by a rapidly thrown-up palisade suggests some experience by the Syracusans. Generally, the Sicilian Greeks were more experienced in siege warfare because of their proximity to Carthage, which was skilled in the art.

In short, the forces were too evenly matched, the logistical problems were too great, the defense was too aggressive, and the Greek siege methods were too limited for the Athenians to invest Syracuse. Their main hope was that bluff and intimidation could bring the desired betrayal within the city. When that failed to happen, they lacked sufficient weight to invest the city and force a surrender. The Athenian failure at Syracuse was not so much a testimony of the incompetence of Nicias as an example of the limitations of Greek siege capabilities during the Peloponnesian War.

Chapter VI

TREATMENT OF
CAPTURED CITIES

Cities

The *Iliad* (c. 750 B.C.) provides us with our earliest images of the treatment of fallen cities in Greek siege warfare. After Achilles rejected Agamemnon's peace offering, his old tutor, Phoenix, turned to the story of a siege to show Achilles the folly of his ways and the sorrows that defeat in war can bring.

Phoenix told Achilles the story of Meleagros, another man who had withdrawn from battle in anger. Meleagros refused to defend his city because he was angry at his mother, just as Achilles was refusing to fight because of his anger at Agamemnon. The city was about to fall when Meleagros's wife made a last plea:

> And then at last his wife, the fair-girdled bride, supplicated
> Meleagros, in tears, and rehearsed in their numbers before him
> all the sorrows that come to men when their city is taken:
> they kill the men, and the fire leaves the city in ashes,
> and strangers lead the children away and the deep-girdled women.[1]

"Evil deed" is the phrase Homer used to describe this terrible event, probably not so much in moral judgment as to find words strong enough to convey the horror of a sack.[2]

Hector had a vision that echoed the story of Meleagros when he spoke to his wife, Andromache, about the fall of Troy:

> For I know this thing well in my heart, and my mind knows it:
> there will come a day when sacred Ilion shall perish,

and Priam, and the people of Priam of the strong ash spear.
But it is not so much the pain to come of the Trojans
that troubles me, not even of Priam the king nor Hekabe,
not the thought of my brothers who in their number and valour
shall drop in the dust under the hands of men who hate them,
as troubles me the thought of you, when some bronze-armoured
Achaian leads you off, taking away your day of liberty,
in tears; and in Argos you must work at the loom of another,
and carry water from the spring of Messeis or Hypereia,
all unwilling, but strong will be the necessity upon you;
and some day seeing you shedding tears a man will say of you:
"This is the wife of Hector, who was ever the bravest fighter
of the Trojans, breaker of horses, in the days when they fought about Ilion."
So will one speak of you; and for you it will be yet a fresh grief,
to be widowed of such a man who could fight off the day of your slavery.
But may I be dead and the piled earth hide me under before I
hear you crying and know by this that they drag you captive.[3]

The fears of Hector and the wife of Meleagros expressed the standard expectation in the *Iliad*: the men will be killed and the women and
children led away into slavery. Some believe that there are vestiges in
the *Iliad* of a more primitive time when conquerors killed the men and
children and claimed the women as captives. They believe the practice
of selling the women and children into slavery reflects a later time, con-
temporary with Homer, when slave markets had developed and it was
profitable to enslave the women and children.[4] As we have seen, however, siege warfare was rare in Greece before the fifth century B.C. and
the scanty evidence we have makes it difficult to tell whether Homer
reflected the reality of Greek siege warfare, much less whether we can
detect stages of development.

Early in the sixth century B.C., the members of the Amphictionic
League took an oath not to destroy any city in the league and may also
have sworn not to reduce cities by starvation or by cutting off the water
supply.[5] These conventions did not establish themselves in Greece, but
they show that the Greeks did not accept the destruction of cities as inevitable.

The earliest sieges in the Greek-speaking world for which we have
even dim evidence took place during the sixth century B.C. in the cut-
throat commercial competition between Greek cities in southern Italy.
Sometime not long before 530 B.C. the city of Siris was defeated and
sacked by its rivals, the most important of which was Sybaris. The circumstances are so obscure that we do not know if a siege was necessary
to take the city. We do hear of an atrocity: the conquerors slaughtered
fifty youths and a priest of Athena who had sought refuge in a temple.
Siris lost its independence, but numismatic evidence shows that it continued to exist after the disaster.[6]

The most famous destruction of a Greek city in the sixth century was that of the wealthy and luxurious city of Sybaris, the old nemesis of Siris, in 511–510 B.C. Sybaris had outraged the people of Croton by killing thirty Crotoniate envoys who had brought a message to Sybaris rejecting a Sybarite demand that Croton surrender some Sybarite exiles who had sought refuge in Croton.[7] Croton determined to avenge this outrage by destroying Sybaris.

The Crotoniates defeated the Sybarites in a field battle in which they gave no quarter, killing most of the Sybarite army. The city itself then fell into their hands, and Diodorus said that they utterly destroyed it.[8] The first-century geographer and historian Strabo reported that the campaign lasted seventy days, a length that suggests a siege was necessary. Strabo also added the detail that the Crotoniates diverted the river over the site of the destroyed city, submerging it completely.[9]

When the news of the destruction of Sybaris reached Miletus, the men of Miletus shaved their heads in mourning, but Miletus had special ties with Sybaris, so we cannot infer that the fate of Sybaris shocked the Greeks in general.[10] There were some survivors, and numismatic evidence indicates they reestablished Sybaris, either on the original site or nearby. But the city never recovered its former preeminence among the Greek cities of southern Italy.[11]

The siege warfare in Sicily during the wars of the Geloan tyrant Hippocrates during the first decade of the fifth century seems to have been less brutal. Hippocrates sold the people of Zancle into slavery; because he needed money to pay his mercenary troops, this may have been his standard practice.[12]

After Gelon took over the leadership from Hippocrates and established himself at Syracuse, he followed a more discriminating policy. Having forced Sicilian Megara to surrender on terms, he surprised the wealthier citizens, who had expected death because they had made war against him. Instead Gelon spared them, transferring them to Syracuse and conferring citizenship upon them. The common people, who had opposed the war and expected to be well treated, Gelon sold into slavery. He followed the same policy after the capture of Sicilian Euboea.

Herodotus attributed Gelon's enslavement of the common people to his dislike of *hoi polloi*.[13] As far as wealthy citizens were concerned, Gelon was trying to enlarge Syracuse and build a base of support, and establishing them as Syracusan citizens served that purpose. However, according to Diodorus, his contemporaries did not take a cynical view:

> since he treated the peoples whom he subdued with fairness and, in general, conducted himself humanely toward all his immediate neighbors, he enjoyed high favor among the Sicilian Greeks.[14]

Herodotus gives us another glimpse of the treatment of captives in siege warfare in the early fifth century in his account of the siege of Sestos in 479 B.C. In this siege, a Greek army under the command of Xanthippus (the father of Pericles) captured Sestos from the Persians. The Persian commander Artaÿctes offered the huge sum of three hundred talents in ransom for himself and his son, but the Greeks rejected his offer in favor of a cruel reprisal. They stoned to death the son before the father's eyes and then crucified Artaÿctes.[15] Artaÿctes had been a brutal and impious ruler, and revenge was undoubtedly the reason for the cruelty of his death.

Herodotus does not tell us what happened to the rest of the Persians in Sestos. But according to Plutarch, Cimon was also a commander at Sestos, and he stripped the Persians of all their ornaments and clothes and told the Athenian allies to choose either the clothes and ornaments or the naked Persians. The allies chose the clothes and ornaments and thought they had the better bargain, until Cimon ransomed the Persians for a sum large enough to maintain his fleet for four months and still have a large amount of gold left over.[16]

About three years after the siege of Sestos, the Athenians besieged another strategic city held by the Persians, Eion on the Strymon River. Cimon, the Athenian commander, offered the Persian governor Boges terms that allowed him to return to Asia. Boges, not wishing to be disloyal to his duty, refused these terms and held out to the end. When food supplies ran out, he cut the throats of his wife, children, concubines, and servants and threw them into a fire. He then threw all the wealth of the town into the river and jumped into the fire himself. These actions earned him great respect, not only among the Persians but with Herodotus as well.[17]

The Athenians enslaved the native inhabitants of Eion. On the same campaign, they also captured the island of Scyros and enslaved its Dolopian population.[18] Some thirty years later the population of the Boeotian town of Chaeronea suffered the same fate when the Athenians captured it in 447 B.C.[19]

Another mass enslavement took place when the Argives captured Mycenae after a siege in 468 B.C. The Argives sold the entire population into slavery and razed the city. Almost five hundred years later, in Diodorus's time, it was still uninhabited. The deed troubled the Argives' conscience, and they dedicated a tenth of the booty to the gods.[20]

A timely surrender could avert death and enslavement. Around the middle of the fifth century, the city of Oeniadae in Acarnania was placed under siege by Messenians from Naupactus. The Spartans had allowed these Messenians to leave the Peloponnese under the terms that ended the long siege of Mount Ithome.[21] Now the Messenians in their turn allowed the people of Oeniadae, who feared death and enslavement if they did not surrender, to evacuate their city under the terms of a truce.[22] The forced evacuation of an entire population was a hard fate but better than massacre and enslavement.

Despite scattered evidence often of questionable reliability, we can know enough about the fate of conquered cities in early Greek siege warfare to see that by the sixth century massacre was no longer the standard practice, as it seems to have been in Homer's day. The general pattern was indeed one of brutality, but there is no clear evidence of any general massacre of a city's population. Enslavement was the most common practice. Cruel reprisals were possible. But no standard practice is evident; as the oligarchs and democrats of Megara discovered, outcomes were unpredictable and depended upon circumstances and even the whims of individual commanders.

The sieges with which the Athenians forged their empire in the middle years of the fifth century B.C. show a further moderating trend in the treatment of captured cities. The first of the sieges that we know of ended in the fall of Carystus, a city of Euboea that had Medized during the Persian War. The town surrendered on terms that left it in control of its internal affairs in return for the payment of an annual tribute to Athens. There is no mention of any reprisals. Sometime later (the chronology is obscure) Naxos suffered the same fate when it tried to break its alliance with Athens and fell after a siege.[23]

About 465 B.C. Thasos revolted from Athens and held out against the Athenian siege for three years. Thucydides gives us the terms on which they surrendered in detail: the destruction of their walls and surrender of their navy, payment of an indemnity and an annual tribute, and economic concessions to Athens.[24] In 450 B.C. Cimon reduced two Cypriot cities by siege, and Diodorus said that he treated the conquered cities "in humane fashion."[25] However, Diodorus provided no details.

In 440–439 B.C. a nine-month siege was necessary to bring Samos to heel. The terms were similar to those imposed on Thasos: destruction of the walls, surrender of the fleet, surrender of hostages, and payment of reparations.[26] The Athenians established the reparations in the amount of 1,300 talents, payable over twenty-six years. Finally, the Athenians imposed a democratic regime on Samos.[27] Plutarch tells us that a Samian historian named Duris claimed that the Athenians crucified the Samian trierarchs and marines, allowing them to suffer for ten days before bashing their heads in with clubs. Plutarch rejected the story as a piece of Samian propaganda, pointing out that no other historian reported this sensational episode.[28]

In the context of siege warfare, the terms the Athenians imposed on Carystus, Naxos, Thasos, and Samos were remarkably light. Even in Thasos and Samos, where the resistance was long and determined, the Athenians limited the terms to political and economic measures. There were no massacres, executions, or enslavements. The Athenians were proud of their moderation. When they defended the empire before the Spartans on the eve of the Peloponnesian War, they boasted:

> And they are to be commended who, yielding to the instinct of human nature to rule over others, have been more observant of justice than they might have been, considering their power. At least, if others should seize our power, they would, we think, exhibit the best proof that we show some moderation. . . .[29]

This pride in moderation did not survive the outbreak of the Peloponnesian War. The harsher circumstances of that war, which began in 431 B.C., manifested themselves early on in the terms the Athenians imposed on Potidaea in 429 B.C. after a two-year siege. The Athenians forced the entire population to abandon the city, leaving behind not only their homes, but their possessions as well. The men could take only the clothes that they wore; the women were allowed two garments. The Athenians allowed the Potidaeans a fixed sum of money, which cannot have been large, with which to pay for the journey into exile. Because the siege had reduced the Potidaeans to the verge of starvation, they must have made a pathetic sight as they marched away. Athenian colonists resettled the city. Although these terms spared the lives of the Potidaeans, they destroyed the polis.

The suffering of the Potidaeans did not satisfy the Athenian people. They were furious with the generals for agreeing to the terms, believing they should have held out for an unconditional surrender.[30] Later events indicate massacre and enslavement were the punishments the Athenian people had in mind.

The harsher standards of siege warfare in the Peloponnesian War pushed the moral implications of total war to the forefront of Greek thought. Thucydides saw this clearly, and his *History* reflected a preoccupation with the moral questions of siege warfare.

Thucydides began his meditation on the moral impact of siege warfare with the description of the Athenian siege of Mytilene in Book III of his *History*. In 428 B.C. Mytilene, one of the few independent allies in the Athenian empire, revolted against Athens. The Spartans supported the revolt by sending a Lacedaemonian officer named Salaethus to Mytilene. A Peloponnesian fleet of forty ships under the command of Alcidas soon followed.[31]

The revolt caught the Athenians at a bad time, for they were exhausted by the lengthy siege of Potidaea and by a plague that had killed around one-fourth of the population of Athens. Alcidas's voyage into Ionian waters, following on the heels of an unexpected Spartan move against Athens' port at Piraeus, deeply disturbed the Athenians. The Mytilenian revolt pressed the Athenians to the limit, but they managed to scrape together a fleet under the command of Paches to send to Mytilene.[32] Thanks to the determination of the Athenians, the dawdling of Alcidas, and a popular rebellion against the Mytilenian leaders of the revolt, Paches was able to force Mytilene to surrender.

The terms of the surrender placed Mytilene under military occupation and reserved the right of Athens to dispose of the Mytilenians in any way it saw fit. However, Paches allowed the Mytilenians to send representatives to Athens to plead for lenient treatment and promised not to take any punitive action until they returned. Under these terms Paches was able to persuade the terrified Mytilenian oligarchs, who had sought refuge in the temples as suppliants, to leave the altars and surrender. These men he interned on the island of Tenedos pending the decision from Athens.[33]

Meanwhile, Alcidas, showing no eagerness to take on the Athenian fleet at Mytilene, kept his distance and satisfied himself with killing a number of prisoners he had taken during his voyage. At the first opportunity he retreated back to the Peloponnese. Against this emotional background of a Peloponnesian fleet penetrating Ionia and massacring prisoners, the Mytilenian representatives arrived in Athens and the Athenians debated the fate of Mytilene.

Salaethus, whom Paches had sent to Athens with the party of Mytilenians, attempted to negotiate for his life, even promising to persuade the Spartans to lift the siege of Plataea, but to no avail. The Athenians killed him, then voted to kill all the adult male Mytilenians, regardless of whether or not they had supported the revolt, and to sell the women and children into slavery.

That was the sort of punishment that the characters in the *Iliad* seemed to take for granted, but Thucydides made clear that the Athenian decision was far from routine in his own day. He emphasized that emotions were running high in Athens over the revolt of an ally and the unexpected threat that Peloponnesian seapower had posed to both Athens and Ionia. He called Cleon, who had introduced the motion, "the most violent of the citizens."[34]

Euripides also was horrified by Cleon's proposal. In *Hecuba*, written a couple of years after the Mytilenian debate, Euripides had the chorus state that the Greeks had decided to kill Hecuba's daughter Polyxena "in full assembly," as if pointedly reminding the Athenians of their dreadful decision.[35] Euripides' description of Odysseus, who persuaded the Greeks to kill Polyxena, as "that hypocrite with honeyed tongue, that demagogue Odysseus" suggested a parallel with Cleon.[36] Thucydides said that the next day the Athenians realized how "cruel and excessive" the punishment was. With this swing in mood, the Mytilenian delegation was able to persuade the Athenians to hold another assembly to reconsider their decision.[37]

Thucydides gives us two long speeches from the ensuing debate, revealing the importance he attached to the issues. The first thing to note about this famous debate is the mere fact that it occurred. The Greeks did not have a god who laid down the law on this matter, as Yahweh had done for the Hebrews in the Book of Deuteronomy, or an emperor whose authority was final, as the Assyrians had; rather, the Greeks had to grope toward a decision in public debate. It was in the context of Athenian democracy that Thucydides was able to probe the moral implications of war.

Cleon urged the Athenians to confirm the original decision to kill the adult male Mytilenians and sell the rest of the population into slavery. He argued that such a punishment would fit the wrong the Mytilenians had done. The revolt threatened the Athenian empire just at the time Athens was engaged in a difficult war and thus threatened Athens' very existence. Therefore the annihilation of Mytilene was just. However, Cleon's argument also rested on expediency. Even if the punishment were unjust, the Athenians should carry it out anyway as a deterrent to further revolts. Leniency would encourage other cities in the empire to revolt in an endless succession that would exhaust Athens and destroy the empire.[38] Cleon did not mention siege warfare in his speech. He set his argument in the context of the empire and the proper policy for a ruling city. That pity and clemency were incompatible with imperial power was the main thrust of his appeal to the people.

In contrast, a keen insight into the nature of siege warfare informed the argument of Diodotus, who opposed Cleon and urged clemency for Mytilene. Even more than Cleon, Diodotus explicitly rejected standards of justice and based his argument entirely on expediency:

> For no matter how guilty I show them [the Mytilenians] to be, I shall not on that account bid you to put them to death, unless it is to our advantage; and if I show that they have some claim for forgiveness, I shall not on that account advise you to spare their lives, if this should prove clearly not to be for the good of the state . . . we are . . . not engaged in a law-suit with them, so as to be concerned about the question of right and wrong, but we are deliberating about them, to determine what policy will make them useful to us.[39]

Diodotus went on to argue that the death penalty had never been an effective deterrent to crime and that it would not deter revolts, because either desperation born of poverty or greed born of wealth would push men into hazardous ventures, hope and desire always serving to banish the dread of failure.[40] Diodotus was keenly aware that the Athenians had brought the siege of Mytilene to an end in the most common way in Greek siege warfare, betrayal from within. If the Athenians now killed the same Mytilenians who betrayed their city to Athens, the Athenians would close off this possibility for the future, thus wrecking chances of an early conclusion to sieges and forcing protracted and economically ruinous sieges on Athens.[41] In this brilliant argument,

Diodotus developed a purely political morality in favor of moderation, an argument that was all the more powerful because it rested on a clear understanding of political and military reality, especially the reality of siege warfare.

Diodotus's reasoning managed to reverse the decision, although by a very slender margin. The Athenians had already dispatched a trireme to Mytilene bearing the original order to kill all the males and sell the women and children into slavery. A second trireme now embarked to overtake the first ship with news of the new decision. The sailors on the second trireme, encouraged by the promise of a large reward by the Mytilenian envoys, rowed around the clock, not even stopping while they were eating. Meanwhile the men on the first ship had no eagerness to hasten along on their distasteful mission. Because of this, the second trireme managed to arrive at Mytilene soon after the first ship, just in time to save the people of Mytilene.[42]

Diodotus's plea, however, had been to spare only the Mytilenians who had supported the surrender to Athens. The Athenians put to death those oligarchs whom Paches had sent to Athens as most responsible for the revolt. Thucydides said they numbered slightly over a thousand men, but some scholars have questioned such a number because it seems too high to number the leaders of the revolt and perhaps even too high to number the entire oligarchic class of Mytilene. Thucydides' other references to Paches' actions concerning the leaders of the revolt do not imply such a large number. Because of these considerations, one emendation of the text, based on the similarity between the Greek numerical symbols for one thousand and for thirty, reduces the number to thirty.[43] This textual uncertainty is unfortunate because there is a large difference between the execution of thirty ringleaders and the massacre of over a thousand men.

Whatever number of Mytilenians met death at the hands of the Athenians, the bulk of the male population survived and no one was enslaved. Thucydides' account made clear that the original motion of Cleon was extraordinary and that a majority of the citizens regretted having passed it. Even though the rationale for rescinding the sentence of doom was purely political, there was great relief that mercy and political expediency coincided. Clearly massacre was not the standard policy for a city that surrendered on terms.

Immediately after the Athenian conquest of Mytilene, Plataea fell to the Spartans. This coincidence gave Thucydides an opportunity to juxtapose the fate of Plataea with that of Mytilene. Even after starvation had weakened the Plataeans so much that they were unable to defend the city any longer, the Spartans patiently refrained from taking the city by force of arms. After four years of inconclusive fighting the authorities in Sparta were already anticipating that the war might end with a compromise peace in which both sides agreed to return all conquests of war. By making a specious distinction between force of arms and the force of starvation the Spartans hoped that Plataea's surrender could be counted as voluntary so that they would not have to return the city to Athens. Accordingly the Spartan commander sent a herald to the helpless defenders and invited them to surrender voluntarily, promising that the Spartans would punish only guilty Plataeans as determined by Spartan judges. The last point was important, because the last thing the Plataeans wanted was to fall under the jurisdiction of their old enemies the Thebans.[44]

Although these terms resemble those the Athenians offered the Mytilenians, their exact meaning is not entirely clear. They certainly implied some sort of trial in which the Plataeans would have the opportunity to defend themselves and in which the Spartan judges would make distinctions between the leaders and the rank and file. Despite this glimmer of hope the Spartans were offering, the Plataeans could hardly have been under any illusions. They had set a precedent by slaughtering Theban prisoners four years earlier. But, dying of starvation, they had no choice, and so they surrendered the city, making a special point that they were surrendering to the Spartans and not the Thebans.

Events justified the worst fears of the Plataeans. The pious Archidamus, who was ill and near death, was no longer in command, and when the judges arrived they included Aristomelidas, a member of the royal Heracleid family and a friend of Thebes.[45] Instead of making specific accusations and then allowing the Plataeans to defend themselves, the Spartan judges put this simple question to the Plataeans: "Have you rendered any good services to the Spartans and their allies in the present war?" The stunned Plataeans realized that the Spartans had already decided their fate. They desperately pleaded for permission to speak on their own behalf. The Thebans stridently objected, but the Spartans, perhaps embarrassed by the promises they had made at the time of surrender, consented to listen to speeches by both the Plataeans and the Thebans.[46]

Thucydides was just as fascinated by this trial and its aftermath as he was by the siege itself. Just as his interest in military technology led him to provide a detailed description of the siege, so his interest in the moral behavior of men at war led him to describe the trial in striking detail.[47] In contrast to the Mytilenian debate, which was a debate among the victors over the fate of Mytilene, the debate at Plataea was between conquered and conqueror. Because of this the tone was quite different. The Plataean and Theban speeches appealed to justice and custom rather than political expediency. The debate revolved around legalities, loyalty, betrayal, and revenge instead of self-interest.

The Plataeans chose two men, Astymachus and Lacon, to make their case. The latter, as his name indicates, had special connections with

Laconia and held an office in which he was responsible for looking after Spartan interests in Plataea. He and Astymachus did their best under difficult circumstances. They repeated the well-known story of Plataean deeds in the Persian Wars, contrasted Plataean Greek patriotism with Theban treachery at that time (the Thebans had fought on the Persian side), and reminded the Spartans of oaths taken after the Greeks had defeated the Persians at the Battle of Plataea in 479 B.C. in which the Greek victors, including the Spartans, swore to preserve Plataean independence.

Astymachus and Lacon claimed for the Plataeans the status of suppliants and appealed to the Greek custom that forbade the killing of suppliants. They argued that the Plataeans acted justly in killing the Theban prisoners at the beginning of the war because the Thebans had attacked them without a declaration of war. This meant that the captured Thebans were not prisoners of war but outlaws, an important distinction in Greek military conventions. For example, after the Battle of Leukimme the Corcyraeans spared the Corinthian prisoners because Corinth had declared war but killed the other prisoners because their cities had not.[48] This was the principle the Plataeans held up in their defense, maintaining that "we were justified in punishing them in accordance with the law which has universal sanction."[49] The Plataean appeal was essentially a religious and legal one, based on the sanctity of oaths and the divinely sanctioned customs governing the treatment of suppliants.[50]

The Thebans based their argument on the right of revenge. They made excuses for fighting with the Persians in the Persian War and belittled Plataea's policy as a mere extension of Athenian policy: when Athens fought Persia, so did the Plataeans; when Athens sought to enslave Greece, so did the Plataeans. Furthermore, any right the Plataeans might have had to defend themselves did not extend to the murder of prisoners whom they had promised not to kill. The Plataeans had violated the natural ties of kinship by opposing Boeotia and then treacherously killed prisoners they had sworn not to harm. Now the Thebans wanted their revenge.

Clearly Thucydides favored the Plataeans. Their speech is moving and arouses our sympathy. In contrast the Theban speech creates an unfavorable impression. Donald Kagan calls it "self-serving, distorted, sophistical, and unconvincing," and Georg Busolt finds it "cold and sophistical, haughty and hateful."[51] However, the Spartans were unmoved. The Spartan judges simply repeated their terrible question to the Plataeans, one by one. As each man gave the only possible answer of no, the Spartans led him away and executed him. They killed all of the two hundred Plataeans who had surrendered as well as twenty-five Athenians who had remained in the city. They sold the Plataean women into slavery.[52]

To preserve appearances the Spartans retained possession of the city that the Plataeans had expressly surrendered to them. But within a year Sparta turned full control over to Thebes. The Thebans razed the city. Theban settlers took over the land.

To placate the vengeance of Zeus, god of justice, the Thebans constructed a large inn out of material taken from the destroyed town, furnished the inn with booty from the town, and dedicated it to Hera, the wife of Zeus. The inn made possible the maintenance of the religious festivals and the many temples in the vicinity, but Zeus can hardly have been satisfied.[53]

Raoul Lonis calls the proceedings a "show trial" and treats the massacre as a clear violation of the surrender terms.[54] The trial drives home one of the central themes of Thucydides' *History*: moral appeals avail for nothing in war, and loyalty and honor are impotent in the face of self-interest. Although Thucydides carefully listed the legal reasons the Spartans put forth as their justification for killing the Plataeans, he rejected them as pretexts. Instead he pointedly emphasized that the real reason was to satisfy Thebes, an indispensable ally.[55] The trial thus fit into the same pattern as the Mytilenian debate between Cleon and Diodotus.[56] It reveals how much Thucydides' *History* is "in its essence, a moral commentary."[57]

In the Mytilenian debate, only Diodotus's appeal to self-interest led the Athenians to spare the Mytilenians. In the Plataean debate, self-interest caused the Spartans to brush aside the Plataean appeal to oaths and mercy and kill all the Plataeans. Thucydides explained why in his analysis of the civil war at Corcyra, the third part of his moral triptych in Book III:

> For in peace and prosperity both states and individuals have gentler feelings, because men are not then forced to face conditions of dire necessity; but war, which robs men of the easy supply of their daily wants, is a rough schoolmaster and creates in most people a temper that matches their condition.[58]

Thucydides was here commenting on the atrocities committed in the course of the Corcyraean civil war, but his insight applies to siege warfare as well. In traditional open-field battles between warriors, codes of honor and military custom limited the violence of war. But civil wars and sieges tended toward total war involving whole populations and frequently divided them against one another. Thucydides' description of the collapse of moral customs, the corruption of language, and the splintering of social cohesion under the "rough schoolmaster" of war in Corcyra showed the weakness of custom and law in the face of such wars. No wonder that Diodotus refrained from moral appeals and that the pleas of the Plataeans went unheeded.

The massacre of the Plataeans was just the first of a series of such massacres in the Peloponnesian War. In 423 B.C. an Athenian army was campaigning in the Chalcidice against towns that had revolted from Athens to support a Spartan invasion of that area. When Athens and Sparta concluded an armistice, the Athenians claimed that the terms of the armistice required the Spartans to return the Chalcidic town of Scione. The Spartan general in the area, Brasidas, refused to turn over Scione, claiming it had joined Sparta before the armistice. Furious, the Athenian assembly passed a motion offered by Cleon to capture the city and kill the men and enslave the women and children. At the same time, the nearby city of Mende also revolted from Athens. Brasidas prepared for siege by sending the women and children of Scione and Mende away to safety.[59]

Once the Athenians besieged the two cities, however, the democratic faction in Mende turned against the pro-Spartan faction and seized control of the city. The situation was chaotic, and someone unwisely opened the gates to the Athenians without any prior negotiations. Probably mindful of the death sentence decreed by the Athenian assembly, the Athenian troops burst in and sacked Mende "as though they had taken it by storm." The Athenian generals only with difficulty prevented their men from slaughtering the inhabitants. The Athenians were determined to place the leaders of the revolt on trial. Most had sought refuge on the acropolis, which the Athenians now walled in. But the rest of the population the Athenians allowed to continue to live and govern themselves in Mende.[60]

Shortly afterward, the Athenians succeeded in forcing the fortifications of another Chalcidic town, Torone, and captured the city. Some of the defenders were slaughtered in the fighting for the city; the rest of the population was captured. The Athenians sent the soldiers to Athens as prisoners and sold the women and children into slavery. The men who had been sent to Athens were later freed in a prisoner exchange after the Peace of Nicias.[61]

The men of Scione were not so lucky. They held out to the bitter end, finally falling to Athens two years later. There was no Diodotus to save the Scionaeans. The Athenians carried out the order to kill the adult males. Thucydides said women and children were sold into slavery, but Brasidas had evacuated the women and children before the siege began.[62] Some women may have stayed in the city as cooks, as happened in Plataea, and some may have given birth to children during the two-year siege, or perhaps Thucydides had simply forgotten about the evacuation. In any case, the Athenians emptied Scione and, in an act that undoubtedly struck them as especially fitting, they settled the town with the Plataeans who had escaped their city before its fall.[63]

Because the timing and the circumstances of the sieges were so similar, the contrasting fates of Mende and Scione are instructive. The city that quickly and voluntarily surrendered received lenient treatment. The one that held out through a two-year siege suffered massacre and enslavement. Diodotus's policy of extending mercy to those who surrendered, saving the Athenians the expense and effort of a long siege, still held.[64]

There were atrocities on the Spartan side as well. Brasidas killed all the Athenian defenders of Lecythus when he captured that place in 424 B.C.[65] In 417 B.C. the Spartans took the town of Hysiae while campaigning in the Argolid and killed all the free adult males whom they captured.[66]

Thucydides related the massacres at Scione, Lecythus, and Hysiae briefly and matter-of-factly. But this does not mean that he considered such bloodletting routine. He had already brought out the moral and political implications of such acts in Book III. This alternation between detailed analysis and bare-bones chronicling was typical of Thucydides' method.

By far the most famous massacre during the Peloponnesian War was the Athenian slaughter of the men of Melos in 415 B.C. The women and children were sold into slavery and Athenian settlers took over the town.[67] Again Thucydides relates these facts briefly and without comment. What makes the case of Melos leap out at us is the debate between the Athenians and Melians before the Athenians placed the town under siege. Thucydides cast it in the form of a dialogue between them, the only time in his *History* that he utilized this technique.

Melos was an island in the southern Aegean. It had been a colony of Sparta, so its people were Dorian with ties to Sparta. Because of this the Melians had remained outside the Athenian empire. In 416 B.C. Athens and its allies sent a force of thirty-eight ships against Melos to bring it into the empire and impose tribute upon it.[68] When the expedition arrived at Melos, it offered terms to the Melians: join the Athenian empire and pay tribute to Athens.[69] In order to hide these mild terms from the people, the oligarchic leaders of Melos insisted on meeting secretly with the Athenian envoys.

The negotiations that followed were remarkable in the naked brutality with which the Athenians stated their case. There would be no "fair phrases . . . that would not be believed"[70] but instead a simple appeal to the power of Athens and the weakness of Melos. This was the reality that Melos must face; appeals to justice, hope, the Spartans, or even the gods would avail nothing. When the Melians rebuffed Athenian *Realpolitik*, the Athenians placed Melos under siege.

Whether Thucydides meant to show the moral decline of Athens under the stress of war or the brutal reality underlying international politics has been a subject of great debate. However, in the context of the history of siege warfare, there was nothing extraordinary about the

Athenian methods. The Athenians attempted to avoid siege by offering terms and tried to intimidate their opponents into accepting them, emphasizing the inevitability of defeat, "for you yourselves are not unaware that the Athenians have never in a single instance withdrawn from a siege through fear of any foe."[71] The abstract, almost philosophical language of the Athenians at Melos seems so artificial that some have suggested that Thucydides created the whole debate to reveal his political and moral philosophy. But understood in terms of the psychological warfare characteristic of sieges, there is nothing artificial about it at all.

The failure of psychological warfare forced the Athenians into a full-scale siege. They circumvallated Melos and left a small garrison to starve the city into submission. The Melians defended themselves aggressively, twice capturing a part of the Athenian wall of circumvallation and eventually forcing the Athenians to reinforce the garrison with a larger force. This sufficed to place Melos under close siege, and by the following summer there was famine in the city.[72]

The hardship of the siege apparently caused resentment among the people against the oligarchs who had decided to resist the Athenians. Faced with famine and internal discord, the Melian leaders unconditionally surrendered the city and the Athenians carried out their terrible punishment.

We know from Xenophon and Plutarch that there were survivors who returned to Melos after the war.[73] Most were probably male children who would have been in their early twenties at the war's end, but it may also be that the Athenians spared the traitors who precipitated the fall of the city.[74] If so, the fate of Melos falls into the pattern of harsh reprisals for those who resisted a long siege and mercy for those who betrayed their city.

A notable exception to this pattern was the fate of Thyrea, a Peloponnesian town inhabited by Aeginetans whom the Spartans had settled there after Athens had expelled the Aeginetans from their island. A Spartan garrison helped protect the town. The Athenians took the town by assault, plundered it, and burned it down. Many of the defenders lost their lives in the battle, but the Athenians took the survivors to Athens.

Thucydides does not tell us the fate of the Spartan garrison or the women and children of Thyrea, but he does tell us that the Athenians imprisoned the Spartan commander and killed the Aeginetan men.[75] Diodorus gave a different account, saying that the Athenians enslaved the Aeginetans, but surely Thucydides was better informed on this.[76]

Thyrea's rapid fall places it outside the category of cities, such as Scione and Melos, that fell only after long sieges. Thucydides attributed the Athenians' brutality to their long-standing hatred of the Aeginetans.

Xenophon reported that the Athenians suffered from a guilty conscience about punishments they had meted out at such places as Scione and Melos and, when they faced defeat at the end of the Peloponnesian War, feared the other Greeks would take revenge.[77] But the fate of Athens at the end of the war dramatically demonstrated how unpredictable siege warfare was and how the issue could hang on the balance of circumstances.

When Athens fell after a several-month siege, Sparta gathered its allies together to decide Athens' fate. The Corinthians and Thebans strongly urged total destruction. The Spartans, however, drew back from the annihilation of a city with such a glorious past and imposed terms that, while depriving Athens of its fleet, empire, and long walls, spared the population and left the city walls intact.[78]

Much depended on the individual commander. In 406 B.C. the Spartans took the Lesbian town of Methymna. Although they thoroughly plundered the place, when their allies urged them to sell the defenders into slavery, the Spartan commander, Callicratidas, refused on the grounds that it was not proper to enslave Greeks. However, he did not extend this consideration to some Athenian soldiers among the captives, whom he sold into slavery.[79] Callicratidas's mercy toward Methymna made his reputation. Xenophon greatly admired him; Diodorus called him "the most just man among the Spartans" and said that "it is agreed by all that also during his period of command he committed no wrong against either a city or a private citizen."[80]

Only a year later, the Spartan commander Lysander followed a quite different policy. He captured a Carian town called Cedreae and sold its entire population, a mixed one of Greeks and Carians, into slavery.[81] According to Diodorus, on the same campaign Lysander took Iasus, which had made the mistake of allying itself with Athens after the Spartans had captured it once before. The Spartans killed all eight hundred males, sold the women and children into slavery, and razed the town.[82] However, some modern historians doubt Diodorus' account because of confusion over who controlled Iasus at the time.[83]

Shortly after his ferocious Carian campaign, Lysander moved to the Hellespont and captured Lampsacus, by assault if we can believe Xenophon. He allowed his soldiers to plunder the city, but he released all the captives and, according to Diodorus, allowed an Athenian garrison there to leave under a truce.[84] Xenophon does not explain why Lysander enslaved one city and freed another. The fact that Lampsacus was a purely Greek city in an area where it was important to the Spartans to win the support of other Greeks was probably the deciding consideration.[85] The contrast again reveals how the fate of a city depended more on specific circumstances than on general practice.

The strongest incentive for leniency was the hope of a quick surrender. In 424 B.C. Brasidas convinced Amphipolis to surrender to him by offering to allow its citizens to remain in their city still in possession of their property and political rights. Those who did not want to remain could leave with all of their property within five days.[86] In 408 B.C. the Athenian Alcibiades captured the Hellespontine town of Selymbria by a combination of treachery and bluff. To reassure the Selymbrians, Alcibiades ordered his Thracian troops, who were notorious for their ferocity, out of the city and promised the Selymbrians there would be no sack. He obtained a quick surrender on the easy terms of an indemnity and the placement of an Athenian garrison in Selymbria.[87] Epigraphical evidence indicates the Selymbrians surrendered hostages until the treaty was ratified in Athens.[88] Similarly Alcibiades facilitated the surrender of Byzantium by promising lenient treatment. He kept his word, returning the city to the Byzantines and contenting himself with sending the die-hard defenders to Athens for trial.[89]

When Calchedon and Byzantium readily surrendered to Lysander at the end of the Peloponnesian War, he allowed the Athenian garrisons to return to Athens. Xenophon said that he did this to hasten the starvation of Athens by swelling the population, but we can suspect that the Athenians would not have lived to starve if they had resisted as the garrison at Methymna had done.[90] When the last Athenian stronghold at Samos offered to surrender to Lysander, sparing the Spartans a costly assault, he agreed to allow the free men to leave with only one cloak, the same terms the Athenians had allowed the Potidaeans in the first siege of the war. Lysander turned the city over to the Samian oligarchs whom the Athenians had expelled, although he appointed the new ruler.[91]

Purely economic considerations often determined the fate of a town. During the Sicilian expedition, Nicias, ordinarily a man of moderation, captured the Sicel town of Hyccara, brought its entire people to Catana, and sold them in the slave market there for 120 badly needed talents.[92] We know that sailors in the Athenian fleet bought some of the slaves, because Nicias complained later that their use as rowers lowered the discipline and effectiveness of the fleet.[93]

There was a story that the famous courtesan Lais, whose breasts were so beautiful that artists clamored to use her as a model and who became the lover of famous philosophers and orators, was a girl of Hyccara who was bought at Catana and taken to Corinth, where she began her renowned career.[94] Despite her success, it was not a field she chose. The story is undoubtedly mostly legendary (there were conflicting versions), and we should not allow the dazzling beauty of Lais to blind us to the reality of the sordid life of slave women. Even in the romantic tale about her, Lais met an unhappy end far from home, a murder victim of a jealous woman.

Viewing the course of Greek siege warfare over a long time, Pierre Ducrey has studied one hundred sieges that took place between the sixth and second centuries B.C. Twenty-five ended in massacres, thirty-four in the enslavement of the population, and forty-one on such terms as the deportation of the population (very common), the destruction of walls, the imposition of a garrison, and the surrender of hostages. Ducrey emphasizes that his data do not permit exact conclusions because of the many gaps in the sources and the significant variations of circumstances. But he does see a tendency toward massacre in cases such as Plataea, Scione, and Melos where cities were defended to the bitter end.[95] Places that surrendered immediately usually received lenient treatment.

The number of massacres rose sharply just before and during the Peloponnesian War in comparison with the preceding two centuries and the fourth century.[96] There were nine cases of the massacre of a city's male population during the Peloponnesian War but only two in the fourth century up to the death of Philip II (336 B.C.).

Franz Kiechle sees a pattern in which the treatment of conquered peoples steadily moderated until an outburst of brutality during the Peloponnesian War. This brutality marked a turning point in Greek thought about war, and the level of cruelty dropped sharply in the fourth century.[97] Andreas Panagopoulos also detects a rising tide of violence during the Peloponnesian War, noting that over half of the cities taken by force suffered either massacre or enslavement. Milder terms, such as deportation, destruction, and the surrender of hostages, usually resulted from capitulation on terms.[98]

Josiah Ober also argues that traditional restraints on war began to break down in Greek society after 450 B.C., but, unlike Kiechle, Ober believes the breakdown was permanent.[99] Ducrey explains this pattern by arguing that Athenian imperialism made the Peloponnesian War a total war which was fought with a ruthlessness more characteristic of a civil war.[100]

W. Kendrick Pritchett, who has compiled a list of massacres and enslavements in Greek warfare, does not find Kiechle's pattern of amelioration convincing and cites Demosthenes' lament in the *Third Philippic* that war had become more brutal in the fourth century than in the past. Pritchett finds no conventions governing the treatment of prisoners but emphasizes that victors regarded prisoners as being entirely at their disposal. He believes the record is too incomplete to allow generalizations.[101]

Raoul Lonis emphasizes that much depended on the mood of the conqueror and his attitude toward the conquered. Like Pritchett, Lonis argues that conquerors considered captured cities at their disposal, so that if mood and circumstance moved them to harsh measures they felt free from guilt.[102]

Although several accounts show that Greeks could feel guilty about harsh treatment of conquered cities—the Athenian guilt and fears at the end of the Peloponnesian War over their treatment of Melos, Scione, and Torone; the Spartan dedication of a temple to Zeus at Plataea; the Argives' dedications to the gods after they sold the people of Mycenae into slavery—Lonis is correct that the attitude of individual commanders was more important than general standards of conduct. But even if no general rules of conduct prevailed, there were some patterns in the practice of siege warfare in Greece.

Certainly Pritchett's caution against inventing moral standards is well taken, but his own list of massacres (not confined to sieges) includes seven up to the events leading to the Peloponnesian War and twenty-four just before and during the war, so it is difficult to avoid the impression that war became more violent during that war.[103] Of these thirty-one massacres, fifteen took place after the capture of a city and two were massacres of unsuccessful besiegers; eleven came after naval battles, two resulted from civil wars, and for one, the massacre of some Athenian colonists by Thracians in 465 B.C., we do not know the exact circumstances.

What is striking here is that not a single massacre took place after a traditional hoplite battle. One explanation of the rising brutality of Greek warfare during the Peloponnesian War could be that the war marked the beginning of a watershed in Greek military technology. Traditional hoplite warfare, fought in the open field to a quick decision, was becoming less dominant. Siege warfare, naval warfare, and the use of light armed troops and the employment of mercenaries were all becoming more prominent.[104] Siege warfare was becoming important in Greek warfare, but siege technology had not been perfected.

The harsh treatment often meted out to cities such as Plataea, Scione, and Melos which forced a siege upon their enemies and held out for a prolonged period contrasted with the conventions of war in set battles between hoplite armies. Hoplites were valuable citizens, and their battles were fought in the open field along conventions designed to bring a quick decision. The end of the battle was clearly defined by the request from the losers for a truce in order to recover their dead from a field they no longer controlled and the erection of a trophy by the victor.[105] These sorts of conventions broke down in city fighting with the whole society caught up in the struggle. Even women and slaves fought in these battles.

The refusal to surrender to a superior force shattered the convention of a quick military decision, and the limited technology of siege warfare prevented the forcing of a quick decision. As a result, besiegers tried to fan factional hatreds and encouraged treachery. Thus siege warfare during the Peloponnesian War brought together a peculiar concatenation of moral and technical circumstances that led to a *guerre à outrance* that previously had not been typical of Greek warfare.

Women and Children

The presence of women in sieges disturbed Greek men. When Andromache pleaded with Hector to defend Troy from its walls, he ordered her to return home because "the men must see to the fighting."[106]

Hector knew that to fight from the city would blur this fundamental distinction between the roles of men and women, and he always preferred to defend the city in open-field combat outside the walls, where only men ventured. When Poulydamas advised him to retreat to the city to defend its walls, Hector showed his distaste for that kind of fighting:

> Poulydamas, these things that you argue please me no longer
> when you tell us to go back again and be cooped up in our city.
> Have you not all had your glut of being fenced in our outworks?[107]

Hector told Andromache that he would be ashamed to come within the walls, and in the end he rejected the entreaties of his father, Priam, and his mother, Hecuba, to come into the city, preferring to die in open combat with Achilles. Hector's ideal battle was single combat between warriors with a trophy to the winner and a proper burial with a memorial grave site for the loser.[108]

This view persisted into the fifth century. In Aeschylus's *Seven against Thebes*, Eteocles becomes very upset at the presence of women in the siege of Thebes, fearing that their fright will demoralize the soldiers. He wishes he did not have to have anything to do with women.[109] Thucydides also was deeply conscious of the disturbing nature of city fighting. In his account of the fighting in Plataea at the beginning of the war when the Thebans attacked Plataea, he described shrieking women and slaves hurling stones and roof tiles from the rooftops at the Thebans. The scene is identical to the one during the civil war in Corcyra when women threw tiles from the rooftops and stood up to the roar of battle "with a courage beyond their sex."[110] These scenes were one way Thucydides emphasized the breakdown of conventional standards.

Aineias Tacticus viewed such scenes with dread. Despite the frequent examples of women pitching in during a siege in Greek warfare, Aineias did not envision this. He believed the proper place for the women and children during a siege was at home. The deployment of troops should enable men to take positions "very close to their homes— close enough to maintain domestic control."[111] Only if there were a severe shortage of manpower could Aineias envision a role for women in a siege and then only to impersonate men on the walls in order to fool the enemy. But these women disguised as men should never throw anything, because that would instantly give them away.[112]

The modest role Aineias assigned to women during a siege, relegating them mostly to house and home, revealed that he believed the ideal situation under siege was one that preserved traditional social relations. To him the breakdown of gender roles signified a fatal social dissolution that would characterize a polis in final collapse.

The representation of siege warfare in Greek epic and tragedy placed women and children in a central position in which their fate represented the pathos and the terror of war. When a bard singing of the Trojan War moved Odysseus to tears, Homer employed the simile of a woman in a fallen city to capture his grief:

> ... As a woman weeps, lying over the body
> of her dear husband, who fell fighting for her city and people
> as he tried to beat off the pitiless day from city and children;
> she sees him dying and gasping for breath, and winding her body
> about him she cries high and shrill, while the men behind her,
> hitting her with their spear butts on the back and the shoulders,
> force her up and lead her away into slavery, to have
> hard work and sorrow, and her cheeks are racked with pitiful weeping.[113]

Such pathetic scenes show that the Greeks did not view the fate of captured women with indifference. But it is the moral chaos inherent in the collapse of a city that strikes us most forcefully in the Greek representation of siege warfare. The nauseating cruelty of ripping pregnant women open and dashing babies to the ground, which were so striking in the Biblical prophets' vision of siege warfare, are not absent from Homer, and they perform the same function: to project an image of utter moral chaos.

During one battle, Menelaos captures a Trojan warrior and is preparing to send him back to the Greek ships to be held for ransom, when Agamemnon runs up and says:

> Dear brother, o Menelaos, are you concerned so tenderly
> with these people? Did you in your house get the best of treatment
> from the Trojans? No, let not one of them go free of sudden
> death and our hands; not the young man child that the mother carries
> still in her body, not even he, but let all of Ilion's
> people perish, utterly blotted out and unanswered for.[114]

Homer then comments:

> The hero spoke like this, and bent the heart of his brother
> since he urged justice.[115]

The moral context here is not siege warfare. Agamemnon makes clear that the morality of his judgment is not based on the right of a conqueror but on the right of revenge for the violation of the hospitality that Menelaos had shown the Trojan Paris. But this scene is suggestive of terrible things to come for Troy.

In another scene, Andromache, mourning for the slain Hector, bewails the fate that awaits Troy and imagines her child, Astyanax, dashed to death from the towers of the city:

> My husband, you were lost young from life, and have left me
> a widow in your house, and the boy is only a baby
> who was born to you and me, the unhappy. I think he will never
> come of age, for before then head to heel this city
> will be sacked, for you, its defender, are gone, you who guarded
> the city, and the grave wives, and the innocent children,
> wives who before long must go away in the hollow ships,
> and among them I shall also go, and you, my child, follow
> where I go, and there do much hard work that is unworthy
> of you, drudgery for a hard master; or else some Achaian
> will take you by hand and hurl you from the tower into horrible
> death . . . [116]

By the fifth century, Greek tragedy presented the fate of Andromache and Astyanax in even starker terms. Here Andromache does not merely anticipate Astyanax's death; she sees her worst nightmare fulfilled.

In Euripides' play about her, Andromache mourns that Astyanax was "hurled headlong down from the highest tower when Troy was taken!"[117] That happened in the heat of battle. Euripides tells a much more cold-blooded story in *The Trojan Women*, written in 415 B.C., perhaps with the massacres at Plataea, Scione, Hysiae, and Melos in mind. In this play the Greeks decide as a matter of policy to kill Astyanax. The herald Talthybius persuades Andromache to give up the child with an argument worthy of the Athenians at Melos:

> Let it happen this way. It will be wiser in the end.
> Do not fight it. Take your grief as you were born to take it,
> give up the struggle where your strength is feebleness
> with no force anywhere to help. Listen to me!
> Your city is gone, your husband. You are in our power.
> How can one woman hope to struggle against the arms
> of Greece? Think then. Give up this passionate contest.
>   This
> will bring you no shame. No man can laugh at your submission.[118]

All Talthybius can offer is the hope that the Greeks will allow Andromache to bury her child.[119] The chilling implication that they might not reinforces the impression of complete moral collapse. Talthybius was ashamed of his ugly mission:

> I am not the man
> to do this. Some other
> without pity, but as I ashamed
> should be herald of messages like this.[120]

After the Greeks hurled the baby Astyanax from the walls of Troy, Hecuba expressed her horror and grief in a moving lamentation over the shattered body, which the Greeks had brought to her lying in the shield of Hector:

> O Darling child, how wretched was this death. You might
> have fallen fighting for your city, grown to man's
> age, and married, and with the king's power like a god's,
> and died happy, if there is any happiness here.
> But no. You grew to where you could see and learn, my child,
> yet your mind was not old enough to win advantage
> of fortune. How wickedly, poor boy, your fathers' walls,
> Apollo's handiwork, have crushed your pitiful head
> tended and trimmed to ringlets by your mother's hand,
> and the face she kissed once, where the brightness now is blood
> shining through torn bones—too horrible to say more.
> O little hands, sweet likenesses of Hector's once,
> now you lie broken at the wrists before my feet;
> and mouth beloved whose words were once so confident,
> you are dead; and all was false, when you would lean across
> my bed and say: "Mother, when you die I will cut
> my long hair in your memory, and at your grave
> bring companies of boys my age, to sing farewell."
> It did not happen; now I, a homeless, childless, old
> woman must bury your poor corpse, which is so young.
> Alas for all the tenderness, my nursing care,
> and all your slumbers gone. What shall the poet say,
> what words will he inscribe upon your monument?
> *Here lies a little child the Argives killed because*
> *they were afraid of him* That? The epitaph of Greek shame.[121]

Talthybius's embarrassment and Hecuba's lamentation make clear that the Greeks considered the massacre of women and children shameful. The genius of Euripides has captured the plight of a fallen city in the small, shattered body of Astyanax. In the midst of the rising tide of violence in the Peloponnesian War, his three plays about the fall of Troy represented an increasingly urgent call for a more humane treatment of captured cities.[122]

Aeschylus also associated the chaos of a sack with the blood of women and children. In *Seven against Thebes*, the chorus of Theban women expressed their horror at the possible sack of Thebes:

> Tumult reigns through the town, against it advances a towering net of ruin. Man encounters man and is laid low by the spear. For the babes at their breast resound the wailing cries of young mothers, all streaming with blood. Kindred are the prey of scattering bands. Pillager encounters pillager; the empty-handed hails the empty-handed, fain to have a partner, all greedy neither for less nor equal share. Good reason is there to surmise the issue of deeds like this.[123]

The image of streams of blood flowing from mothers with suckling babes at their breasts against the wild background of pillage again uses women and children to capture the horror of siege war.

Perhaps the dominant theme in the representation of siege warfare in Greek literature is rape. Rarely mentioned by historians, it is a constant theme of Homer and the tragic poets. Siege warfare in the *Iliad* is very much a battle for sexual rights. Paris's violation of Menelaos's sexual rights began the war, and there is no reticence by the Greeks about how they intend to avenge this insult. Nestor exhorts the Greeks:

> Therefore let no man be urgent to take the way homeward
> until he has lain in bed with the wife of a Trojan
> to avenge Helen's longing to escape and her lamentations . . . [124]

The Trojan defenders did not fail to understand what was at stake. Agenor tells Achilles:

> . . . there are many of us inside, and men who are fighters,
> who will stand before our beloved parents, over wives and our
>   children,
> to defend Ilion . . . [125]

Poulydamas vainly advised Hector and the Trojans to retreat to the city and defend the walls, because if Achilles returns to the battle "the fight will be for the sake of our city and women."[126] Priam also begs Hector to come inside the walls "so that you can rescue the Trojans and the women of Troy."[127]

Victory to the Greeks and defeat to the Trojans meant the absolute power of the Greeks to claim the Trojan women, parcel them out among themselves (Agamemnon promised twenty to Achilles),[128] and carry them away to their beds. This image of victory reinforces the image of a chaotic climax to siege warfare. Agamemnon boasted that

> Vultures shall feed upon the delicate skin of their bodies,
> while we lead away their beloved wives and innocent
> children, in our ships, after we have stormed their citadel.[129]

This boast yokes together two themes of chaos, the antifuneral and sexual violation.[130] Both represent impurity, the former in the image of rotting flesh and carrion birds, the latter in the violation of sexual taboos.

The sexual violation of women was largely a characteristic of siege warfare. Unburied corpses could be a part of any warfare, to be sure, but the slaughter of the men and enslavement of the women and children in sieges made the funeral rituals much less likely than in the formalities of armistice and body recovery in conventional battles. When Thucydides described the complete degradation of the Athenian expedition to Sicily when it attempted to escape from Syracuse, he emphasized the shame the Athenians felt at leaving behind unburied corpses. The image that came naturally to his mind to make the terrible scene comprehensible was that of siege warfare: "For indeed they looked like nothing else than a city in secret flight after a siege."[131]

Sexual mores and funerals are uniquely human constructs (the gods of the *Iliad* notoriously lacked sexual mores) that define human culture by separating human procreation and death from that of the animals. Only humans surround these elemental acts with ceremony. The moral chaos inherent in the breakdown of these fundamental elements of human culture represents the real terror of siege warfare in the *Iliad*.

The theme of sexual violation became even more explicit and pronounced in Greek tragedy. The chorus in Aeschylus's *Seven against Thebes*, lamenting the fate of a fallen city, summons up visions of virgin women, clothes torn from their bodies, being dragged away by the hair:

> For piteous it were thus to hurl to destruction a city of olden time, made slave and booty of the spear, in dust and ashes laid by Heaven's decree and ignominious ravage of Achaean men. Piteous, too, for her captive daughters (ah me, ah me!), young and old, to be haled by their hair, like horses, while their raiment is rent about them. A city made desolate waileth as the captive spoil is borne to its doom and mingled cries. Grievous in truth is the fate my fear forebodes.
>
> Woeful it is for modest maidens, plucked all unripe, before the nuptial rite, to pass on a detested journey from their homes. Nay, the dead, I trow, have a happier fate than they. Aye, for many and wretched are the miseries (alas, alas!) when a city is taken.[132]

There is no doubt among the Theban maidens that they will be raped if the besiegers break into the city and in their minds the enemy's phallus is a weapon:

> Perish the braggart who vaunteth loud against the city! May the thunder's bolt stay him ere ever he burst into my home and with o'erweening spear despoil my maiden bower![133]

The chorus also raises images of pollution and defilement of a fallen city:

... Man drags off man, or slays, or carries fire; the whole city is befouled with smoke. Mad, inspiring to frenzy, slaying the people, defiling holiness in war.[134]

Such defilement called into question the very existence of the gods. When Talthybius sees the wretchedness of Hecuba after the fall of Troy, he cries out:

> O Zeus what can I say?
> That you look on man
> and care?
> Or do we, holding that the gods exist,
> deceive ourselves with unsubstantial dreams
> and lies, while random careless chance and change
> alone control the world?[135]

A general air of religious desecration hangs over a sacked city. In *Seven against Thebes*, Eteocles comments that "a captured city is forsaken by its gods."[136] In Euripides' *Trojan Women*, Poseidon announces that he must abandon his altars in Troy because "once a city sinks into sad desolation / the gods' state sickens also, and their worship fades."[137] He ominously warns of the danger of sacking a city:

> That mortal who sacks fallen cities is a fool,
> who gives the temples and the tombs, the hallowed places
> of the dead to desolation. His turn must come.[138]

Euripides presented *The Trojan Women* in 415 B.C., the year of the Melian massacre. Surely he intended to disturb the conscience of Athens over its treatment of captured cities.[139]

The image of captured women in the *Iliad* is ambiguous. At times they appear as mere objects to be handed out as prizes, and not especially valuable ones at that. Homer values a slave woman at four oxen, an iron tripod at twelve oxen.[140] But he presents Agamemnon's and Achilles' captives, Chryseis and Briseis, as human beings and treats them with pity. Both Agamemnon and Achilles confess that they felt love for their captives, Agamemnon saying that he liked Chryseis "better than Klytaimestra, my own wife"[141] and Achilles calling Briseis "the bride of his heart" and saying "I now loved this one from my heart, though it was my spear who won her."[142] Briseis, in her lamentations over the dead body of Patroclus, even says that Patroclus had promised her that she could become Achilles' lawful wife when they returned to Phthia.[143] But here we reach the limit of the prospects for captured women, because the promise echoes hollowly off Briseis's description of the mangled body of her husband and her own enslavement.[144] Homer meant to show the kindness of Patroclus rather than the prospects for Briseis.

The image of spear brides in Greek tragedy was much grimmer. In Euripides' *Hecuba*, the chorus of captive Trojan women poignantly expresses the horror of capture.

> My husband lay asleep,
> his spear upon the wall,
> forgetting for a while
> the ships drawn up on Ilium's shore.
>
> I was setting my hair
> in the soft folds of the net,
> gazing at the endless light
> deep in the golden mirror,
> preparing myself for bed,
> when tumult broke the air
> and shouts and cries
> shattered the empty streets:—
> *Onward, onward you Greeks!*
> *Sack the city of Troy*
> *and see your homes once more!*
> Dressed only in a gown
> like a girl of Sparta,
> I left the bed of love
> and prayed to Artemis.
> But no answer came.
> I saw my husband lying dead,
> and they took me over sea.
> Backward I looked at Troy,
> but the ship sped on
> and Ilium slipped away,
> and I was dumb with grief.[145]

Artemis was the goddess of virgins, so the chorus leaves us in no doubt for what it prayed. "But no answer came."

In the same play, Hecuba's daughter Polyxena graphically describes the peripety of a captive woman:

> I had a father once,
> king of Phrygia. And so I started life,
> a princess of the blood, nourished on lovely hopes
> to be a bride for kings. And suitors came
> competing for the honor of my hand, while over the girls
> and women of Troy, I stood acknowledged mistress,
> courted and envied by all, all but a goddess,
> though bound by death.
> And now I am a slave.
> It is that name of slave, so ugly, so strange,
> that makes me want to die. Or should I live

to be knocked down to a bidder, sold to a master
for cash? Sister of Hector, sister of princes,
doing the work of a drudge, kneading the bread
and scrubbing the floors, compelled to drag out
endless weary days? And the bride of kings,
forced by some low slave from god knows where
to share his filthy bed?[146]

Polyxena shares none of Briseis's illusions about what awaits her:
drudgery and forced sex with some loathsome creature.[147]

Even though Hector's wife, Andromache, became the mistress of no
less than Achilles' son Neoptolemus, her life in Euripides' play about
her is little better than Polyxena's nightmare. Neoptolemus married
another woman, who persecutes Andromache, causing her to reflect bitterly that Neoptolemus's bed was

A bed that from the first I never wanted
and now reject for good. The gods are witness
that was a bed I never crept in gladly.[148]

The chorus of Trojan women in Euripides' play of that name sees
nothing in the future but an endless succession of bitter nights of forced
sex:

Shall I be a drudge besides
or be forced to the bed of Greek masters?
Night is a queen, but I curse her.[149]

What Homer and the Greek tragic playwrights leave us with most
of all is a sense of pity for the victims of siege warfare. Like the Hebrew
prophets, Homer revealed a prophetic vision of moral chaos so complete and frightening that we can scarcely bear to look. The tragic playwrights presented terrible scenes that must have been all too real to their
audiences. Neither Homer nor the tragic playwrights drew didactic conclusions, but their deep sense of humanity made the moral questions of
slaughter and enslavement implicit in their work.

Chapter VII

DIONYSIUS I

The New Siege Warfare

As the Peloponnesian War approached its dreary conclusion, a new era of Greek siege warfare was beginning in Sicily. The Syracusan tyrant Dionysius emerged as the first Greek commander to master all the methods of siege warfare. The shocking catastrophe of the Athenian defeat in the siege of Syracuse had been the great event of his youth and undoubtedly impressed him with the pitfalls of passive siege methods.

As a tyrant, Dionysius had the air of an eastern monarch about him, and some even compared him to the Persian emperor.[1] From the fringes of the Greek-speaking world, he had more in common with Macedonian princes than with the classical Greek polis.[2] A transitional figure between the polis and Hellenistic monarchs, he moved away from traditional agonal battles between citizen soldiers toward the military methods of the Hellenistic age.[3] His methods of warfare—the use of mercenaries, the efficient mobilization of labor on a large scale, the utilization of the full range of siege machinery, and the transportation of populations—either looked back to the eastern empires or forward to the Macedonian empire. They were part of a general military revolution in Greece after the Peloponnesian War that saw the development of integrated armies of heavy and light infantry, skirmishers, heavy and light cavalry, more sophisticated logistics, and greater siege capabilities.[4]

Four years after the Syracusans repulsed the Athenian invasion, Sicily suffered another, much more devastating invasion by the Carthaginians, a Phoenician people who knew eastern siege methods. Threatened his entire life by the Carthaginians, Dionysius learned the art of siege in this hard school. Ephorus numbered the Carthaginian invading force at 204,000, Timaeus at 100,000. But even the smaller figure is probably too large; 40–50,000 is a more realistic estimate.[5] A fleet containing sixty warships and 1,500 transports supported the army.[6]

That was a much more powerful force than the Athenians had mustered for their invasion of Sicily. Although the Athenians sent a larger battle fleet (134 triremes), their army of 6,400 men was much smaller.[7] Perhaps the most significant difference was the greater logistical support the Carthaginians enjoyed. Although the 1,500 transports may be an overestimation based on an inflated estimate of the army's size, it seems likely that the Carthaginian fleet constituted a logistical support system far more organized than that of the Athenians, who took only 130 transports to Sicily, relying primarily on a trailing fleet of private boats for logistical support.[8] Also the Carthaginian manpower was much more adequate for the labor-intensive task of siege warfare. In addition to the manpower, however, the Carthaginians brought a full range of siege equipment, including battering rams and siege towers. The Phoenicians had an old tradition of skill in engineering, an important asset in siege operations.[9]

The Carthaginian general Hannibal quickly demonstrated the devastating effect of his siege equipment at Selinus, a town in western Sicily that was his first objective. Hannibal first secured his logistical position by capturing a trading station at the mouth of the Mazarus River. This trading station was probably the main port of Selinus; it certainly provided good landing facilities for ships from Africa.[10]

Having secured a supply line, Hannibal then divided his army. One contingent probably took a position east of the city to guard against relief forces from the Greek cities east of Selinus. This will have been the smaller force. The larger Carthaginian force immediately began assaulting the walls, probably on the northern side where the terrain was most favorable.[11] Diodorus said Hannibal invested the city first, but we should not take this literally, because investment was not usually a part of Carthaginian siege tactics, which were much more aggressive than Greek tactics and relied almost entirely on assault.[12] Hannibal deployed six huge wooden siege towers and a like number of battering rams. Iron plating covered the rams to protect them from fire.

The Selinuntians, who had been allies of Carthage in the past, had not expected a Carthaginian campaign against them, and they had no recent experience in siege warfare. As a result, they had neglected their walls. The scarcity of archeological evidence makes it impossible to trace the course of the Selinuntian walls. Most likely a common wall enclosed both the city and the acropolis.[13]

In any case, the Carthaginian siege towers far exceeded the walls in height, and the Carthaginians employed large numbers of slingers and archers to drive the Selinuntians from the battlements. Protected by the withering covering fire, the rams were able to breach the walls on the first day of the attack, a remarkably short time that revealed the weakness of the walls and the inexperience of the Selinuntians in defending against siege machinery.[14]

The size and power of the Carthaginian siege machines terrified the Selinuntians, but the entire population fought back fiercely. The young men manned the walls, while the older men and the women carried food and supplies to them. Diodorus said that the women forgot "the modesty and sense of shame which they cherished in time of peace" as they fought for their city. He explained such extraordinary behavior by "the magnitude of the emergency [that] called for even the aid of their women."[15]

After the rams broke through the Selinuntian wall, the Carthaginians sent Campanian mercenaries, whom Diodorus considered Hannibal's best soldiers, into the breach. They attacked in relays, a tactic that greatly increased the intensity of the assault.[16] These mercenaries had come to Sicily three years earlier to serve in the Athenian army, but the Athenian defeat left them without an employer. We do not know what they had been doing, probably nothing good. They may have fought for the Greek enemies of Syracuse and then crossed over to Africa.[17] From there the Carthaginians had sent them back to Sicily to fight for Egesta against Selinus.

The Carthaginians had provided them horses so they could serve as cavalry.[18] Perhaps we should infer from this that the Campanians were knights who had come to Sicily to serve the Athenians as cavalry, for which the Athenians had a great need. Why would cavalrymen accept the role of infantry shock troops? Cavalry were of little use in siege warfare. Diodorus tells us that later, when Dionysius's enemies had him cooped up in Syracuse, they sent home their cavalry because they were of no use in a siege.[19] Perhaps the Carthaginians, knowing that they would be besieging cities, made it a term of employment that the Campanians be willing to climb off their horses to assault cities. In any case, it is a good example of the flexibility mercenary troops could afford.

The important point is that the assault was carried out by mercenaries and not Carthaginian citizens, for whom attacking cities would have been too hazardous. Later, in the siege of Syracuse, Himilco's deal with Dionysius, which secured the withdrawal of the Carthaginian citizen troops but left the mercenaries to the mercy of Dionysius, showed how important citizens were and how expendable mercenaries. This assault on Selinus proved the point, because it ended in a bloody failure.

The effort to exploit the breach was premature. The Carthaginians had not cleared away the rubble, and in the face of determined Selinuntian resistance, the Campanians soon bogged down in the debris and could not force the breach. Indeed, at the end of the campaign the Campanians felt used and complained bitterly that Hannibal had not rewarded them sufficiently for their hard service.[20] The battle of the walls of Selinus lasted all day, and darkness forced the Carthaginians to break off their assaults.[21]

The Selinuntians took advantage of this respite to send for help to Acragas, Gela, and Syracuse. The Acragantini and Geloans did not want to move without the support of Syracuse. The Syracusans had their own problems, being engaged in a local war, but they broke off that fighting and began gathering a relief army. Even though the walls of Selinus were breached, the Syracusans believed that the Carthaginians would be unable to take Selinus by assault before the relief army arrived to rescue the city.[22] This confidence that a passive siege would be necessary to reduce Selinus showed that the Syracusans were still living in the limited world of Greek siege warfare.

The next day, Hannibal proceeded more systematically by attacking the walls in several places in order to draw defenders away from the breach so that engineers could enlarge the breach with their rams and clear away the debris. Nine days of hard work and hard fighting were necessary before Iberian mercenaries succeeded in breaking into the city. But the fighting was far from over.

The Iberians suffered heavy casualties as the Selinuntians defended their city street by street and house by house. As in the street fighting at Plataea and Corcyra, the women and children took to the rooftops and threw stones and tiles at the enemy. This grim battle of attrition lasted into the afternoon, but, attacking in relays, the larger Carthaginian force finally exhausted the Selinuntian defenders.

By the day's end, Hannibal's troop had fought their way from the north wall all the way across the city to the acropolis and secured the city.[23] Although the construction of the siege machinery may have taken as long as a month, once the Carthaginian assault began the city had fallen in just ten days.[24] A new era had arrived in Sicily.

Hannibal next turned his army against Himera, a town on the left bank of the Himera River (probably the modern Fiume Grande) on the northern coast of Sicily.[25] He had a grudge against Himera; it had played a role in the exile of his father, and his grandfather, Hamilcar, suffered a terrible defeat there in which he lost his life, either in battle or by suicide.

Hannibal again divided his army, leaving about a third of his troops in a camp he established on some hills south of the city to guard against the approach of relief forces. The rest of his army, reinforced by thousands of native troops who had joined him against the Greeks, moved against the city. Again Diodorus said that the army invested the city, but later events make clear that this is a formulaic phrase of Diodorus that reveals a lack of understanding of Carthaginian siege methods. As he had at Selinus, Hannibal moved directly to assault the walls with his siege engines without establishing a full-scale investment of the city.

Apparently Himera's walls were stronger than those at Selinus because Hannibal's battering rams failed to breach them. Hannibal had appreciated the greater strength of the Himeraean walls and set sappers to work to undermine them, a method he had not used at Selinus. His sappers dug under the wall and then collapsed the tunnel by burning the wooden support beams, causing a large section of the wall to fall. However the Himeraeans repulsed the attack through the breach and quickly threw up a make-shift wall out of the rubble of the old wall. The Syracusan force that had been too late to save Selinus now hurried to Himera with other allies, the reinforcements amounting to about four thousand men. All this forced the Carthaginians to withdraw to regroup.[26]

There could now be no illusions about Carthaginian siege capabilities. The Himeraeans decided not to submit to siege as the Selinuntians had done but to take the offensive. That their attack caught the Carthaginians completely by surprise revealed the novelty of aggressive defense tactics by Greeks. Carthaginian confusion and panic was so great that their mercenaries fought each other more than the Himeraeans. However, when the mercenary troops fled in disorder to the Carthaginian camp in the hills, the Himeraeans pressed too hard and far, enabling Hannibal to catch them in disorder with a successful counterattack by fresh troops. The counterattack drove the Himeraeans into retreat, but three thousand of them took a stand and fought to the last man so the rest could make it back to the city.[27] Despite the failure of the sortie, its near success demonstrated the potential of surprise attacks against besiegers, and this soon became the standard defensive tactic of the Sicilian Greeks.[28]

At this point, the defenders of Himera were further reinforced by the arrival of twenty-five Syracusan triremes. Hannibal's navy was still at its main base at Motya, a Carthaginian port on the western tip of Sicily, so Syracusan naval supremacy was complete. But when a rumor reached Himera that the entire Syracusan army was on its way to Himera and that Hannibal was preparing to embark troops on his triremes to surprise a Syracuse stripped of its defenders, the Syracusan commander at Himera, Diocles, decided to rush back to Syracuse as quickly as possible. The Himeraeans protested in vain.

The triremes could accommodate about half the population of Himera, so Diocles planned to evacuate that many to safety while the other half protected the city until the triremes returned. Diocles himself set out with his troops toward Syracuse. He departed in such haste that he left behind the bodies of those Syracusans who had died in the fighting, a sacrilege for which he was later exiled. A throng of women and children followed because they feared there would not be room for them in the triremes.[29] This partial evacuation of Himera reveals that Hannibal had not invested the city.

The next day Hannibal resumed his assault on the city. The reduced garrison held out for a day, but the next day, just as the triremes that were returning to evacuate the rest of the population came into sight, Hannibal's battering rams knocked down a section of the wall and Iberian mercenaries poured through the breach. The Iberians gained control of the wall on both sides of the breach and from that vantage point were able to keep the Himeraeans at a distance while the rest of the Iberians rushed through the breach.[30] Once they had gained entrance into the city, the Iberians made short work of Himera.

The Carthaginians had captured two of the most important Greek cities in Sicily in a single season. When Hannibal returned to Carthage laden with booty, the Carthaginians greeted him "as one who in a brief time had performed greater deeds than any general before him."[31]

The Carthaginians invaded Sicily again in 406 B.C. They again appointed Hannibal commander, but he pleaded old age and so the Carthaginians also made his cousin Himilco general.[32] Having raised large numbers of mercenary troops from Iberia and Italy in addition to their African troops, Hannibal and Himilco crossed over to Sicily with a force of 120,000 men. Over a thousand transport ships supported the army.[33] Again Diodorus's sources probably exaggerated the numbers, perhaps by a factor of two, but this force was almost certainly larger than the expedition of 409 B.C.[34] Their objective was Acragas on the southern coast of western Sicily not far east of Selinus.

The wealth of Acragas was famous, and Diodorus made a long digression to describe it.[35] So great was the luxury of Acragas there was some question about the hardiness of its men. Diodorus tells us somewhat scornfully that when under siege the Acragantini passed a decree that the night guards should be limited to one mattress, one cover, one sheepskin, and two pillows.[36]

But there was no sign of decadence in the defense by the Acragantini of their city against the Carthaginians. Unlike their neighbors the Selinuntians, whom the Carthaginians had caught by surprise three years earlier, the Acragantini had realized that their location made them the most likely first target of the Carthaginian invasion. They had prepared by bringing all their crops and possessions within the walls of the city, as well as the entire population of perhaps 200,000 people.[37]

When the Carthaginians arrived at Acragas, they followed their now familiar practice of dividing the army, establishing one camp in the hills east of the city, where they left about a third of their troops to guard against the approach of relief forces from Gela and Syracuse, and a main camp with the rest of the troops on the right bank of the Hypsas River, which ran just west of the city. They fortified the latter with a palisade and a deep trench.[38] Having established this formidable position, they offered terms giving the Acragantini the choice of joining Carthage as allies or remaining neutral in the Carthaginian war of conquest in Sicily.

The Acragantini, being well prepared for a siege and reinforced by fifteen hundred mercenaries under the command of a Spartan named Dexippus as well as the eight hundred Campanians whom Hannibal had left in the lurch after the earlier campaign, rejected the Carthaginian offer.[39] They armed all citizens of military age to defend the walls, keeping some in reserve. The creation of this reserve was a response to the aggressive assault methods of the Carthaginians. In the face of repeated assaults supported by siege machinery, defenders would quickly tire. The creation of a reserve would enable a troop rotation to keep fresh defenders on the wall at the place of attack.[40] The Acragantini relied heavily on the crack mercenary troops, who took a position on the Hill of Athena, the most strategic point guarding the access to the city.[41]

Difficult terrain made the access to the Acragantini walls poor, but Hannibal and Himilco found one place where they were able to move two very tall siege towers forward toward a gate. The Carthaginians inflicted many casualties from these towers, but when nightfall forced them to withdraw, the Acragantini made a daring nighttime sortie that succeeded in burning the siege towers. Stung by this setback, Hannibal and Himilco decided to throw up siege ramps across unfavorable terrain to make possible multiple assaults. Large numbers of troops turned to the task. They tore down the monuments and tombs in the area and, using the debris for fill, rapidly completed the ramps.[42]

Tearing down tombs made the soldiers uneasy, however, and there was alarm when lightning struck one of the tombs. The soothsayers interpreted the lightning strike as a sign of divine displeasure. A plague that broke out immediately afterward made the anger of the gods even more evident, and guards on night duty reported seeing the spirits of the dead. When the plague struck down Hannibal, the troops were in a state of panic.[43]

The death of Hannibal left Himilco in sole command. He moved decisively and terribly to end the panic. First he stopped the tearing down of monuments and tombs. Then he placated the anger of the gods by sacrificing a young boy to Cronus and a large number of cattle to Poseidon by dumping them into the sea. Thus reassured, the troops continued building the siege works, filling in the Hypsas River, which flowed close enough to the city at this point to serve as a moat. That enabled Himilco to press on with the siege.[44] Diodorus does not tell us what happened to the flow of the Hypsas when the Carthaginians filled it in. They may have laid a base of underbrush and covered it over with soil, thus allowing the river to flow through. Or perhaps the stream of the Hypsas was so slight in the summer that damming it had little effect.[45]

The Carthaginians had pressed the siege with great energy, but Acragas's walls were still intact. Now the Carthaginian position became precarious. The Syracusans, determined that Acragas would not suffer the same fate as Selinus and Himera, rushed a large relief force, consisting of thirty thousand foot soldiers and five thousand cavalry supported by thirty triremes, toward Acragas. Diodorus's figure of five thousand cavalry is probably too large, but the cavalry force must have been substantial.[46] The numbers may not include light armed troops, so the total force may have been considerably larger than thirty-five thousand.[47]

Himilco sent his mercenaries to intercept them, but after a hard-fought battle somewhere west of the Himera River, the Syracusans defeated Himilco's mercenaries, who fled in disorder.[48] The Syracusan general Daphnaeus, mindful of the disaster that had befallen the Himeraeans when they pursued too eagerly after defeating the Carthaginians, regrouped his army before following the fleeing enemy.

When the Acragantini saw the Carthaginian army in retreat, they begged their generals to lead them out to deliver a coup de grâce. This the generals refused to do. They too may have remembered the ill-fated pursuit by the Himeraeans and, with winter setting in and the sea from Syracuse to Acragas still open, decided that caution was in order. The Acragantini, however, were so angry that a rumor that the generals had been bribed found easy belief. Whatever their motives, the Acragantini generals died for their caution. The people were so enraged by the escape of the defeated enemy that they stoned to death four of the five generals, an outburst of violence that starkly revealed the terrible stress in a city under siege.[49]

Daphnaeus made a reconnaissance in force of the Carthaginian camp, but when he found it well fortified he contented himself with cutting the roads into the camp and harassing Carthaginian foragers with cavalry. His reluctance to attack the camp showed how far the Greeks were behind the Carthaginians in siege warfare.[50] However, Daphnaeus maintained the harassment all summer and this tactic had the advantage of maintaining an aggressive defense without risking a crushing defeat in an all-out attack on the enemy. Eventually it caused severe food shortages in the Carthaginian camp. Many died of starvation, and the mercenaries were on the brink of mutiny.

The Greeks must have established complete naval supremacy, because the huge flotilla of transport ships that had brought the Carthaginians to Sicily was certainly large enough to supply the army. Himilco was learning the hard way the logistical vulnerability of siege armies. In this case, however, the incredible complacency of the Greek defenders of Acragas, or treason by the generals, saved the Carthaginians from defeat.[51]

Himilco, finding himself bottled up on land, decided to attack by sea. Good intelligence had brought him the information that a large shipment of grain from Syracuse was approaching Acragas. He persuaded his mercenaries to hold on a few more days while he made an effort to capture this grain. The Syracusans had unchallenged naval supremacy and it was winter, so they had become careless about the security of their grain convoys. When Himilco gathered forty triremes and attacked a grain convoy, the surprise was so complete that he sank eight Greek warships and forced the rest to the beach. The entire convoy of transports fell into his hands, and for the first time in weeks he was able adequately to feed his army.[52]

Because of the siege, the Acragantini probably had not been able to bring in that year's harvest, and so they were heavily dependent on the grain the Syracusans were bringing in by sea.[53] As a result, this single success collapsed the morale in Acragas. The Campanian mercenaries in the city, encouraged by the offer of fifteen talents from Himilco, immediately switched sides and joined the Carthaginians. The Acragantini had been so confident of their grain supply that they had not bothered to ration it. The swollen population of the city was rapidly exhausting the food supply, causing the Italian Greek allies to leave. Their departure may have been encouraged by Dexippus in return for a fifteen-talent bribe from Himilco, but Diodorus does not seem confident in the story.

Whatever the mercenaries' motives, the episode revealed the dangers of hiring them for the defense of a city. By the fourth century the practice had become universal in the Greek-speaking world. For example, Aineias Tacticus assumed that part of a city's defenders would be mercenaries, and he viewed them with suspicion. He believed a city should never force mercenaries to endure a siege. Instead it should gather the mercenaries together and make the following announcement:

> Anyone dissatisfied with the conditions here and wishing to leave is allowed a discharge—but in future [any malcontents] shall be sold into slavery. For lesser offenses than these the penalty is imprisonment, in accordance with the prevailing law. Any manifest harm to military operations, however, by undermining morale in camp, is a capital offence.[54]

Aineias went on to caution that a city should not rely too heavily on mercenaries. Citizen troops should always outnumber mercenary troops.[55] Mercenaries should never command but instead be under the command of "the most trustworthy citizens."[56] Wealthy citizens should pay the mercenaries, but the state should contribute to their pay as well, there being nothing more dangerous than mercenaries dissatisfied with their pay. Mercenaries should be quartered in the houses of those same wealthy citizens to keep them isolated in small units.[57] Acragas had learned to its cost the soundness of Aineias Tacticus's warnings.

An inventory revealed too little food to sustain the city, and the generals ordered an evacuation.[58] It is a story of incredible negligence, and later Daphnaeus was executed by the Syracusans. But this seems to have been more because he opposed the tyranny of Dionysius than for his role at Acragas, so although an air of suspicion hangs over the generals there is no proof of treason.[59]

The following night—it was mid-December—a pitiable throng of men, women, and children departed the city on the road to Gela, forty miles away. The army accompanied them to provide protection, but there is no word of any effort by the besiegers to stop them. The march was difficult for the women and children, but they did reach Gela safely. Those who were too old or sick to march forty miles were left behind. Diodorus said that some Acragantini committed suicide rather than face life in a foreign place. Diodorus's imperfect account makes it difficult to reconstruct the chronology of the siege, but it had taken the Carthaginians eight months to reduce Acragas.[60]

In the summer of 405 B.C., after destroying Acragas, Himilco moved against Gela, which lay along the coast east of Acragas. He spent some time plundering the countryside to obtain provisions for his large army and arrived before Gela in July.[61] He established a camp along the sea to the west of the city. Certain that Syracuse would try to relieve Gela, the Carthaginians fortified their camp with a palisade and a trench, leaving only the seaward side open. The Geloans prepared to evacuate their women and children to Syracuse; but the women insisted on staying, and they would play an important role in the events that followed.

Although the Geloans made life miserable for the Carthaginians by harassing Carthaginian foragers, Himilco was still able to launch a series of assaults with his battering rams against the western walls of Gela. The Geloan walls were not strongly fortified and the rams made some breaches, but the Carthaginians were unable to exploit them because of the spirited defense and because the Geloans repaired the breaches during the night. This would not have been possible without the help of the women and children who had chosen to stay, because the men were fully occupied in the fighting.[62]

The determined defense of Gela kept the Carthaginians at bay until a large army arrived from Syracuse, including Italian Greeks and mercenaries as well as most of the Syracusan army. The force numbered at least thirty thousand men, probably not counting the light armed troops, three to four thousand cavalry, and fifty ships.[63]

This army was under the command of Dionysius. Dionysius had fought bravely at Acragas and had risen to power by attacking the Syracusan generals who had abandoned Acragas. Now he was in the process of establishing his tyranny. Only in his mid-twenties, he gained his first opportunity to show his mettle as a commander.

Dionysius camped at the mouth of the Gela River on the side of the city opposite from the Carthaginian camp, a position from which he could coordinate both land and sea operations. For three weeks he used his light armed troops to prevent the Carthaginians from foraging and his ships to cut the Carthaginian supply line from Africa. This was the same tactic that had almost defeated the Carthaginians at Acragas, but Diodorus indicated that Dionysius's patrols were not entirely successful.[64]

Dionysius may have simply been biding his time while he prepared and organized a complicated plan of attack.[65] But it seems more likely that his own troops forced his hand. Dionysius's army consisted mainly of the citizens of Sicilian Greek cities. As we know, such troops did not like protracted campaigns. They were accustomed to ending campaigns by a decisive battle, and they may have become impatient with Dionysius's tactic of attrition.[66] In any case, Dionysius now moved to attack the Carthaginian camp.

He developed an imaginative plan of attack.[67] The Carthaginians had fortified their camp and possessed numerical superiority. To offset these disadvantages, Dionysius devised a coordinated three-pronged attack involving heavy and light infantry, cavalry, mercenaries, naval support, and an amphibious landing. The plan revealed a deep insight into the weakness of the traditional massed phalanx against a fortified position, but it failed because the coordination of the rather complicated plan broke down. Dionysius attempted to lead a group of mercenaries through the city so they could sally forth through the western gate at the critical moment. But with the besieged city packed with the rural population, the mercenaries became so bogged down in the narrow, crowded streets that they never reached their start line. The Carthaginians were able to defeat the other two parts of Dionysius's army in detail. [68]

This defeat left Dionysius in an unenviable military position. The failure to break the siege meant that Gela faced the doubtful prospect of enduring a siege through the winter. The defeat also endangered Dionysius's political position in Syracuse. If he brought his troops into Gela to help defend the city over the winter, his absence from Syracuse would give his political enemies too good an opportunity. These considerations caused Dionysius to decide on an immediate evacuation of Gela.[69] He carried out the evacuation that very night. His departure was so precipitous that he did not even recover the dead from the battlefield.[70] The next day the Carthaginians plundered the city.

Dionysius withdrew to Camarina, but even though his army still had fight left in it, the same logic that had driven him from Gela led him immediately to decide to evacuate Camarina also. These decisions showed Dionysius's good military and political judgment, but the surrender of two Greek cities in rapid succession was an inauspicious beginning for a man trying to establish a military reputation.

By now the countryside was teeming with refugees, most of them women, children, and old people.[71] The Carthaginian reputation for cruelty added terror to their flight, as they had heard of crucifixions and other tortures inflicted by the Carthaginians on their captives. The plight of the refugees was so pitiable that opposition against Dionysius grew into a plot to overthrow him. Dionysius had to hurry back to Syracuse. He did so with such speed that he caught his enemies so much by surprise that the gate to the outer city was unguarded. He forced his way in by burning it and thwarted the uprising.[72]

Himilco was unable to exploit Dionysius's embarrassment because the plague had continued to ravage his army. Both sides finding themselves in difficulty, they negotiated a peace. The terms left Syracuse to Dionysius and formally recognized him as its ruler. But he paid a high price. The treaty required Selinus, Acragas, Himera (nothing was left of Himera; the treaty must refer to a new settlement at Therma),[73] Gela, and Camarina to pay tribute to Carthage and forbade them to rebuild their fortifications, leaving the Carthaginians in control of much of the Sicilian coast.[74]

These terms freed Dionysius from the threat of Carthage, but his position was far from secure. He soon found himself under attack by his Greek enemies, who, like the Athenians, seized Epipolae and placed Syracuse under siege, bringing Dionysius to the verge of despair.[75] Only the incompetence of the besiegers and some Campanian mercenaries the Carthaginians made available to Dionysius bailed him out.[76]

This narrow escape, as well as the threat of Carthage's devastating siege prowess, convinced Dionysius to strengthen the fortifications of Syracuse. He had already strongly fortified the island of Ortygia with a well-built, powerful wall strengthened by high towers at close intervals. The most recent archeologist to study the towers calls them "unusually costly and well-built."[77] He had also enclosed the dockyards at Laccium, creating a fortified haven large enough to accommodate sixty triremes.[78] Now, sometime between 401 and 399 B.C., he turned to the task of preventing future enemies from threatening Syracuse from Epipolae, a danger that had twice recently brought Syracuse to the brink of defeat.[79] He therefore began a crash project to build a wall along the northern edge of Epipolae.

Diodorus provided a detailed description of how Dionysius organized his labor to build this wall. He recruited sixty thousand peasants from the countryside and divided them about equally between wall building and working the quarries that were providing stone for the wall. The status of these peasants is not clear. They were not landowners; most were probably free, but a good number of them may have been serfs.[80] Six thousand yoke of oxen were employed to haul the stones to the construction site. Dionysius divided the men assigned to wall construction into labor gangs of two hundred men each and assigned a one-hundred-foot section to each labor gang. Master builders, one per six hundred feet of wall, supervised the labor. Each labor gang had one mason to provide the necessary skilled labor. Dionysius offered prizes to the master builders, masons, and labor gangs to inspire them to work quickly. To Diodorus's great surprise, Dionysius even joined in the work himself, laboring alongside the peasants to inspire them to greater effort. These methods enabled Dionysius to complete six thousand yards of wall in just twenty days.

Diodorus called the wall impregnable to assault. Dionysius's masons had carefully joined together four-foot-long hewn stones to form the curtain. High towers at frequent intervals protected the curtain.[81] Because of the abundance of stone on Epipolae, it seems likely that the wall was built entirely of stone. The engineers designed a casemate wall with an innovative chain masonry technique in which every ten to twenty feet rectangular stones alternated as headers and stretchers. The headers reached through the fill to the opposite facing wall. This connected the two walls, holding them together and counteracting the tendency of the walls to fall outward. As a result, the wall was able to reach a significantly greater height.[82]

The keystone of Dionysius's fortification of Epipolae was a powerful fortress at Euryalus on the western point of Epipolae (see map 2). Inexplicably, Diodorus, who provided such a detailed account of the northern wall, did not mention this fortress nor the wall on the southern edge of Epipolae until his history reached the year 384 B.C.[83] That has led some scholars to conclude that there was as much as a fifteen-year delay before Dionysius completed the wall around Epipolae.[84] But such a delay would have left the northern wall vulnerable to outflanking. Such compelling military logic strongly suggests that Dionysius must have pushed on to complete the wall and that Diodorus's silence resulted from his sometimes sloppy use of sources rather than a long delay by Dionysius in the completion of the wall. The best guess is that the entire wall was completed by 398–397 B.C.[85] The fortress had underground chambers and galleries and guarded the main point of access to Epipolae. Impressive remains of this fort still exist today, although some of them probably date from a later time.[86]

The southern wall continued to the fortifications of the city, thus circumvallating Syracuse.[87] According to Diodorus, "its circuit was the greatest possessed by any Greek city."[88] Such a long wall posed problems to the defenders because there was so much more wall to defend, but by following the natural line of defense along the height of Epipolae it made the advance of heavy siege machinery almost impossible, an essential advantage against the Carthaginians in light of the fate of Selinus and Himera.

The problem of defending long circuits apparently became common in the fourth century, because Aineias Tacticus advised how to solve it. He suggested raising the wall in places where it was most accessible to such a height that it would be impossible for intruders to jump down from it. Guards stationed on both sides of the built-up stretches could

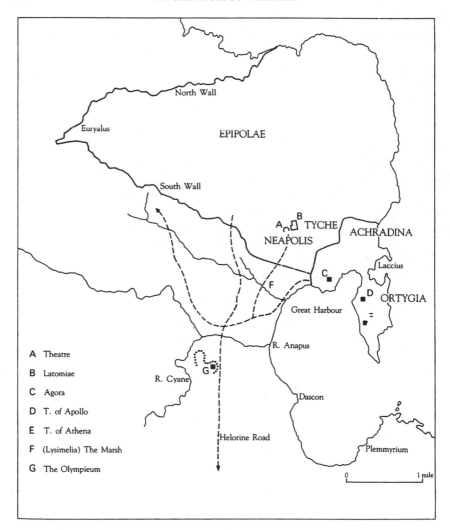

Map 2. Syracuse at the time of Dionysius I. *From Brian Caven,* Dionysius I,
Warlord of Sicily. *Copyright 1990 by Brian Caven. Used with permission of Yale
University Press.*

kill any who dared to jump down.[89] We do not know, however, what
methods Dionysius employed to guard his wall.

Having provided Syracuse with the most advanced defenses in the
ancient world, Dionysius turned to another problem in his preparation
for war with Carthage: Greek inferiority in armaments and siege equipment. In doing so, he showed the same thoroughness with which he had
prepared his defenses.

Dionysius brought large numbers of skilled craftsmen from Italy,

Greece, and Sicily—even from Carthaginian territory—to Syracuse, divided them into groups according to their particular skills, and set them to work making a huge supply of arms. He kept morale high with good wages and prizes for the most productive workers. Dionysius accomplished two things by such a systematic organization of labor. First, by producing arms, instead of relying on his soldiers to provide their own arms, he transcended one of the primary limits on the size of Greek armies by opening his ranks to all social classes. Second, he stimulated technological innovation.

In these conducive circumstances, some unknown armsmaker invented the catapult (*katapeltikon*).[90] This is the first time we have a description of how an innovation in military technology was achieved. Most interesting is the relationship between social organization and invention. A systematic division of labor and the stimulation of interaction between skilled craftsmen from widely scattered areas, not serendipity, produced the first catapult.

Diodorus did not describe these first catapults, although it is clear that they shot arrows. They may have used torsion power generated by tightly twisting bundles of animal hairs, but more likely they were a type of crossbow called the *gastraphetes*, or "belly bow" (see figure 13). Bracing the bow against the stomach, archers could use both hands to push a slider on the composite bow with more strength than could be produced by pulling back with one hand. Although these first catapults did not have winches, they were mechanical devices equipped with a slider, ratchet, and trigger.[91] These catapults shot arrows with devastat-

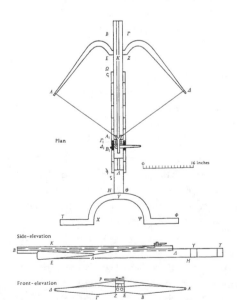

Figure 13. Dionysius I's *gastraphetes*, or belly bow. *From E. W. Marsden, Greek and Roman Artillery. Oxford: Clarendon Press, 1969. By permission of Oxford University Press.*

ing effect up to a range of 200 to 250 yards, around 50 yards greater than the best hand-bows could obtain. Greater striking power and accuracy added to the advantage of the belly bow.[92]

After two years of preparation, in 398 B.C. Dionysius launched a campaign to wrest all of Sicily from Carthage. His army consisted of forty thousand infantry and three thousand cavalry.[93] As many as half, certainly at least ten thousand, of these troops were mercenaries, mostly Greek, who could provide the professionalism and flexibility necessary for siege warfare. Especially important was a corps of engineers to direct the disassembly, transport, and assembly of siege machinery and to take charge of the construction of siege ramps and other siege works.[94]

This "new model" army could match the capabilities of the largely mercenary Carthaginian army.[95] Indeed, as time went on, an increasingly higher percentage of Dionysius's troops consisted of mercenaries. The growing sophistication of Greek armies after the Peloponnesian War demanded specialized troops such as the traditional citizen-soldier could not provide. Mercenary troops were more important to Dionysius's siege capabilities than his catapults. They could develop specialized skills, they could stay in the field as long as their commanders could pay them, and they were expendable. When one considers that mercenaries were indispensable to the maintenance of Dionysius's power as tyrant of Syracuse, it becomes apparent that there was a symbiotic relation between tyranny and the new demands of warfare in the fourth century.[96]

A large fleet of two hundred warships and five hundred transports supported the army. The transports carried "great numbers of engines of war and all the other supplies needed."[97] This army was better equipped for siege warfare than any previous Greek army. We would like to know more about Dionysius's logistics, but Diodorus tells us relatively little. He required his troops to carry rations to last thirty days, a much longer time than Greek armies normally provided for.[98] The fleet may well have been a necessary support for sieges. All of the sieges in the war between Carthage and the Sicilian Greeks were against coastal cities, suggesting that neither Dionysius nor the Carthaginians could move their siege trains overland.[99]

Hiring mercenaries, providing them with weapons, constructing fortifications and fleets cost much more money than Greek polises were used to raising. Dionysius paid his soldiers and his armament workers well and handed out lavish rewards for meritorious service. But he seems to have stuck with traditional Greek financing methods, relying on payments from tribute-paying allies, harbor dues, taxes on alien residents, plunder, and the sale of slaves and booty to pay his mercenaries. He sometimes paid off mercenaries by giving them land of defeated enemies. In times of financial crisis he resorted to borrowing from

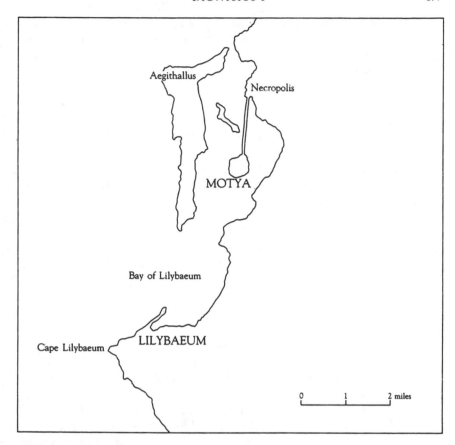

Map 3. Environs of Motya and Lilybaeum. *From Brian Caven,* Dionysius I, Warlord of Sicily. *Copyright 1990 by Brian Caven. Used with permission of Yale University Press.*

temple treasuries, special levies on citizens, and debasing the coinage to make ends meet. There is some evidence that Dionysius imposed a regular direct tax on the citizens of Syracuse, but, if so, that is the only financial expedient he employed that was not common to Greek cities. He commanded substantial financial resources for the Greek world, but he could not match the wealth of the eastern empires, or even of Carthage.[100]

Dionysius's first target was Motya, which was located on an island off the western coast of Sicily and was the main Carthaginian base of operations for Sicily (see map 3). Motya posed a formidable problem to besiegers, who would have to traverse a channel over a mile wide to reach the island. A narrow causeway connected Motya to the mainland, but the Motyans had breached it in preparation for the siege.[101]

A stone wall surrounded the island.[102] The circumference of the wall was a little over a mile and a half and there were about twenty towers, which meant the towers were about 125 meters apart. But they were not at a uniform distance, falling at more frequent intervals near the gates. The wall lay very close to the shore, in some places rising from the edge of the sea. Only along the northern and northeastern coasts was there some space between the wall and the sea, and because of this the Motyans had reinforced their wall in these more vulnerable places. In these reinforced areas the wall was about six meters thick. We cannot know with certainty the height of the wall; one estimate is eight to nine meters high. Thick stone battlements protected the defenders on the top of the wall.

The main gate of Motya was on the northeastern coast and opened to the mole connecting the island to the mainland. Two powerful towers protected it. Their design followed the old tradition of making troops trying to force the gate approach with their exposed right side open to enemy fire. The passage into the city from the outside gate ran through a series of inner gates.

There was a smaller gate on the southern side of the city, which apparently served as an entrance to the city from boats and ships. Archeological evidence suggests the Motyans barricaded this gate with a stone wall during the siege. There also were smaller gates on the western and eastern sides of the city. At least one of them appears to have been bricked up during the siege.

Near the southern gateway, the Motyans constructed a small artificial harbor that was connected to the island with a drawbridge. The presence of three large stones in the channel leading into the harbor indicates that the Motyans blocked the channel during the siege.

There is evidence that a lead pipe from the mainland supplied Motya with water. The Greeks would have cut such a pipe. But there were also cisterns and wells on the island. These must have been adequate, for we do not hear of any water shortages during the siege.[103]

Dionysius began the siege by rebuilding the mole from the mainland to the island. Dionysius himself led his infantry against neighboring cities and left his brother, Leptines, who commanded the fleet, in command of the construction.[104] Leptines beached his ships to free his sailors for the task. The crews of the five hundred transports probably provided most of the labor. Diodorus said the Greeks completed the mole "by employing a large force of laborers."[105] The mole was fifteen hundred meters long and ten meters wide. The water around Motya was quite shallow, which made the task easier.[106]

The scale of the project recalls the ability of eastern emperors to mobilize and organize labor.[107] The capabilities of Dionysius, however, did not match those of the east. His campaign against the neighboring cities met with little success. The Assyrians had been able to reduce

small cities in the area while carrying on a siege, but Dionysius's resources were too small for simultaneous siege operations, especially
when they required moving his siege train overland, a logistical capability he lacked.[108] He soon returned to Motya and put his infantry to
work to accelerate the construction of the mole.

Beaching the warships proved risky, as it gave a Carthaginian naval
force an opportunity to attack them. With most of his crews on land,
Dionysius was in a precarious position, but his new invention and his
own ingenuity saved the situation. Although his ships were trapped on
the beaches of the harbor, Dionysius got them into the open sea by dragging them across a narrow strip of land that separated the harbor from
the sea.

Himilco hastened to attack the Greek ships at their most vulnerable moment, as they were being launched one by one, but the Greeks
drove his ships away with a hail of missile fire from archers on the Syracusan ships and from Dionysius's land-based catapults, whose greater
range and accuracy had a great psychological effect on the Carthaginians.[109] It may be that Dionysius's craftsmen had equipped their belly
bows with winches, which would have made them still more powerful.[110] If Polyaenus's report is true that Dionysius was able to haul eighty
triremes across land in one day, we have an impressive example of Dionysius's ability to mobilize and organize labor.[111] Once Himilco's surprise attack failed, the Carthaginian fleet, which the Greek fleet outnumbered, fled to Carthage, freeing Dionysius to resume the siege of
Motya.[112]

When his men had completed the mole, Dionysius advanced his
siege machinery against the walls in the vicinity of the northern gate.
The citizens of Motya, cramped for space on their island city, had built
their houses to a great height, reaching well above the walls. Dionysius
built six-story siege towers to reach beyond the top of the walls and
equal the height of the houses. Such tall siege towers probably exceeded
the height of Carthaginian towers.[113]

The Greeks rolled these towers forward on wheels. Presumably
teams of oxen pulled the towers. Xenophon tells us that Cyrus developed a means of hitching eight yoke of oxen to the first story of his movable towers, which enabled him to pull the siege tower with twenty soldiers in it, a load he estimated to be more than six thousand pounds.
Cyrus yoked the oxen to a single pole so that they could advance in file,
greatly reducing their vulnerability to enemy fire. This must have been
the universal method for moving siege towers.[114]

Catapults helped drive the Motyan defenders off the wall while battering rams attacked it. The abundance of arrowheads among the ruins
of the northern walls of Motya bear witness today to the intensity of
their fire.[115] The Motyans attempted to set the siege machinery on fire
by throwing burning pitch from crow's nests which they suspended on

yard arms reaching out from the highest places along the wall. They managed to set some of Dionysius's machinery on fire, but fire squads quickly doused the flames with water and soon the battering rams had opened a breach in the wall.[116]

The presence of fire squads in Dionysius's army revealed how much better prepared he was for siege warfare than Greek armies in the Peloponnesian War. When Demosthenes had attacked the Syracusan counterwall with siege machinery during the Athenian siege of Syracuse, the Syracusans had burned his machinery. Evidently the Athenians were not prepared for this very common threat to siege machinery. By the fourth century, matters were quite different.

Aineias Tacticus emphasized the necessity of fire prevention in siege warfare. His advice included using fire-resistant material, such as felt or hides, to cover any wooden towers or wooden components in the walls and smearing bird-lime on flammable materials to fireproof them. He recommended vinegar as an especially effective fire quencher.[117] If the enemy succeeded in setting a gate on fire, he advised the defenders to augment the fire to hold the attackers at bay to gain time to dig a trench and erect a palisade as a second line of defense. When the enemy advanced its siege machinery to the wall, the defenders should drop flammable material on it and set it on fire. Aineias described a special device with small spikes that would attach itself to wooden siege machinery when dropped on it. Highly combustible material could be attached to it so that it would set the machinery on fire. Aineias also gave a recipe for a combustible combination that he claimed would be impossible to extinguish: pitch, sulphur, tow, powdered frankincensegum, and pine sawdust.[118] Fire could also create smokescreens and hinder siege operations at the foot of a wall. If countermining against enemy mining opened up the enemy's tunnel, smoke could drive the troops out of the tunnel or even suffocate them.[119] Aineias made clear that in city fighting, fire was a prime weapon, and Dionysius's ability to defend against it showed how well he had mastered the art of siegecraft.[120]

Breaching Motya's wall did not end the battle. The Motyans barricaded themselves in their high houses, which were contiguous so that they formed a wall themselves. From these heights the Motyans were able to direct withering missile fire down on Dionysius's troops. These tactics forced Dionysius's men to push their high siege towers into the city up to the houses from which the Motyans were fighting. The towers had gangways which could bridge the gap between the towers and the houses, enabling Dionysius's troops to force their way into the houses.

With the safety of their women and children at stake and no hope for escape from their island home, the Motyans fought with a grim determination. The knowledge that they could expect no mercy from the Greeks, who were eager to avenge the cruelty the Carthaginians had inflicted on the towns they had conquered, added desperation to the Motyan resistance. Advancing across narrow gangways against such desperate opposition was no easy task, and Greek casualties were heavy.[121]

This bloody fighting continued for several days before Dionysius brought it to an end with a stratagem. Each evening he recalled his troops and rested them during the night. When the Motyans had grown accustomed to this pattern, Dionysius sent some picked troops under the command of a Thurian named Archylus into the city during the night. They carried ladders with them and under the cover of darkness were able to climb to an advantageous position before the Motyans had discovered their presence. Archylus's troops held their position long enough to open the way to more Greek troops. The Motyans fought bravely, but superior numbers now overwhelmed them.[122] Dionysius rewarded Archylus with one hundred minas for his feat.[123] Archylus and his men were mercenaries; they exemplified the critical role mercenaries and mercenary commanders played in Dionysius's army.[124]

The fall of Motya shattered any complacency the Carthaginians may have felt about the war in Sicily. Redoubling their effort the next year (397 B.C.), they raised a large force, probably numbering around fifty thousand men, in both Africa and Iberia and equipped a large fleet of about a thousand ships, four hundred of which were war ships, to transport the army with its extensive siege machinery and keep it supplied in the field. Himilco took command of this formidable force.[125]

Dionysius had stationed a fleet under the command of his brother, Leptines, at Motya to prevent the Carthaginians from reinforcing Sicily, but the Carthaginian fleet somehow eluded the Syracusans, except for about fifty stragglers that Leptines managed to sink with a squadron of thirty huge quinqueremes. Most of the Carthaginian fleet put in safely on the northern coast of Sicily at Panormus, not far from Motya. The numbers we have are too unreliable to determine the relative size of the fleets, but probably the Carthaginians outnumbered Leptines' ships by a wide margin. Dionysius had taken his army to besiege Egesta, enabling Himilco to march to Motya and quickly reduce it by siege.[126]

Diodorus does not tell us why Dionysius made no effort to hold Motya after he had expended so much effort to capture it, but it seems most likely that the arrival of such large Carthaginian forces made Dionysius's position in western Sicily precarious and he preferred to withdraw rather than risk a battle in unfavorable circumstances.[127] The Carthaginians were apparently dissatisfied with the site, for they abandoned it in favor of a new town on the nearby promontory of Lilybaeum. Motya was never rebuilt.

Himilco then set his sights on Messene, located on the narrow strait between Sicily and Italy. It had a harbor large enough to serve as a base for Himilco's fleet of over six hundred ships, and its strategic position would enable Himilco to block reinforcements arriving from either Italy or Greece. The possession of Messene would also pose a dire threat to the city of Syracuse.

Messene was in a poor situation to defend itself. Incredibly, the walls of Messene were in a state of complete disrepair (Diodorus said they had fallen down). The cavalry was away in Syracuse. When the Messenians realized the Carthaginians were moving against them, despair set in. Nevertheless, they prepared to defend themselves as best they could. They evacuated their women and children and their valuables.

With their walls unfit for a siege, it seemed best to the Messenians to confront the Carthaginian army before it invested the city. Accordingly, they sent a force to the Carthaginian encampment, hoping to stop the Carthaginian army before it reached Messene.[128] These moves left the city almost undefended, and when Himilco dispatched a fleet of two hundred ships against Messene, it had no difficulty capturing the city. It apparently landed without opposition. The absence of the Messenian fleet is a puzzle Diodorus did not explain.

Most of the Messenian population escaped to fortresses in the countryside. This system of fortresses, reminiscent of the defensive system developed by the Kings of Judah in the Judean hill country, proved effective. Himilco quickly realized the futility of trying to reduce one small fortress after another and contented himself with the capture of Messene.[129] There he prepared his army for an advance against Syracuse, and when he was ready, he razed Messene to the ground so thoroughly that "no one would have known that the site had been occupied."[130] Apparently Himilco doubted his ability to hold a city in eastern Sicily and desired to make it useless to his enemies.

The threat to Syracuse was clear to Dionysius, and he took drastic action to secure the city, freeing the slaves to man sixty more ships and hiring over a thousand Lacedaemonian mercenaries. But Dionysius devoted most of his efforts to strengthening the system of fortifications in both Syracuse and Leontini. The difficulty Himilco had encountered with the Messenian fortresses had not escaped Dionysius, and he made sure his fortresses were well fortified and fully provisioned.[131] The second-century Macedonian author of a book on military strategy, Polyaenus, said that Dionysius intended these fortresses to surrender quickly, thus luring the Carthaginian army away from Syracuse, but this may have been mere speculation on Polyaenus's part.[132] More likely, Dionysius hoped the fortresses could protect the harvest and serve as bases from which the Syracusans could harass Carthaginian foragers.[133]

Dionysius's first line of defense was to attack the Carthaginian fleet at Catana, to which it had advanced, but the Greeks suffered a crushing defeat that left the way to Syracuse open to the Carthaginians. Dionysius then faced the decision of whether to meet the Carthaginian army in the field or to accept a siege at Syracuse. The dread of a siege was so great that Dionysius's men urged him to lead them against Himilco. Some of Dionysius's advisers, however, reminded him that Himilco had captured Messene by sea when the Messenian army marched out of the city to block the Carthaginian army at the frontier and convinced Dionysius that he should go to Syracuse. This decision to accept a siege was so unpopular that most of Dionysius's Sicilian Greek allies deserted him, either returning home or dispersing to the fortresses in the countryside.[134]

These desertions left Dionysius deeply shaken, and he desperately sent to Italy and Greece for mercenaries. But he was helpless to prevent the Carthaginians from sailing into the Great Harbor with a fleet of 250 battleships supported by perhaps as many as three thousand merchant ships.[135] The Carthaginian army advanced at the same time. Because of the fortification of Epipolae, Himilco had to march around the plateau and encamp on the low ground to the south, just south of the Anapos River, less than one-and-a-half miles from the city. From there Himilco challenged the Syracusans to battle but refrained from assaulting the walls when the Syracusans remained in the city. Instead he devoted the next thirty days to ravaging the countryside in the hopes of intimidating Syracuse into surrender before winter set in.[136]

The long-range prospects for Himilco were not promising. The fortifications of Epipolae made circumvallation a virtual impossibility, and an assault on the southern side of Syracuse would leave the Carthaginian flank vulnerable to an attack from Epipolae. Under these circumstances, Himilco adopted the strategy of Nicias, maintaining the pressure and hoping for an internal upheaval in Syracuse.[137] Like Nicias, Himilco made preparations for a long siege by building three forts, including one at Plemmyrium, to provide secure landing points for his supply ships and establishing a regular supply line with Africa.[138]

Only in the spring did he move against the walls of Syracuse, assaulting the part of Syracuse known as Achradina and seizing a walled suburb that contained several temples, which he immediately plundered. In response, Dionysius increased the aggressiveness of his defense, successfully engaging the Carthaginians in some small skirmishes.

According to Diodorus, the first significant counterstroke by the Syracusans took place when Dionysius was absent from the city. He had left with a naval squadron to escort a critical supply convoy to Syracuse. While he was gone, the Syracusans launched their ships and won a notable naval victory in the Great Harbor.[139] The contrast with Dionysius's generally passive defense was so great that he faced a serious threat of overthrow from the Syracusans, which was thwarted only by the failure of the Lacedaemonian allies to support it.[140] Diodorus's story of this battle is implausible and has the tone of an antityranny tract. The leading modern historian of Dionysius rejects Diodorus's account and believes that no such battle ever occurred.[141]

Himilco's cautious strategy assumed that time was on his side, but the opposite proved to be the case. Almost as soon as he encamped before Syracuse, a plague broke out in his army. Diodorus attributed the plague to divine retribution for the plundering of the temples, but he also pointed to other causes: the crowded conditions in an army encampment, the hot summer weather, and the wet, marshy terrain. He pointed out that the Athenians also had suffered from the plague when they besieged Syracuse under the same conditions.[142] Soon there were so many corpses that the burial details, overwhelmed by the number of deaths and demoralized by the spread of the plague to those who came into contact with it, could not keep up with their task and the stench of putrefying bodies hung over the Carthaginian army.[143]

The extent of this catastrophe emboldened Dionysius to attack the Carthaginians by land and sea. His attacks were entirely successful. He captured two of the fortified naval stations, and fire destroyed a large part of the Carthaginian fleet. He assigned mercenaries to attack the rear of the Carthaginian camp, an exceptionally hazardous task. Diodorus tells us Dionysius did this because they were mutinous and therefore expendable, another example of commanders' willingness to sacrifice mercenary troops.[144]

Himilco did not want to suffer the fate of Nicias, and he immediately entered into secret negotiations with Dionysius. Dionysius agreed to allow Himilco to embark for Carthage with his citizen troops in return for a payment of three hundred talents. Diodorus attributed his leniency to Dionysius's belief that the only reason the Syracusans tolerated his tyranny was that they feared the Carthaginians more than they feared him, but the political and military advantages for Dionysius of an end to the siege are evident. Breaking the siege was a glorious victory that undoubtedly took its place alongside the victory over Athens in Syracusan military lore.

Himilco left his mercenaries and allies in the lurch and fled to Africa. Some of the allies escaped, but Dionysius destroyed or captured the rest.[145] Himilco never recovered from his disgrace. He became obsessed with the idea that his defeat was divine retribution and took to wandering from temple to temple, dressed in rags, confessing to impiety. In the end, he starved himself to death.[146]

Having broken the Carthaginian siege of Syracuse, Dionysius seemed content to leave Carthage in control of western Sicily for the time being. He set out to secure the eastern portion. The destruction of Messene left a vacuum on the strait that Dionysius hastened to fill. He rebuilt Messene, fortified it, and resettled it. He also settled some Messenians from the Peloponnese on a site on the northern coast of Sicily and powerfully fortified it. The Messenians named the place Tyndaris.[147]

The fortification of Messene, whose walls had been in a state of collapse in the years before the city fell to Himilco, deeply alarmed the city of Rhegium, which lay just across the strait in Italy. Convinced that Dionysius intended to use Messene as a base for an invasion of Italy, the Rhegians in 394 B.C. raised an army of Sicilian Greek exiles and sent it to besiege Messene. The defenders defeated the would-be besiegers and followed this victory up with a quick march to Mycale, on the northern coast of Sicily a little west of Messene, planted there by Rhegium as a rival to Tyndaris. They captured the town, apparently by offering to allow its inhabitants to leave as free people.

Having secured the territory west of Messene, Dionysius proposed to gain firm control of the coast stretching south of the city along the east coast of Sicily. This required the capture of Tauromenium, a city that Himilco had built and settled with Sicels.[148] Camping on the southern side of the city, Dionysius laid siege to it in 394 B.C.[149] The Siceli defended Tauromenium stubbornly, forcing Dionysius to extend the siege into winter.

Tauromenium was perched on high mountain peaks and presented a formidable challenge to the besieger. The acropolis was fortified with high walls atop steep cliffs that were inaccessible to siege machinery. The Siceli did not even feel the need to keep it well guarded, and this carelessness encouraged Dionysius to attempt a daring nighttime attack in the depth of winter.

Choosing a moonless, stormy night, Dionysius led his men against the steepest, and therefore least guarded, approach to the acropolis. Deep snow made the approach even more arduous, but even though he suffered so much from the cold that his vision became blurred, Dionysius and his men managed to scale one of the peaks of the acropolis, from whence they broke into the city.

The Siceli quickly recovered from their surprise and counterattacked, and Dionysius's men, perhaps because they were weary from their arduous climb, got the worst of the fighting. Dionysius himself barely escaped. Six hundred of his men died and most of the rest lost their armor as they scrambled down the mountain. Dionysius managed to emerge from the fray with his corselet still on, a feat in which he took some pride.[150]

After the assault failed, Dionysius broke off the siege. Two years later, however, the Carthaginians sold out Tauromenium in a treaty with Dionysius that handed Tauromenium over to him, enabling him to occupy the city and exile the Sicel population.[151]

The year after his failure to take Tauromenium, 393 B.C., Dionysius began a campaign against Rhegium, located on the toe of Italy just across the strait from Sicily. Rhegium had shown itself to be a threat to

Dionysius's control of the strait as well as a knife pointed at his back in the war with Carthage. So Dionysius determined to capture the city.

Gathering a force of one hundred triremes, Dionysius attempted to surprise the Rhegians with a night attack. His troops set fire to the gates of Rhegium and brought ladders forward with which to scale the walls. At first the Rhegians tried to put out the fire, but then, on the advice of their general, an exiled Syracusan named Heloris, they decided to fight fire with fire. They brought up combustible materials to feed the fire and make it much larger. This conflagration drove back Dionysius's troops and gave the Rhegians time to assemble their army.

Faced now with determined opposition, Dionysius decided to spend his time ravaging the countryside. The threat from Carthage prevented a protracted siege. In the end he concluded a truce that at least prevented the Rhegians from allying with the Carthaginians and then returned to Syracuse.[152]

After another attempt to take Rhegium failed in late 390 B.C. because a storm severely damaged his fleet before he had even reached Rhegium,[153] Dionysius tried yet again in 389 B.C. His force consisted of twenty thousand infantry, five thousand cavalry, seventy warships, and three hundred transports. This time he landed at Caulonia, a city on the eastern coast of the toe of Italy, and placed it under an aggressive siege, assaulting the walls with siege engines, although apparently without much success.[154] The Italian Greeks hastened to rescue Caulonia, mobilizing an army of twenty-five thousand infantry and two thousand cavalry.

A besieging army is dangerously vulnerable to attack from behind, and Dionysius wisely disengaged from the siege of Caulonia to face the rescue army. We would like more detail about how he accomplished this difficult maneuver. How did he divide his forces? To what degree was he able to maintain the siege, or did he lift it altogether? How did he protect his divided forces against a sally from the city? But Diodorus provided the answers to none of these questions. Perhaps Dionysius disembarked the crews of his fleet and deployed them as light infantry to maintain the siege.[155]

In any event, Dionysius defeated the rescue army. He drove the main body of the Italian Greek army to a hill where he was able to cut them off without water. They endured the rest of the day and night without water, but in the morning they offered to surrender if Dionysius would accept ransom. This he refused, but they were able to hold out only a few more hours until the desperation of their situation forced them to surrender unconditionally, clear testimony to how short a time troops can last with no water at all.

The defeat of the Italian Greek army left the way open from Caulonia to Rhegium. The citizens of Rhegium did not believe they could resist Dionysius, and so they hastened to negotiate a surrender under the conditions that the Rhegians pay three thousand talents, turn over their entire fleet of seventy ships, and hand over one hundred hostages. The Rhegian surrender doomed Caulonia, which quickly surrendered.[156]

Dionysius had a deep-seated hatred of Rhegium over a rebuff he had received when he tried to marry a Rhegian woman; and besides, the possession of Rhegium was essential to his control of the straits. He considered the peace he had made a mere truce and was still determined to take the city. Accordingly, in 387 B.C., Dionysius launched another campaign against Rhegium.

This time he approached Rhegium in feigned friendliness, requesting food for his troops and promising to repay the Rhegians once he had returned to Syracuse. He thought in this way to empty the city of food or, if they refused, to gain a pretext for breaking the truce. At first the Rhegians complied with Dionysius's request, but when he lingered on with a variety of excuses, his intentions became apparent and the Rhegians ceased providing him with supplies. Thereupon Dionysius returned the hostages, declared the truce at an end, and besieged the city.[157]

Dionysius pressed this siege aggressively. He assaulted the walls with huge siege engines that Diodorus said were of "unbelievable size." But the defense by the Rhegians was determined. They made a successful sortie and burned the siege engines. The failure of Dionysius's siege engines led to a grim battle of attrition. Dionysius was seriously wounded by a lance that struck him in the groin.[158]

The siege dragged on for eleven months and ended only when starvation forced the Rhegians to surrender. Diodorus reported that the price of wheat reached five minas a bushel and that the Rhegians first ate their draft animals, then boiled leather, and finally ate grass. Dionysius supposedly brought in livestock to crop the grass in order to deny the Rhegians even this source of food. When the Rhegians finally surrendered, the Syracusans found heaps of corpses piled up in the streets.[159]

Dionysius lived another twenty years, during which he extended his power into the Adriatic Sea and up the west coast of Italy, but we know little of his military operations in these years. He began the construction of a wall across the toe of Italy to guard against incursions by Italian barbarians but failed to complete it.[160] He took the city of Croton, perhaps by betrayal, but we do not know the details of that siege.[161] At the end of his life, he tried to repeat the greatest success of his military career, the siege of Motya, by besieging Lilybaeum, the port on the western tip of Sicily that the Carthaginians had built to replace Motya (see map 3). All Diodorus tells us about this siege is that "there were so many soldiers in the place that he [Dionysius] abandoned the siege."[162] The next year, 367 B.C., Dionysius died.

Dionysius represented a watershed in the history of Greek siege warfare. Before him, Greeks, if they could not capture a city by betrayal, either abandoned the effort or resorted to the slow methods of passive siege. Dionysius changed this by introducing the coordinated use of the full range of siege machinery, and even invented a new machine, the catapult.

Just how wide-ranging an effect Dionysius exerted on Greek siege warfare is reflected by Aineias Tacticus, who made clear that in contrast to Greek cities in the fifth century, those in the fourth century could expect assaults on their walls. His treatise on how to defend against besiegers revealed that siege machinery was a familiar sight in mid-fourth century Greece. Indeed, we learn from him that Greeks were familiar with a wider variety of siege machinery than we would suppose from reading the historians.[163] Aineias told how to defend against battering rams, drills, catapults, and towers; how to set fire to siege machinery; how to counter undermining; how to prevent escalade; how to extinguish fires; and how to protect against arrows and other missiles.[164]

In fact, Aineias said nothing about how to defend against investment, leaving us with the impression that assault had become the dominant siege tactic in Greece by the middle of the fourth century. However, Aineias may have discussed investment in some other work and, in any case, fourth-century historians leave us with a very different impression. Xenophon especially made it evident that Greeks had little taste for assaulting walls and frequently failed when they did so. Only men who were extraordinarily motivated or professional mercenaries who were highly rewarded willingly assaulted fortified positions.[165] Not surprisingly, Dionysius, the first great Greek siege captain, relied on mercenary troops.

Although Aineias occasionally drew from historical examples and although we can find historical examples for several of his defensive tactics, his work had a theoretical character that presages a new genre of military writing: the military manual. It was no accident that the first military manual concerned siege warfare, the most technical branch of warfare in the ancient world. Aineias Tacticus's treatise reflected the age of Dionysius, who first began the professionalization of siege warfare among the Greeks.

The professionalization of siege warfare under Dionysius resulted in one of the most impressive records of successful sieges in Greek military history. His most recent biographer justifiably calls him "Dionysius the Besieger."[166] His methods pointed the way for the greatest Greek conqueror of them all, Alexander the Great.

Treatment of Captured Cities

The new siege warfare in Sicily was no less cruel than the old. Indeed, it was often crueler, especially when barbarians were involved. No other war between Greeks and barbarians was fought with such ferocity.[167] The Carthaginians, from whom the Greeks learned so much about siege warfare, set an unhappy example. The Carthaginians had a reputation for cruelty. They crucified captives and inflicted other tortures on them that may have included the old Assyrian practice of impalement.[168]

The fall of Selinus starkly revealed the ferocity of Carthaginian siege warfare. The Carthaginians plundered the city, burning some people alive in their homes and forcing others into the street, where "without distinction of sex or age but whether infant children or women or old men, they put them to the sword, showing no signs of compassion."[169] Diodorus reported that "according to their custom" the Carthaginians cut off the hands and heads of their victims, stringing the hands around their necks and carrying the heads aloft on swords and javelins. Sixteen hundred Selinuntians died in the carnage; some five thousand survived as captives. Only 2,600 people escaped.[170]

The Carthaginian rampage in Selinus deeply shocked Diodorus. He found the looting of temples especially barbaric. The plight of the women "deprived now of the pampered life they had enjoyed" also aroused his pity. Diodorus could not bring himself to describe plainly what had happened, resorting to a burst of euphemisms:

The women . . . spent the nights in the very midst of the enemies' lasciviousness, enduring terrible indignities, and some were obliged to see their daughters of marriageable age suffering treatment improper for their years.[171]

These women now faced a life of slavery under masters who "used an unintelligible speech and had a bestial character."[172]

After the fall of Selinus, the Syracusans and Selinuntian fugitives sent ambassadors to Hannibal offering ransom for the prisoners and urging him to spare the temples. Diodorus said that Hannibal refused these requests, except that he released the relatives of one Selinuntian who had been a special friend of Carthage. As for the gods, Hannibal said that they had already left Selinus. He ordered his army to tear down the walls of Selinus, but he allowed those Selinuntians who had escaped to return to the city under the condition that they pay tribute to Carthage.[173] In light of this, it seems likely that Hannibal did, in fact, agree to ransom the prisoners, despite Diodorus's report to the contrary.[174] Diodorus, who always gave great importance to sacrilege, does not tell us that Hannibal destroyed the temples; and so we can assume that despite his comment about the desertion of the gods, he spared them.[175]

At Himera, Carthage's Iberian mercenaries pillaged the city and slaughtered its defenders. They even dragged suppliants from the temples and despoiled them. Hannibal finally ordered a halt to the massacre, but he enslaved the surviving women and children and distributed them to his troops. Diodorus leaves it to us to imagine their fate. The captured Himeraean soldiers he led to the same spot where his grandfather had fallen and ordered his troops to torture and kill them. The city of Himera was razed to the ground.[176]

Another Carthaginian violation of temples took place at Acragas. When Himilco occupied the nearly empty city after most of the population had fled, he killed all those who remained behind, even dragging suppliants from the temples. The Carthaginians plundered the city thoroughly, finding great wealth. Because it was mid-December, Himilco wintered in Acragas, but when summer came he razed the city.[177]

Dionysius's record was better than the Carthaginians, but he too was capable of cruelty. A calculating and rational man, he rarely let his emotions control his actions. Even after the arduous siege of Motya, he hoped to recover his expenses by selling the population into slavery. However, his troops were enraged and he lost control of them. They killed every person they found, men, women, children. Only those Motyans who managed to seek refuge in the temples survived the butchery. Respecting the inviolability of the sanctuaries, Dionysius's troops turned to plunder and thoroughly sacked the city. There were some survivors whom Dionysius was able to enslave. Diodorus later mentioned some Motyan survivors, so many may have been ransomed by neighboring cities rather than enslaved.[178] Among the prisoners at Motya were some Greeks who had fought for the Carthaginians. These Dionysius crucified, presumably as a warning to his Greek enemies who might hate him enough to collaborate with the Carthaginians.[179]

Dionysius was capable of acts of pure vengeance. He had a special hatred of Rhegium, which had not only resisted him a long time but had insulted him. Only six thousand Rhegians survived the ordeal of Dionysius's siege. The number was shockingly small for a city as large as Rhegium. Perhaps it referred to the adult males or perhaps a large part of the population had managed to escape during the long winter months. In any case, Dionysius sent them to Syracuse and ransomed them for one mina apiece. Those who could not pay he sold into slavery. But it was the Rhegian general Phyton on whom Dionysius especially took out his hatred. First he drowned Phyton's son and strung Phyton up on a high siege engine. He then tortured Phyton by flogging him through a gauntlet of the entire army until even the sympathy of Dionysius's troops was aroused. Finally Dionysius drowned Phyton and his family.[180]

Dionysius's most common treatment of captured peoples was enslavement or resettlement. In his campaign against the Chalcidic cities on the east coast of Sicily at the end of the fifth century, he captured three cities: Catana, Naxos, and Leontini. Catana and Naxos fell

through betrayal, and Diodorus tells us that Dionysius sold their populations into slavery.[181] But it seems likely that he spared his supporters,
who had opened the gates to him.[182] He allowed his soldiers to plunder
Naxos and then razed the city. He settled Catana with his Campanian
mercenaries. Leontini surrendered voluntarily. Its citizens Dionysius resettled at Syracuse; they probably retained possession of their land.[183]

The Leontini were not the only people Dionysius resettled as free
men in Syracuse. After he took Caulonia in Italy, he transported the
people to Syracuse, where he granted them citizenship and tax-free
status for five years. He razed Caulonia.[184] Dionysius imposed the same
conditions of transportation to Syracuse on the people of Hipponion
when he captured that city the following year.[185]

In one case Dionysius freed his prisoners unconditionally. The ten
thousand men who came to relieve Caulonia had surrendered to him
with great dread. They knew he could be ruthless and cruel. To their
surprise, Dionysius treated them well, releasing them unconditionally
to return home.[186] He undoubtedly had his political reasons, but the
siege of the hill had been brief and was the culmination of a field battle
such as traditionally ended with an honorable surrender. Nevertheless,
the general reaction was one of astonishment and gratitude, and Diodorus said that "men believed that this would probably be the finest act
of his life."[187]

PART FOUR

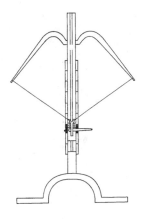

THE
MACEDONIANS

Chapter VIII

PHILIP II AND
ALEXANDER THE GREAT

Philip II

The siege methods that Dionysius had employed with such telling effect on the western periphery of the Greek-speaking world had an even greater impact in the hands of the Macedonians on the northern periphery. The empire of Philip II and Alexander would have been impossible without their ability to reduce cities.

It was Philip who developed Macedonian siege capabilities. This was a remarkable achievement in light of the primitive conditions in which he found the Macedonians. As Alexander boasted in his passionate speech to mutinous troops at Opis deep in Persia in 324 B.C.:

> Philip took you over when you were helpless vagabonds, mostly clothed in skins, feeding a few animals on the mountains and engaged in their defence in unsuccessful fighting with Illyrians, Triballians and the neighboring Thracians. He gave you cloaks to wear instead of skins, he brought you down from the mountains to the plains; . . . He made you city dwellers and established the order that comes from good laws and customs.[1]

Among the arts of civilization that Philip brought to the Macedonians was the art of siege warfare, which went hand in hand with city living. Under his organizing genius, the Macedonians were able to mobilize the resources necessary for siege warfare. By the end of Philip's reign (359–336 B.C.), he could raise an army of twenty-four thousand infantry, four thousand cavalry, and a large number of support troops—the latter especially important in siege warfare—from Macedonia alone.

Moreover, Macedonia and Thrace were rich mining areas that provided Philip with revenues much greater than even the Athenian empire had produced. These vast financial revenues meant that Philip, in an age of mercenaries, "could afford as many he needed."[2]

Philip streamlined the Macedonian army by forbidding women to follow its march and drastically reducing the number of servants and pack animals by training his soldiers to carry their own supplies.[3] That greatly improved the army's logistical ability to sustain a siege. And one of the first things Philip did when he came to the throne was create a corps of engineers, relying mostly on Thessalian Greeks.

Under Philip, Greek siege warfare began to become more systematic and technical, a development that would culminate in the highly technical and professional siege methods of the Hellenistic age.[4] Clearly he fully appreciated the importance of siege warfare.[5]

Philip mastered the full range of siege methods. Although he fought many sieges, we have a detailed account of only one of them, the siege of Perinthus in 340 B.C., and so it must serve as our main example of his siege capabilities.[6] Perinthus was a Hellespontine town located on a high peninsula in the Sea of Marmara, a naturally strong position that presented a formidable challenge to a besieger. The siege began in July, and Philip pressed it with great determination. His greatest strength was an ability to coordinate a variety of siege methods. He used rams, siege towers, catapults, escalade, and undermining in a coordinated way to bring maximum pressure on the city. He divided his army of thirty thousand men into divisions and assaulted the walls in relays, rotating the divisions frequently. This tactic not only prevented fatigue but also helped maintain morale in the face of the extreme hazards of fighting at the foot of a wall. It also allowed the Macedonians to keep up the assaults day and night.

Philip did not send his men against the walls empty-handed. They brought with them the entire array of siege machinery. The Macedonians broke down their machines for transportation in the siege train. Philip's engineers could quickly assemble them in the field. The Macedonian siege towers were exceptionally high; according to Dionysius, they reached 120 feet into the air, possibly twice as high as any previous towers. Philip's catapults used torsion power, which made them much more powerful than Dionysius's "belly bows." Although the evidence is circumstantial, E. W. Marsden believes Philip's engineers invented the torsion catapult.[7] If we can believe Polyaenus, Philip's troops had been badly mauled in a field battle against the Phocians by nontorsion stone-throwing catapults.[8] Marsden believes this so impressed Philip that he ordered his engineers to give special attention to the improvement of catapults. Chief among these engineers was a Thassalian named Polyidus, who became famous for his ingenuity. He wrote a technical treatise, *On Machines*, and first revealed his talent at the siege of Byzantium.[9] Philip's interest in military innovation was apparent, he possessed the requisite resources, and, like Dionysius, he attracted skilled craftsmen from the outside. This created an atmosphere similar to that in which Dionysius's engineers invented the catapult. However, Philip's torsion catapults did not throw stones. Rather they applied torsion power to powerful composite bows that shot arrows.

As impressive as these machines were, Philip's real strength was their tactical integration. The siege towers that rose so high above the walls of Perinthus drove the defenders off their battlements while the rams battered the walls. Philip employed "numerous and varied" catapults to sweep the walls of their defenders and then sent troops with scaling ladders over them. Philip also knew the effectiveness of slingers against defenders on top of a wall. Excavations at Olynthus show that he armed his slingers with lead balls, whose velocity and accuracy were much better than stones. Philip's sappers were able to collapse a long section of wall at Perinthus by undermining. He aggressively exploited breaches opened by rams and undermining by sending assault troops through them. The development of new machines and, even more, their skillful tactical deployment reveals the importance of the highly trained corps of engineers that Philip had developed at the beginning of his reign.[10]

Despite the destruction of their walls by Philip's sappers and machines, the Perinthians kept the Macedonians at bay and, in the end, Philip broke off the siege without capturing the city. There were two reasons for this.

First, the Perinthians were able to buy time by defending their city in savage house to house combat. Their first line of defense within the city was a second wall they threw up behind the breach the Macedonians had created. When the Macedonians destroyed the second wall, the Perinthians took to their houses, which rose up the hill in closely packed rows like the seats of a theater. Each tier of houses formed a sort of wall from which the Perinthians could defend themselves. It took much hard fighting for the Macedonians to dislodge the Perinthians from a row of houses, and when the Macedonians finally succeeded, the Perinthians simply took up the defense again on the next tier.

Second, the Perinthians were not isolated. The Macedonians lacked the naval forces to blockade the city on its seaward side. The Persians in Asia Minor were anxious that Perinthus not fall to Philip, and they sent mercenaries, money, food, and war materiel to the besieged city. Nearby Byzantium also came to the aid of Perinthus, most importantly by sending catapults. That a city as important as Perinthus did not possess catapults shows how new they were. Now the Perinthians had an answer to Philip's catapults. Indeed, this is our first example of the use of catapults to defend a city against siege.[11]

This forced Philip in September, after he had besieged Perinthus for two months, to split his army and march to Byzantium to place it under a close siege also. It was no small feat to besiege two strong cities simultaneously. Only a commander who knew how to deploy his forces with superb efficiency could have managed it. Because his siege train was at Perinthus, Philip's engineers improvised at Byzantium, building the necessary machines out of timbers from Athenian corn ships Philip had captured. However, Byzantium was the stronger and more important city, and it seems likely, given Philip's personal presence there, that it now became the focus of the Macedonians' main effort. Philip caught the Byzantines by surprise. They had sent so much help to Perinthus, secure in the belief that Philip had his hands full there, that they were inadequately prepared for his unexpected move and now found themselves in a precarious position. Macedonian sappers collapsed a section of wall and a daring nighttime assault almost succeeded. But the Byzantines managed to repair their breached wall, and as winter came on the city still held out.[12] At this point, Athens sent a large fleet to support Perinthus. Other Greek cities followed Athens' lead. Philip realized that his position had become untenable and, with the shrewd strategic sense that characterized him, broke off the siege, probably in December, in order to bring Athens into a reliable alliance that would free him to cross the Hellespont and invade the Persian Empire. This turn of events led to the Battle of Chaeronea.

Despite Philip's failure to take Perinthus, the siege displayed his formidable siege capabilities. It showed that he was a masterful organizer and that his machinery and sappers could make short work of walls. His troops showed they were capable of very tough city fighting. Philip showed considerable staying power and resourcefulness against difficult odds. Only outside help on a scale against which no army could have maintained a siege saved Perinthus.

Earlier sieges about which we know less show that Philip was fully capable of capturing cities with stunning speed. His first great siege led to the fall of Amphipolis in 357 B.C. after he breached its wall with rams.[13] The rapid fall of Amphipolis is especially impressive in view of its long and successful resistance against Athens during the Peloponnesian War. In 354 B.C., Philip took Methone by assault. He took an active role in this battle and lost an eye when he was struck by an arrow.[14] In 348 B.C., Olynthus fell in three months, a victory that should be compared with the two years Athens needed to capture nearby Potidaea at the beginning of the Peloponnesian War.[15] Slingshot stones found at Olynthus and dated from this siege suggest that the Macedonians may have driven the Olynthians back with intense fire using heavier than average stones.[16]

Like all siege commanders, Philip preferred finding someone to betray a city to the arduous and expensive work of a siege. Diodorus related the anecdote that upon being informed that a city's walls were impregnable, Philip asked if gold could not scale the walls.[17] His arch-enemy, the Athenian orator Demosthenes, painted Philip as a great cor-rupter rather than a great commander.[18] But Demosthenes was stuck in a past of hoplite warfare and did not realize that in siege warfare the rules were different. Philip must be counted as a worthy successor to Dionysius in the history of Greek siege warfare.

Alexander the Great

Philip had created powerful siege capabilities in the Macedonian army; his son Alexander stretched those capabilities to their fullest extent. Robin Lane Fox is surely right to say that "it was as a stormer of cities that [Alexander] left his most vigorous impression" on his contempo-raries.[19] During the brief course of his conquests, Alexander conducted more than twenty sieges.[20]

Without the ability to take fortified cities, Alexander could not have sustained his spectacular campaigns. The farther he marched from Macedonia, the longer his line of communication became, making it impossible to bypass powerful cities that could pose a threat to his tenu-ous connection with his home base. Alexander recognized this and so, although he generally did not use wagons because they slowed his marches, the siege train was an integral part of his army.[21] And yet siege warfare did not fit easily into Alexander's personality and style of com-mand. He was impatient, a man in a hurry, who preferred the dramatic stroke to the plodding work of siege warfare. This inclination led him to take extraordinary risks as a siege commander, for which he paid dearly.

The nature of Alexander's army also determined his role as a siege commander. Although the Macedonians employed mercenaries, they did not play a central role in their army. The heart of the Macedonian army consisted of a heavy cavalry force and a heavy infantry force, known as the Companion Cavalry and the Foot Companions respec-tively, both drawn from the Macedonian people. Among the Foot Com-panions was an even more elite group called the hypaspists (shield-bearers), who had a special relation to the king. They were famous for their endurance and toughness. The hypaspists were a remarkably flexible unit and could operate with light infantry, especially the elite Agrianian peltasts, as well as heavy.[22] According to circumstances, they could arm themselves differently and they undoubtedly lightened their armament in sieges.[23] Toughness and flexibility were invaluable in siege warfare. These elite troops provided the shock forces for the Mace-donian army, both in the field and in sieges. These men were more like Homeric warriors than citizens of a Greek polis and, unlike the Greek polises that were always reluctant to expose their citizen hoplites to the dangers of assaulting a wall, preferring to employ mercenaries for such hazardous duty, Alexander did not hesitate to deploy the hypaspists and other Foot Companions against fortified positions. The personal bond between the king and these troops required that Alexander take his place at their head. Not even the Assyrian emperors, the greatest siege commanders before Alexander, had done that. In order for Alexander and his elite Macedonian troops to take the lead in assaulting fortified positions, a whole new attitude toward such fighting had to develop. Gone was the traditional Greek distaste for city fighting. Assaulting a wall became a test of valor, perhaps the supreme test, that brought great honor to a soldier.[24]

Alexander's first experience as a siege commander came at the beginning of his reign in 335 B.C. during a campaign in the Danube region that he undertook to pacify his northern frontier before embarking for Asia; it was not an auspicious beginning. He had pursued Clitus, a local Illyrian ruler who was in a state of rebellion, into the fortress of Pellium, which Arrian, the second-century A.D. historian who is the most reliable source on Alexander, described as "the strongest in the country."[25] Without hesitation, Alexander determined to assault the walls immediately. The Illyrians sacrificed three boys and three girls and three black rams to appease the gods in preparation for the siege.[26] This may have had some effect, because Arrian tells us that Alexander, instead of carrying out the assault, prepared to circumvallate the place. Circumvallation suggests the more passive siege methods of the fifth century, and it is a surprise to see Alexander resorting to it. Perhaps his full siege train had not arrived or perhaps Arrian simply meant that Alexander was positioning his troops around the fortress. J. F. C. Fuller calls Alexander's tactic a contravallation, suggesting that what Alexander had in mind was something less than a complete circumvallation.[27]

In any case, whether Alexander was attempting to circumvallate or contravallate Pellium, he had to abort the siege when an ally of Clitus arrived on the scene, leaving Alexander's army in too vulnerable a position should it assault the walls.[28] Indeed, Alexander had to fight his way out of an embarrassing situation in which his army was caught between a fortified city and an enemy force occupying higher ground to the rear. After a series of field battles, in one of which the Macedonians used catapults and in which Alexander outwitted the enemy twice,[29] Clitus fled Pellium, setting fire to the fortress as he left.[30] Alexander had brilliantly extricated his army from a precarious situation and won a smashing victory; but despite this success, Pellium had demonstrated the hazards of siege warfare rather than Alexander's ability to conduct a siege.

Any doubts that the close call at Pellium had raised about Alexander's ability to capture cities were quickly erased by the destruction of one of Greece's greatest cities, Thebes. Encouraged by a rumor of Alexander's death on the northern frontier, the Thebans had risen in rebellion and besieged the Macedonian garrison stationed in the citadel. Luckily for Alexander, he had just defeated Clitus when word reached him of the Theban revolt, and he was free to march to Greece. He covered 240 miles in the amazing time of thirteen days and caught the Thebans completely by surprise. Alexander waited a couple of days in hopes that political divisions within Thebes would produce a surrender. The Thebans responded by sending a sally against the Macedonians, so Alexander's patience did not last long.[31]

Alexander prepared to attack by moving his army around to the south of the city as near to the gates as possible to lend support to the beleaguered garrison in the citadel. The Thebans had constructed a double palisade to protect themselves while they besieged the citadel. Alexander divided his troops into three forces: one, under the command of Perdiccas, to attack the outworks; a second to counter any Theban sally; and heavy infantry held in reserve to strike at the decisive moment.[32] Arrian said that Alexander still hesitated, hoping yet for a capitulation, but that Perdiccas, acting without orders, forced the action by attacking the outer palisade.[33] Diodorus believed that Alexander himself ordered the attack, having become enraged by Theban insults.[34]

Whoever precipitated the attack, the Thebans fought so stoutly that Perdiccas's troops were soon in trouble and Perdiccas himself was seriously wounded. Alexander had to throw in his reserve heavy infantry to tip the balance in favor of the Macedonians. He struck at just the right moment, when the Thebans had broken formation to pursue the Macedonians. That routed the Thebans. Such was the confusion that the Thebans failed to shut their gates and Alexander's troops burst into the city. Stubborn fighting in which few Theban soldiers surrendered soon gave way to a general massacre. Six thousand Thebans died and thirty thousand became Alexander's prisoners. The battle demonstrated Alexander's effective use of reserves, a tactic that apparently caught the Thebans by surprise.[35]

The capture of Thebes demonstrated to Alexander that he could take a powerful city by assault if the psychological and material conditions were favorable, and that remained his preferred siege tactic the rest of his career. The severity of Perdiccas's wound showed the peril in assaulting a fortified position, but events were to prove that Alexander was not unwilling to lead such assaults.

However, siege conditions did not always allow an assault. Alexander was brave but not a fool, and in Asia he soon showed that he could conduct a classic siege battle. For example, in 334 B.C. at Miletus on the shore of Asia Minor, Alexander faced a strongly fortified coastal city supported by a Persian fleet. Logistical problems inherent in the location of Miletus made it impossible to deploy a large force against it.[36]

Thus an immediate assault with overwhelming force was not possible. Alexander first neutralized the Persian fleet by blocking the narrow entrance into the Milesian harbor with his own ships. The Milesians abandoned their outer works to seek the protection of the inner city. Having isolated the city and driven the Milesians within their innermost walls, Alexander then deployed siege machinery. As far as we know, this is the first time he had overseen the use of siege machinery. He attached such importance to the machinery that he personally commanded the battering rams. Under his direction, they soon knocked down a long section of the Milesian wall. Cut off from the Persian fleet and their wall in ruins, Milesian resistance collapsed.[37] Many tried to flee, but few escaped. The Macedonians caught most of the Milesians in the city and killed many of them.[38]

Alexander's success at Miletus inspired in him an unconventional strategy. The Persian fleet was stronger than his and he lacked the money to strengthen his navy. But the fall of Miletus so encouraged him that he hit upon the idea of capturing all the cities along the coast of Asia Minor, thus depriving the Persian fleet of its harbors. The conquest of Miletus was so intimidating that the coastal cities fell like dominoes as he marched along the coast.[39] Generous treatment encouraged the cities to open their gates to him. He did not even have to unload his siege equipment from the ships that carried it.[40]

Alexander's main objective when his campaign reached the Carian coast was Halicarnassus, the main naval base of the Persians in the southern Aegean. Halicarnassus presented a formidable challenge (map 4). It was located on a naturally strong site; it had strong fortifications including walls six feet thick, three citadels, and a moat forty-five feet wide and twenty-two feet deep; and a determined and experienced military commander, a Greek from Rhodes named Memnon whom Diodorus called "outstanding in courage and strategic grasp."[41] Memnon had prepared carefully for the siege, and a large force of mercenaries and Persians defended the city. They possessed arrow-shooting catapults. A fleet of triremes guarded the harbor; their sailors could provide additional manpower to the city.[42] The Persians controlled the sea and could supply the city with whatever it needed.

In contrast, Alexander's logistical situation was difficult. The peninsula on which Halicarnassus sat was barren and without adequate sources of water. This problem determined Alexander's choice of campsite, which was a little over half a mile east of the city. This location positioned him near a bay that could serve as a harbor for supply ships bringing both food and water.[43] Nevertheless, his supply line was precarious enough to encourage Alexander to attack even without his siege machines. His first approach to the city was against the nearest gate.[44] He quickly learned that Memnon intended to defend the city aggressively when Memnon sent a sortie against the approaching Macedonian

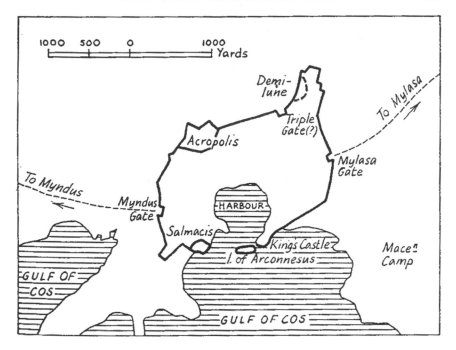

Map 4. Alexander's siege of Halicarnassus. *From J. F. C. Fuller*, The General-ship of Alexander the Great. *Eyre and Spottiswoode, 1958. Used with permission.*

troops, backing it up with withering fire from catapults. The Macedoni-ans beat back the sortie, but it showed the defenders of Halicarnassus were not going to rely solely on the strength of their walls.[45]

A nearby fortified coastal town named Myndus threatened Alexan-der's approach to the western side of Halicarnassus. He believed that the capture of that town would leave Halicarnassus much more vulner-able and, even more important, Myndus's port could help relieve his weak logistical position. He especially needed the port to land his heavy siege equipment.[46] Encouraged by a report that the town was ready to surrender, he led a strong force, including his elite foot soldiers, the hypaspists, and his elite cavalry, the Companions, as well as several infantry battalions, archers, and the Agrianian javelin throwers, against Myndus. His informants within Myndus promised to surrender the town if he arrived at night. Alexander personally led his men to Myn-dus around midnight, but the traitors were nowhere to be found. Alex-ander had not brought any siege equipment, but nevertheless he or-dered the Macedonian phalanx to undermine the wall. They toppled one tower, but the wall remained intact and the citizens put up a spir-ited defense. When reinforcements from Halicarnassus arrived by sea, Alexander broke off the siege and returned to his camp.[47]

Having failed to take Myndus, Alexander turned to the siege of Halicarnassus in earnest. His ships had finally eluded the Persian squadrons and managed to land his siege equipment at the bay near his camp.[48] The failure to take Myndus decided him to assault the northern side of the city, which was farthest from the sea. He began by filling up the moat to gain access to the walls for his siege engines. Arrian said he accomplished this "without difficulty."[49] Diodorus said that the Macedonians pushed siege sheds into the ditch so that the shed's flat roofs formed a bridge over the ditch. These sheds may have been designed expressly for this purpose.[50]

However, when the Macedonians began moving their siege machinery forward over the ramp, the defenders again made a sally against them, attempting to set fire to the siege towers and other machinery. The attack came at night and resulted in a frightening battle in the dark in which the defenders lost 170 men. Only 16 Macedonians died, but there were over 300 casualties because they had been caught without their armor. However, the Macedonians drove the defenders back into the city before they were able to disable any of the siege machines.[51]

The Macedonians now set to work against the walls. Sheds protected the sappers from missiles, and they soon collapsed two towers as well as the curtain in between. A third tower was severely damaged. The defenders built a crescent-shaped brick wall behind the damaged portion of their fortifications, a task they completed quickly owing to the large amount of manpower available in the heavily defended city.[52] According to Diodorus, it was stronger than the original wall.[53]

At this point a curious and perhaps revealing incident occurred. Two Macedonian hoplites were in their camp drinking one night when an argument arose over who was braver. Boastful words soon led to rash action when the two men armed themselves and charged the wall of Halicarnassus. When the defenders saw two soldiers coming against the wall, they hastened out to kill them. According to Arrian the two Macedonians were apparently sober enough to repulse the first sally against them and soon more Macedonians hurried to their aid. A stiff fight followed in which the Macedonians drove the defenders back inside their walls.[54] Diodorus had a somewhat different account that is more plausible. He said Memnon's men killed the drunken soldiers and got the best of the battle that ensued. Alexander himself had to rush forward to stiffen the Macedonians, and even he found it necessary to request a truce to recover the Macedonian dead, an admission of defeat.[55] The commander of the drunken troops was Perdiccas, who, whatever his role at Thebes had been, was a hothead, and we might suspect the whole thing was staged by him in the hope of a rich reward if the attack led to the fall of the city.[56] But everyone agrees the soldiers were drunk, and that may have reflected a more common reality: that soldiers fortified themselves with drink before assaulting a wall. Drinking was an important part of Macedonian culture, and it is hard to believe that their soldiers went into battle fully sober.[57]

After this distraction, Alexander again brought his siege machinery forward. He deployed his machines against the emergency wall the besieged had built behind the breach. Once again, Memnon did not allow the Macedonians uncontested access to the wall and sent out a sortie against their machinery. The commander of the Greek mercenaries, Ephialtes, was especially anxious not to wait passively behind the walls and in a council of war urged an attack against the Macedonians.[58] The Macedonians had set out mantlets to protect their engines and crews. Memnon's men succeeded in burning some of the mantlets, as well as one of the siege towers, despite the efforts of a special light armed unit whose job it was to protect the siege machinery. Intensive fire from arrow-shooting catapults mounted on a high wooden tower on the secondary wall drove the Macedonians back. The situation was serious enough for Alexander to come forward personally to stiffen his troops. His action had the desired effect, and the Macedonians drove the attackers back into the city.[59]

However, Alexander's machinery was unable to breach the inner wall because of the fierce resistance of the defenders. Especially troublesome was the fire from the towers on each side of the breach, which raked the Macedonians not only from the flank, but also from behind because they had moved past the towers to the new wall. Diodorus said the defenders had built a tower forty-six meters high, a remarkable height probably necessitated by the lofty height of Alexander's towers. As a result, this initial attack failed.[60]

Chastened by this failure, Alexander led the next assault himself. Again Memnon sent a sally against the Macedonian machinery, but this time with disastrous results. The sally struck from two places, one from the breach as was to be expected, the other from a gate where the Macedonians least expected it. Ephialtes commanded this attack on the Macedonian flank. Memnon stayed in reserve, waiting for the right moment to lead a third force out of the city. But the Macedonians were well prepared for a sally this time. Alexander had positioned strong reserves of heavy infantry to counterattack any sally against his machines and had also placed catapults in the siege towers to rain stones down on the enemy.[61] This is our first example of stone-throwing catapults. They may have been nontorsion catapults, but more likely they were the first experiment with torsion power for stone-throwing catapults. Although Arrian said these catapults hurled "large stones" at the enemy, this seems unlikely, especially because the catapults were mounted on wooden towers. In all probability, these first stone-throwing catapults were small and were strictly antipersonnel devices.[62]

Despite these preparations, Ephialtes' flank attack caught the Macedonians by surprise. The light-armed troops who were covering the siege engines were young and inexperienced, and they faltered when struck on the flank. But the royal bodyguard, supported by some heavy battalions and some light infantry, quickly moved forward and repulsed it. Diodorus said that old veterans in reserve rebuked the young Macedonians who were fleeing and moved forward themselves in a disciplined phalanx against the enemy.[63] This stroke proved decisive. The attacking force was caught beyond the moat and retreated in such disorder that there was a jam at the narrow bridge that spanned the moat. The bridge gave way under the weight of too many soldiers crowding onto it and casualties were heavy. The rout turned into a slaughter when the city's defenders, fearing that the Macedonians might burst through the gate on the heels of the fleeing troops, shut the gate before all their soldiers had gotten into the city. Ephialtes was one of the dead. Arrian claimed one thousand of the defenders of Halicarnassus died in this sally, but this is probably an exaggeration. Only forty Macedonians died, although several important officers were among the dead.[64]

Having dealt the besieged a crushing defeat, Alexander called off the assault. Arrian said he hoped that Halicarnassus would surrender, but it seems more likely that the fierce battle had left Alexander's troops in too much confusion to mount an assault into the city, especially because darkness had fallen. Indeed, Memnon and the Persian governor Orontobates decided that the city could not withstand the siege much longer when so many of the defenders were casualties. They may well have feared a revolt by an anti-Persian faction within the city now that Alexander seemed about to break through the walls.[65] Accordingly, during the night they set a fire in the city to cover a retreat and withdrew to the two citadels on the seaward side of the city. Memnon left by sea, leaving Orontobates in command of the citadels. When the Macedonians saw what was happening, they rushed into the city to kill those setting the fire.[66] We do not know how long Halicarnassus had held out. Because of his unfavorable logistical position, Alexander had clearly pressed the siege vigorously, and Donald Engels estimates he took Halicarnassus in just two weeks. Robin Lane Fox believes the siege lasted two months. My own estimate, based on an analysis of Arrian's rather vague chronology (how many days are "a few days" or "not many days later?"), is three to four weeks.[67]

Had Memnon and Ephialtes erred in staging costly sorties against an enemy superior in numbers? They could have held out longer with a defensive strategy. Their dilemma was how such a strategy could lead to victory. In Alexander they faced a strong-willed opponent unlikely to quit. So they bled their troops white with offensive maneuvers. The fates of Tyre and Gaza would soon vindicate their judgment that no city could hold out against Alexander the Great with a passive siege defense.

Alexander now possessed all of Halicarnassus except the citadels. Because they posed no strategic threat to him and he was anxious to get on his way, he decided not to assault the citadels. He sent his siege machinery and their crews to Tralles, a city in northern Caria, closed in the citadels with a wall and a ditch, and left three thousand mercenaries to guard those in the citadels.[68] Because they still had access to the sea, the defenders were able to hold out for a year before the citadels finally fell.[69]

Alexander had demonstrated at Halicarnassus that no city could withstand him if he was determined to take it. The city had strong fortifications, the terrain around it was difficult for a besieger, the city had plenty of time to prepare itself for the siege and had not wasted that time, there was ample manpower in the city, and its access to the sea was never cut off. Despite all these strengths, the siege ended with Halicarnassus in Alexander's possession, stark evidence that under Alexander the advantage in siege warfare had passed to the offense.

One of the most striking characteristics of Alexander's siege methods at Halicarnassus was his personal leadership. He personally led the foray against Myndus. When a sally threatened his siege machinery, he personally led the relief force. And he personally led the final assault against the city. This personal brand of leadership was characteristic of the Macedonian army, but it exposed the commander to a level of hazard unprecedented in siege warfare, the most dangerous form of battle in the ancient world.[70] So far Alexander had avoided injury, but he was trusting a great deal to luck.

THE SIEGE OF TYRE

Halicarnassus was preparatory for Alexander's greatest siege, the siege of Tyre. Its capture was his greatest accomplishment as a siege commander. Located on an island two and three quarters miles in circumference and surrounded by high walls reaching 150 feet on the eastern, landward side, the city had long been impregnable. The walls rose from the edge of the sea, so there was no land available for siege operations.[71] The Assyrians had failed to take it, settling for terms; and according to Menander of Ephesus, the Babylonians, in spite of a thirteen-year siege under Nebuchadnezzar, had done no better. The Persians were still a threat at sea and the Tyrians themselves possessed a formidable fleet and two harbors. Even though a part of the fleet was with the Persians in the Aegean, no less than eighty triremes remained at Tyre. A population of fifty thousand provided ample manpower for the defense of the city. Alexander's decision to besiege the city was so serious that he felt it necessary to justify it with a speech to the army, and Arrian tells us that an omen helped Alexander overcome his own doubts about the decision.[72]

The siege began in January 332 B.C. (map 5). Alexander faced formidable logistical problems. There was an adequate water supply, but the small coastal plain of Tyre could not supply Alexander's army with

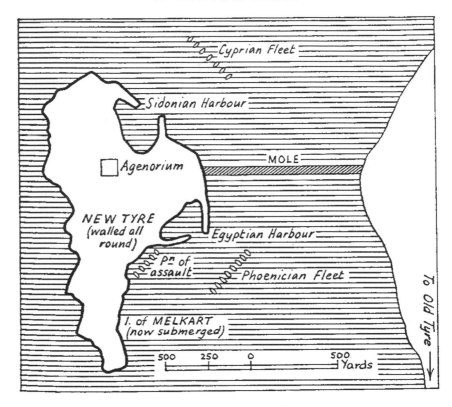

Map 5. Alexander's siege of Tyre. *From J. F. C. Fuller,* The Generalship of Alexander the Great. *Eyre and Spottiswoode, 1958. Used with permission.*

grain. Moreover, the siege took place in the worst possible time of the year in relation to harvest time, which was in June in the Tyre area. This meant that the granaries of nearby cities would not be full. Alexander had to bring grain from some distance. He made an alliance with Palestine to gain access to the rich plain of Jezreel; but that could only partially alleviate the Macedonian supply problem, so Alexander brought in considerable amounts of grain by sea. That undoubtedly accounted for the regular arrival of ships from nearby coastal cities.[73]

Tyre also presented formidable technical problems to Alexander's siege engineers. Fortunately, he employed two famous engineers, Charias and Diades, both trained in the Thassalian school of Philip's favorite engineer, Polyidus. Diades, and perhaps Charias also, wrote a treatise on machines. The siege of Tyre made them famous. Alexander's engineering corps was so large and famous that the names of two other of his engineers, Poseidonios and Philippos, also come down to us. In addition to these siege specialists, Alexander also had in his employ civil engineers who could apply their skills to military problems if need be. The most famous of them were a mining engineer named Gorgos, a hydraulic engineer named Crates, and two architects named Aristoboulos and Deinocrates. That the names of so many of Alexander's engineers have been preserved reflects their importance to his conquests.[74] Alexander valued them highly. One story that comes down to us has Philippos banqueting with Alexander, a sign of his high rank and an indication that Alexander may have even made him one of his Companions.[75]

Alexander's first task was to gain access to the island city by building a mole to the island, which lay one-half mile off the mainland, as Dionysius had done at Motya when confronted with the same problem. The construction of gigantic siege works presents a commander with two problems: finding the manpower to complete the task and keeping up morale among soldiers forced to lay aside their weapons and work with common tools under exceptionally dangerous conditions and under the taunts of the enemy. (According to Quintus Curtius, the Tyrians jeered at the Macedonian soldiers for performing manual labor instead of fighting.)[76] Alexander solved the former by pressing local inhabitants into labor gangs to help the soldiers build the mole.[77] He solved the latter by personally directing the work (although Arrian did not say that he actually worked alongside his men as Dionysius did during the construction of the walls on Epipolae) and awarded large gifts as incentives to the workers.[78]

Alexander chose a place where there were shallows and mudflats near the mainland. The Macedonians drove stakes into the mud and then filled in the mole with mud, wood, and stones. This was no easy task because prevailing southwest winds blew pounding waves against Alexander's works.[79] The Macedonians tore down Old Tyre for stones and felled trees in the famed forests of Lebanon for wood.[80]

At first the work went easily enough, but as the mole neared the city the water became much deeper, three fathoms (eighteen feet) at its deepest point, so the work became more difficult at the same time the workers were coming within the range of catapult fire from the Tyrian walls. The Tyrians were so abundantly supplied with catapults (probably nontorsion) they could cover the entire circuit of their wall with them. The area close to the mole fairly bristled with catapults.[81] Armor was too heavy and confining for the workers to wear, so they were excruciatingly vulnerable. To make matters worse, the deeper water allowed Tyrian ships to approach the mole and fire on the Macedonians.[82] The Tyrian ships carried both light arrow-shooting catapults and stone-throwing ones. Alexander tried to cut them off, but the Tyrian ships beat him back to their harbor.[83] Both sides were disturbed by a portent in the form of some sort of huge fish that washed up against the mole and remained there for a long time before swimming back out to sea. Seers on both sides provided favorable interpretations of the omen.[84]

The Macedonians defended their workers with two siege towers placed at the end of the mole. To be effective against the towering Tyrian walls, these towers reached 150 feet into the air in order to equal Tyre's high walls; they may have been the highest siege towers ever constructed.[85] The men in the towers used arrow-shooting catapults to sweep the walls of their defenders and repulse any approach by sea. Leather hides protected the towers from flaming arrows. The Macedonians placed the towers so as to screen the workers as much as possible and built a palisade in front of them.[86]

The Tyrians countered these defensive measures with a fire ship. They modified a cavalry transport ship by increasing its capacity with bulwarks built all around the sides of the ship. They filled this expanded ship with dry boughs and other combustible materials, including sulphur and pitch. Two masts held yardarms from which hung cauldrons full of more flammable materials. The Tyrians weighted the stern of the ship to lift the bow out of the water so the ship would go as far as possible up on land when it hit the mole. On a day when the wind was favorable, the Tyrians fixed ropes to the stern of the fire ship and towed it toward the mole with triremes. As the ship neared the mole, its crew set it ablaze and dove into the sea. The ship hit the mole with great momentum and, as the yardarms burned through, the cauldrons poured still more sulphur and pitch on the flames. This conflagration set the Macedonian siege towers on fire and missiles from the Tyrian ships kept the Macedonian fire squads from the burning towers. Under cover of this inferno, Tyrian soldiers reached the mole in small boats and tore down the palisade and set fire to any siege machinery that had escaped the flames of the fire ship.[87]

Quintus Curtius said the Tyrians also sent divers against the mole. It is difficult to understand how they could have done so, but he claimed the divers used hooks to pull tree branches out of the foundation, thus undermining it.[88] Diodorus's story that nature dealt the Macedonians another blow in the form of a storm that washed away a large section of the mole is more credible.

This setback forced Alexander to revise his plans. He ordered his men to widen the mole, which was about two hundred feet wide, so he could bring more siege machinery against the city. They also reinforced the sides with breakwaters to protect it against storms.[89]

While his engineers began constructing new machinery, a new threat presented itself. The Tyrian fleet was returning from the Aegean to lend powerful support to the besieged. Other contingents from the Persian fleet were also returning to Cyprus and Sidon. Together they amounted to more than the Tyrian fleet, and if they joined the Tyrians Alexander would face an overwhelming naval force.

Fortunately for Alexander, his great victory at Issus had made a deep impression on the Phoenicians, Greeks, and Cypriots. He hastened to Sidon and soon had gathered a fleet of 223 ships.[90] The size of this fleet stunned the Tyrians when it approached their city. Their consternation was so great they abandoned their plan to meet any naval threat on the open sea and instead withdrew into their harbors, barricading the mouths with triremes. That left Alexander free to impose a naval blockade on Tyre.[91] Perhaps even more important, domination of the sea secured his logistical position, which, as we saw, relied heavily on sea transport.

The Phoenicians and Cypriots contributed not only ships but also engineers in great numbers, so the work of rebuilding the siege machinery went swiftly. Protected by Alexander's new fleet, the Macedonians completed the expansion of the mole and finally reached the walls of Tyre. The mole was so well built that it still stands today, forming the core of an isthmus that connects Tyre to the mainland.

Arrian said the Tyrian wall was 150 feet high and correspondingly thick at this point. It was built with huge blocks of stone set in mortar. The Tyrians gained extra height by building wooden towers on their battlements from which they shot missiles and flaming arrows at the siege machinery and at the Macedonian ships that were bringing the machinery forward.

Alexander wanted his ships to sail right up to the wall, but they could not because the Tyrians had placed boulders in the sea in front of their walls. Alexander ordered these boulders cleared away, no easy task. First the Macedonians tried to pull these boulders out of the way with ships, but the Tyrians sent special armored ships against the anchors of the Macedonian triremes snapping their cables. Alexander covered some of his ships with armor and placed them in front of the anchors to block the approach of the Tyrian ships. But the Tyrians sent divers under water to cut the cables, and it was not until the Macedonians substituted chains for their anchor cables that these ships were able to get on with their work. They now lassoed the boulders and pulled them up on the mole. From there they used engines to lift them and drop them into deep water. These engines may have been catapults capable of throwing large stones, evidence that Alexander's engineers had succeeded in building catapults more powerful than any previously employed (figure 14). Thus they succeeded in clearing an area where their ships could anchor right alongside the wall.[92]

This success deeply alarmed the Tyrians, who now decided that they must break Alexander's naval blockade. They stretched sails across the mouth of their harbor opposite Sidon to screen their preparations. Alexander's headquarters was on the other side of the mole from Sidon. The Tyrians hoped to surprise the Cyprian ships at Sidon at a time when Alexander would be unable to rush to their aid. They had

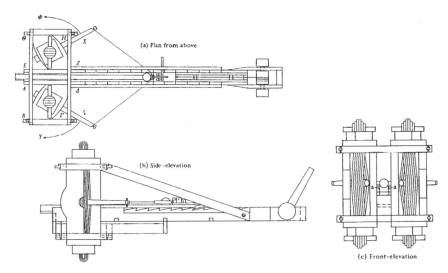

Figure 14. Alexander's stone-throwing torsion catapult. *From E. W. Marsden,* Greek and Roman Artillery. *Oxford: Clarendon Press, 1969. By permission of Oxford University Press.*

observed that Alexander usually rested in his tent at midday and so planned their sortie at that time of day.

Behind the sails they readied three of their huge quinqueremes, three quadriremes, and seven triremes and manned them with elite crews and crack marines. Paddling as quietly as possible, these ships glided out of the harbor toward the Cyprian ships at Sidon. When they were almost in view of the Cyprian ships, the Tyrians raised a great shout and surged forward. The surprise was complete and the Cyprian ships were at anchor. The Tyrians sunk several of them and drove many of the rest into shore, where they broke up on impact.[93]

Unfortunately for the Tyrians, Alexander had not taken his usual rest in his tent that day and was with his ships when word came of the Tyrian sortie. Alexander reacted with characteristic decisiveness. He ordered some of his ships to guard the Tyrian harbor on his side of the mole to prevent reinforcements from sailing out. He led a squadron around the city to attack the Tyrian ships that had ventured out. Tyrian lookouts in the city observed Alexander's ships and tried to signal the Tyrian ships, but it was too late. Only a few of them made it back into harbor; Alexander sank or captured the rest.[94]

Alexander's control of the sea was now (midsummer) complete, and he was able to deploy his siege machinery against Tyre unhindered. Diodorus said that among the machines were stone-throwing torsion catapults powerful enough to propel large stones capable of damaging walls.[95] These are the first such catapults we know of, but it is difficult to say if Alexander's engineers had devised a new type catapult or simply modified an arrow-shooting catapult.[96]

Such was the desperation of the Tyrians that they considered resuming an ancient practice of human sacrifice to appease the gods. Only the opposition of the city elders prevented the sacrifice of a freeborn boy.[97] Diodorus described curious countermeasures by the Tyrians, such as a spinning spoked wheel that deflected arrows and soft materials that cushioned the blows of the stones, that are otherwise unknown in the ancient world.[98] There is suspicion that Diodorus may have inserted these devices into his story from some theoretical military manual. But the report of the heavy stone-throwing catapults remains credible, because they represented a logical extension of the development of catapults by the Macedonians.[99]

The rams that Alexander moved forward on the mole had no success owing to the remarkable strength of the wall there and the intense fire along a frontage that was only two hundred feet wide. The Tyrians had built wooden towers on their walls to regain their height advantage over the huge Macedonian siege towers. They also built a fifteen-foot-thick wall seven feet behind their outer wall and filled the inner space with stones and earth. The Macedonians tried to cross the wall over bridges they dropped from siege towers. The Tyrians fought back by spearing the Macedonian shields with tridents attached to ropes. The Macedonian soldiers had to let go of their shields or be pulled off the bridge. The Tyrians also threw nets over the Macedonians and poured hot sand over them. The sand filtered down the necks of the Macedonians and got under their shirts and armor, causing excruciating pain. The Tyrians also used long grappling hooks to pull men off the towers and bridges. Long poles with blades were used to cut the ropes of the rams. The Tyrians had some sort of fire thrower that propelled hot metal at the besiegers.[100]

Diodorus reported that Alexander was discouraged enough at this point to consider breaking off the siege.[101] The constricted space of the mole made it too easy for the Tyrians to defend against assaults on such a narrow front. But Alexander solved this problem by mounting rams on ships, the only case we know of ship-borne rams.

Bringing these rams against the powerful walls of Tyre was an extraordinarily difficult task. The ships would have to be firmly anchored and the rams unusually large to reach the walls. J. F. C. Fuller speculates that the ships would have to be anchored at all four corners and that two ships may have been lashed together to provide the necessary space to operate a long ram. He also believes that the ships must have been roofed over with fire-resistant material to protect the crews from missiles.[102] Alexander first brought these ships against the wall opposite

Sidon, but here too the wall frustrated his rams. Finally, after considerable probing, he found a weak spot on the southern side of the city and
achieved a partial breach with ship-mounted rams.

The Macedonians dropped gangplanks across the breach and assault troops tested the defense, but they found it too strong to try a full-
scale exploitation of the breach. Here Alexander showed he could be
patient. He waited two days for calm wind conditions and then sent
his ships equipped with siege machinery to complete and expand the
breach.

Then the siege ships backed off and two transport ships came
forward. One of these ships carried Alexander's elite heavy infantry,
the hypaspists; the other carried a battalion of Macedonian heavy infantry not quite so elite as the hypaspists but still crack troops. Alexander intended to lead the hypaspists, who were a royal bodyguard, into
the breach himself.[103] This was an excellent example of Alexander's
method of personally leading the troops most closely tied to him, the
hypaspists, on hazardous missions.

Alexander's sense of timing rarely failed him. Also the use of
reserves and attacking in relays organized in columns, tactics that the
Carthaginians had perfected, may have been safer.[104] As a result, Macedonian casualties in siege warfare were low. Twenty hypaspists died in
this action, and Macedonian losses for the entire seven-month siege
amounted to four hundred.[105] This undoubtedly goes far to account for
his soldiers' willingness to follow him into battle. The other battalion
he brought with him may have come from the mountainous area of
upper Macedonia. If so, they will have had some of the wild barbarian
spirit that characterized such peoples as the Thracians, Illyrians, and
Campanians from whom the Greeks recruited their best mercenaries.

Alexander coordinated the assault on the breach with seaborne attacks on the city. He sent triremes with archers and arrow-shooting
catapults sailing around Tyre, harassing the defenders and watching for
an opportunity to attack by sea, presumably by escalade, should the
Tyrians become too preoccupied defending the breach. Once these
ships had started threatening Tyre from all sides, the two transport
ships moved forward to the breach and dropped gangplanks across it.
The hypaspists crossed the gangplanks first, led by their commander,
Admetus, and Alexander. Arrian reported that Alexander took "a
strenuous part in the action."[106] Admetus reached the wall first and was
killed as he urged his men through the breach. Alexander was right behind him and succeeded in seizing a section of the wall, including not
only the breach but several of the towers and the curtains in between.[107]

Now the defense of Tyre began to collapse. Alexander led the hypaspists into the city, closely followed by the other battalion of heavy
infantry. At the same time Phoenicians and Cyprian ships broke into

the harbors of Tyre and began destroying the Tyrian fleet. The Cypriots, attacking from Sidon on the opposite side of the city from Alexander's assault, captured the wall there. The Tyrians abandoned their walls and retreated into the city. However, they were unable to organize an effective fighting force and a sharp attack by the Macedonian hypaspists scattered them in flight.[108]

The siege of Tyre had taken seven months, but the fall of a city that was considered impregnable enhanced the reputation of invincibility that Alexander was gaining. His new powerful stone-throwing catapults surely made a deep impression, and they gave Alexander an advantage that Philip had not possessed.[109] But, like Philip and all great siege commanders, the key to Alexander's success was his ability to coordinate naval and ground forces, organize a gigantic construction project, and deploy his technology to sound tactical purpose. His willingness personally to lead his best troops into the city procured the quick victory once his patient campaign against the walls had finally achieved a breach. After Tyre, no city could consider itself safe from the Macedonian army.

THE SIEGE OF GAZA

The lesson of Tyre was not learned by Batis, the Persian governor of Gaza. When Alexander's route to Egypt brought him to Gaza, Batis did not surrender the city to him, believing it was impregnable. He based his decision on the elaborate preparations he had made for a siege and on the strong position of his city. For a long time he had been stockpiling grain, and he had hired a force of Arab mercenaries to help defend the city.

The approach to the city was difficult. It lay a little over two miles from the sea. Shoals prevented ships from coming even that close. This was the beginning of desert country, and deep sand made a march to the city a hard slog for men, animals, and machinery. There were few wells in the area, completely inadequate to provide the 100,000 gallons of water a day the Macedonian army needed. Gaza's preparation for the siege would have left no grain in the surrounding country. Therefore Alexander would have to make arrangements to bring great amounts of water and grain by sea.[110]

Gaza sat on a high tel that formed a natural glacis for the strong wall that surrounded the city. So formidable were these obstacles that Alexander's engineers advised him against a siege. Alexander, however, thought that the stronger Gaza was, the more it was necessary to capture it. To allow it to stand would encourage resistance by other cities and might even breathe new life into Darius.[111]

Gaza was the sort of city that the Assyrian emperors had regularly reduced, and Alexander adopted their methods. The Macedonians began to build a siege ramp against the city so they could bring their machines against its walls. At first they worked on the southern side of the city; when the mound reached the wall, they brought siege machines forward. Alexander's engineers must have constructed these machines in the field, because the siege train had not yet arrived from Tyre. Quintus Curtius tells us one of the towers was smeared with bitumen and sulphur, so apparently it was designed to be pushed up to the gate and set afire to burn down the gate.[112]

Moving the siege towers in the sand proved impossible. The wheels sunk in the sand so deeply the floor of the towers broke. The Macedonians incurred heavy casualties in this futile effort.[113] Not surprisingly, an Arab sally was able to drive the Macedonians back, and even though Alexander's seer had warned him not to take risks, Alexander felt compelled to lead his hypaspists forward to relieve the siege party. His counterattack was successful, but the seer's prophecy was true: an arrow shot by a catapult penetrated Alexander's shield and corselet and seriously wounded him in the shoulder.[114]

This mishap convinced Alexander that he had to attack on a broader front, and he ordered that the mound be extended completely around the city. If true, this was a truly gigantic siege work, over 1,200 feet wide and 250 feet high. Hammond envisions the Macedonians working "with Chinese concentration" to complete the mound in less than two months,[115] but it may be that Alexander pressed labor gangs into service from the local population to help with the labor, as he had done at Tyre.

None of our sources provide any details about this ramp. Some modern scholars have doubted that it extended entirely around the city. Johann Droysen thought the Macedonians more likely constructed several separate ramps around the city in order to mount simultaneous assaults, but Grote believed Arrian's language did not support such an interpretation.[116] Fuller, who finds such a project "totally impossible" in a two-month siege, suggests that perhaps Alexander surrounded the city with works of contravallation to prevent sorties while he completed the ramp.

Making a ramp from sand presented special difficulties because of the instability of such loose soil. Sand's flat angle of repose might account for the unusual length and width of the ramp, although wooden walls probably shored up the sides. Some sort of cover, probably wooden, undoubtedly covered the surface of the ramp, because it would have been impossible to move siege engines on a sand surface.[117]

By the time the Macedonians completed the expanded mound, the siege train from Tyre had arrived by sea. The engineers reassembled the machinery and hauled it up onto the mound, but when the Macedonians attempted to bring their engines against the wall, the Gazaeans drove them away three times. Finally, Macedonian firepower drove the defenders from their towers and cleared a large area of enemy fire. On the fourth assault, Alexander deployed his Macedonian heavy infantry and they provided the push necessary for the rams to reach the wall. The rams soon battered down a large part of the wall. Sappers also undermined the wall in several places.[118] Here the sandy soil was an advantage, making the digging easier and allowing the work to go quickly because of the absence of stone.[119]

Once they had leveled the wall, the Macedonians used ladders to cross the rubble. The elite shock troops vied with one another for distinction, and Arrian recorded the name of Neoptolemus from the family of the Aeacidae as the first man over the wall. This suggests there was an official dispatch honoring those soldiers who distinguished themselves the most and reflects the new prestige attached to assaults on walls.

The first troops across the wall threw open the gates, allowing the entire army to pour into the city. Unlike the Tyrians, whose resistance had collapsed once their walls were breached, the Gazaeans were able to organize a tenacious defense within the city. They fought to the last man.[120] The siege had taken two months.[121] Batis's underestimation of Alexander's siege capabilities had doomed his city.[122]

Although Alexander's engineers had advised him against the siege, their prowess made possible the reduction of Gaza once he had decided to besiege it. His main engineers were Diades and Charias, protégés of Philip's great engineer, Polyidus of Thessaly. The mole at Tyre and the massive ramp at Gaza attested to their ability to complete great construction projects. The height of their towers, the strength of their rams, and the development of stone-throwing torsion catapults attested to their ingenuity.[123] The increasing mechanization of siege warfare meant that the number and skill of a commander's engineers was decisive.[124]

DISTANT SIEGES

If Alexander was master of the large set-piece siege, he could also take smaller places in far-reaching campaigns over a variety of terrain. In this, too, he reminds us of the great Assyrian armies. Indeed, Alexander captured fortified positions at the farthest reach of his campaigning, in Sogdiana along the Jaxartes River.

Barbarians who had risen against him seized seven forts the Persians had built to guard their northeastern frontier. The largest of these was Cyropolis, and in 329 B.C. Alexander sent Craterus there to besiege it.

Craterus circumvallated Cyropolis with a ditch and a palisade, and his engineers built siege machinery. It is not clear how much, if any, of his siege train Alexander had gotten across the Hindu Kush, so we do not know if they were assembling engines from the siege train or if they were building them in the field. At the very least, they must have brought along the metal parts for their machines and the tools necessary to build machines in the field.[125]

Alexander wanted to tie down the defenders of Cyropolis while he led another force against the smaller forts. The first fort Alexander approached was named, ironically, Gaza. But this Gaza was fortified only with low earthen walls. Alexander moved directly to assault, clearing the walls of its defenders with intense fire from slingers, archers, javelin throwers, and catapults. The Macedonians quickly crossed the wall with ladders and slaughtered the defenders on Alexander's orders. The fort was put to the sack, and the women and children were seized as booty. Alexander quickly took the next two forts by the same method. When the inhabitants of two more forts tried to flee, Alexander sent his cavalry to cut them down. Arrian tells us that Alexander captured these five forts in just two days.[126]

Cyropolis, which, as its name indicates, had been built by Cyrus, another great siege general, had stronger walls. When Alexander arrived he intended to move the siege machinery against the walls. But then he discovered an easier way. A river ran through the city, and as it was summer the river was low enough for men to follow its course under the wall of the city. Alexander took his bodyguard, hypaspists, archers, and javelin throwers and led a small number of them through this opening into the town. It seems incredible that the defenders did not guard so obvious an entry point, but Arrian explained that they were preoccupied with the threat the Macedonian siege engines posed to their walls. Perhaps they were inexperienced in siege warfare. Fuller speculates that Alexander may have slipped into the city at night or just at daybreak.[127]

In any case, once inside the town, the Macedonians threw open the gates to the rest of the army. The barbarians fought bravely and wounded both Craterus and Alexander. A stone hit the latter in the head and neck. Nevertheless, the Macedonians succeeded in driving the defenders into a citadel. More than half of the fifteen thousand defenders died in the battle. The rest surrendered the next day because their citadel had no water supply.[128]

The fall of Cyropolis left only one more fortress. It fell quickly, although Arrian's sources gave conflicting accounts. One said the fort surrendered and that Alexander held the defenders captive until he left Sogdiana. The other said the fort fell by assault and Alexander killed all the defenders.[129] These seven forts were not so formidable as such cities as Tyre or Gaza. They were designed to provide protection against nomadic raiders. The barbarians had counted on distance to protect them from more sophisticated armies such as the Macedonian army. Alexander, with a tremendous feat of logistics and arms, had shown that distance was no protection against his siege capabilities.

Alexander also proved adept at capturing seemingly impregnable mountain strongholds during his pacification campaign on the northeastern frontier. One of his most remarkable strokes was the capture of the Rock of Sogdiana in 328 B.C. It was a peak high in the mountains protected by sheer cliffs all the way around, so impregnable that the Bactrian king Oxyartes had sent his wife and daughters there for safety. The place had sufficient provisions to hold out indefinitely against a siege. Deep snow provided a plentiful supply of water while making the going difficult for the Macedonians.

Faced with seemingly insurmountable obstacles, Alexander offered easy terms. So confident were the defenders that they not only refused the terms but stung Alexander with a derisory taunt that he would have to find soldiers with wings to reach their redoubt. Alexander offered the incredible prize of twelve talents to the first man to scale the mountain and lesser prizes to all who reached the top, three hundred gold darics going to the last. Arrian tells us that about three hundred men had practiced rock climbing in previous sieges, and they now eagerly came forward. This shows how highly specialized the Macedonians had become in siege warfare.[130]

These troops used iron tent pegs for pitons and, driving these pegs into the ice and crevices, hauled themselves up the sheer rock under the cover of darkness. Climbing a cliff at night was a terrifying task, and thirty men fell to their death. Their bodies plunged into snowbanks so deep it was impossible to recover them. The rest reached the top. Their appearance so astonished and demoralized the Bactrians that they overestimated the strength of the party on top and surrendered.[131]

An even more difficult stronghold that Alexander captured in the same year was the Rock of Chorienes, so called after the king who had taken refuge there. The Rock of Chorienes was a sheer peak of great height. Arrian claimed it was twenty stades high (about twelve thousand feet), but that cannot be correct. Perhaps he was giving the length of the narrow and difficult path that wound to the top.[132] The circumference of the peak at its base was almost seven miles, and a deep ravine ran all the way around this distance.

The ravine was the first obstacle Alexander had to overcome to capture the place. Although Arrian's description of the way Alexander crossed the ravine is difficult to understand, it seems that he built a hanging bridge over the ravine and then piled earth upon it to create a level crossing. The work was extremely arduous and progressed at a rate of only about thirty feet a day and even less at night. It was winter, and heavy snow fell continuously. Alexander divided his army in half to keep the work going around the clock, supervising the daytime labor himself. The work was strenuous and the work area constricted, so the soldiers worked in relays. Screens protected them from missile fire.[133]

At first the barbarians on the rock laughed at the project, but when the bridge brought the Macedonians within arrow shot, King Chorienes' nerves snapped. Despite the huge stores he had accumulated on the rock and the precipitous height the Macedonians had yet to scale, he opened negotiations. Alexander was anxious to treat because his army was dangerously low in supplies and was suffering greatly from the winter. When it became apparent that Alexander intended to treat him in a friendly way, Chorienes surrendered. When Alexander saw the enormous supplies on the rock, he concluded that Chorienes had submitted of his own free will and treated him as an ally rather than a conquered enemy. The first benefit of the new alliance was that Chorienes supplied the Macedonian army for two months.[134]

As Alexander led his army into India in 327 B.C., he drove many Indians to seek refuge on the Rock of Heracles (Rock of Aornos). The Rock of Heracles, so called because legend had it that Heracles himself had failed to capture it, was a huge peak in India about two and a third miles in circumference. The way to the lowest part of the top was over a one-mile climb. There was arable land on top, enough, according to Arrian, for a thousand men to till. A year-round spring provided plenty of water.[135]

Alexander determined to take the Rock of Heracles, even if it required a long siege. He established a powerful base at a near-by city and ordered Craterus to stockpile grain and any other supplies there to sustain a long siege. He himself gathered a force of archers, javelin throwers, and heavy infantry, including "the most nimble but at the same time the best armed men."[136] Two hundred of the Companion Cavalry and a hundred mounted archers also accompanied him.

Alexander approached the rock cautiously and established a camp nearby. According to Arrian, some deserters told Alexander that they could lead him to a part of the rock where an assault was feasible and which would give him a strategic advantage. Diodorus and Curtius said that it was an old man who lived in a cave in the area and knew the territory intimately. They reported that Alexander paid him the fabulous sum of eighty talents for his invaluable information.[137]

Alexander sent his general Ptolemy forward with light troops, as well as some hypaspists, to seize this spot. Ptolemy somehow got there undetected by the enemy and occupied the place without opposition. He immediately girded his position with a ditch and a stockade. Alexander saw the signal that Ptolemy had secured his position and moved the rest of his force toward the rock, but the terrain was so difficult that he got nowhere.

When the Indians saw that Alexander could not threaten them, they turned against Ptolemy's position and made an all-out effort to recapture it. A stiff fight resulted in which Ptolemy succeeded in holding on.

Alexander thought Ptolemy had been too passive in the assault, and that night he sent an order to attack the Indians from his position when Alexander moved forward the next day.

After the previous day's fiasco, Alexander decided to move up the rock by the same route Ptolemy had taken. This time there was no surprise and Alexander's men had to fight their way up against fierce opposition. They attacked in relays and slowly forced their way up to Ptolemy's camp. They had left at dawn and did not reach Ptolemy until afternoon. Despite the exhaustion his men must have felt, Alexander, having united his forces, moved immediately to assault the entire rock. Not surprisingly, the effort failed.[138]

Alexander now decided to build a siege mound from the hill the Macedonians held up to the top of the rock. He thought this would enable him to bring his siege engines close enough to fire missiles at the defenders. Apparently there was a plentiful supply of wood and soil for the project. Alexander had each man cut a hundred wooden stakes and put the entire army to work on the mound. Alexander personally supervised the work and punished those who failed to make progress. The first day they completed about six hundred feet, and that brought the slingers and catapults near enough to keep the Indians from sallying forth from their position. The Macedonians worked for three days on the mound, bringing it to a height where they were able to seize a small hill on a level equal to the Indian position. Alexander immediately began extending the mound to this new position.

The Indians watched the steady ascent of the Macedonians with alarm, and when Alexander's mound reached the new hill they opened negotiations. Their intent was to drag out the negotiations for the rest of the day and then flee from the rock during the night. Alexander's intelligence learned of the plan, and he then planned some treachery himself. He agreed to allow the Indians to leave the rock if they would remove their sentries. When the Indians started to leave the top, Alexander led seven hundred bodyguards and hypaspists to the evacuated area. When they reached the top, they suddenly attacked the retreating Indians, killing many of them. Many others fell to their death fleeing in panic.[139]

Siege warfare was crucial to Alexander's Indian campaign. Cities such as Bazira and Ora were too strong to take by quick assault, and Alexander had to resort to such methods as counter forts and circumvallation. Bazira was the stronger of these two cities, which Alexander simultaneously besieged in 327–6 B.C. He had a counter fort constructed at Bazira to deny the Bazirans the use of their countryside. It was the weaker city of Ora that he ordered circumvallated. However, the circumvallation was apparently not completed, because Arrian tells us that the place served as a haven for Indians fleeing the countryside. Alexander probably ordered the circumvallation to stop the flow of refugees into the city.[140]

Alexander also circumvallated the Indian city of Sangala in 326 B.C. His purpose was to prevent the inhabitants from escaping. He built a double wall all around the city except where a lake next to the city prevented its continuation. Alexander plugged this gap with guard posts and wagons. When the Indians tried to break out, the Macedonians drove them back, killing five hundred of them. If Alexander circumvallated some cities, he never satisfied himself with a passive siege. In the case of Sangala, he attacked the city with engines, undermining, and escalade and captured it by assault.[141] He also had no trouble capturing Ora. He attacked its walls without any preliminaries and took it on the first assault.[142]

When Alexander moved onward into the territory of the fierce Mallians, he began to encounter morale problems among his troops. The campaign began smoothly enough when Alexander drove many Mallians into a fortress that Arrian described as "strong." But that cannot have been true, because a detachment of Macedonian troops and cavalry marched there and took the place without a pause. Fuller conjectures that these Indian cities "were probably much like many present-day Indian villages—conglomerations of mud-huts, the circumference of which formed a protective wall of no great height or thickness."[143] The Macedonians enslaved the survivors.[144]

Other Mallians fled to the city of the Brachmanes, whose walls were much stronger. Characteristically, Alexander boldly led his heavy infantry right up to the wall; under the protection of missile fire, they undermined the wall. The defenders deserted their wall and took their stand in the citadel. Some Macedonians broke in on their heels, but the defenders were able to drive them from the citadel. Alexander then assaulted the citadel by escalade and undermining. Apparently the escalade was tentative, because when the undermining collapsed one of the towers Alexander mounted the wall ahead of the rest of his troops. The rest followed him "out of shame." This is unmistakable evidence that Macedonian morale was flagging as Alexander's ambition pushed them into ever remoter territories. The Indians resisted bravely, and in the end some set fire to their houses and burned themselves to death. Out of five thousand, few survived.[145]

The Macedonians continued to pursue the Mallians and drove them across the Hydroates River. Alexander crossed the river, defeated the Mallians and chased them into a nearby fortified town called Multan. He surrounded this town with cavalry. By the time the infantry arrived it was late in the day and Alexander allowed his exhausted army to rest over night before assaulting the town.[146]

The next day, Alexander divided his army in half in order to as-

sault Multan from two sides. He personally commanded one of the assaults. Once again the Mallians abandoned the walls in favor of making
their stand in the citadel. Alexander's troops broke through a postern
gate and got into the city far ahead of the other force, which had some
trouble crossing the wall even though the Mallians had deserted it. Alexander's force immediately began undermining the citadel, and although most of the ladders had still not been brought up, some tried to
enter the citadel by escalade. Alexander was again dissatisfied with the
efforts at escalade, and he impetuously snatched a ladder from a soldier
and brought it forward to the wall himself. Covering himself with his
shield, he mounted the wall. Peucestas, his shieldbearer, and Leonnatus, his bodyguard, followed Alexander up the ladder. An officer
named Abreas reached the top of the wall on another ladder. Alexander
was engaged in hand-to-hand combat on the top of the wall, and the
hypaspists rushed to support him. But in their hurry, too many
mounted the ladder and it broke under their weight.[147]

Standing on top of the wall exposed Alexander to intense missile
fire, so he jumped down to a mound on the inside of the wall. Backing up against the wall, he fought off the Mallians. Peucestas, Abreas,
and Leonnatus immediately jumped after him. At this point an arrow
struck Alexander in the chest, penetrating his lung, causing a sucking
wound that sprayed blood with every breath. Abreas had fallen with an
arrow in the face as soon as he jumped, but Peucestas and Leonnatus
covered the prostrate Alexander with their shields and fought desperately to hold off the Mallians.

The ladders had still not arrived, and the hypaspists were having
great difficulty crossing the wall. Their alarm had turned to desperation when they saw Alexander leap into the citadel. Some inched up the
earthen wall by driving pegs into it. Others tried to reach the top by
climbing onto their comrades' shoulders. In these ways a few made it
over the wall and cut the bar holding the gate to the citadel. Once they
broke through the gate, the Macedonians slaughtered the Indians, killing all including the women and children.[148]

Alexander's wound was extremely grave. The arrow had to be cut
from his body, and he lost a great deal of blood.[149] The seriousness of
the wound gave rise to the rumor that he was dead, and it was necessary
for him to make an agonizing parade before his army to reassure it that
he still lived. His boldness had been foolhardy, and Arrian made a rare
critical comment: "And yet his rage in battle and passion for glory made
him like men overcome by any other form of pleasure, and he was not
strong-minded enough to keep out of dangers."[150]

Arrian failed to appreciate the dilemma Alexander faced in motivating his troops to storm yet another fortified position in a seemingly
endless succession of cities. As John Keegan has shown, Alexander's

leadership was personal and heroic. In the face of sagging morale, he could capture cities only by increasingly personal and heroic leadership.[151] This necessity reached its limit when Alexander mounted the wall of Multan alone. He never fully recovered from his wound, his troops refused to go farther, and Alexander had to turn back to Persia.

Chapter IX

TREATMENT OF
CAPTURED CITIES

Alexander was a brutal siege commander. He learned the harsh treatment of captured cities from his father, who was a much feared siege commander. Philip had deported more than 10,000 inhabitants of the Illyrian town of Sarnus and regularly sold entire populations into slavery, as he had at Potidaea, Stageira, Olynthus, and undoubtedly a number of other places where the sources do not reveal the exact fate of the inhabitants.[1]

The stark fate of Thebes showed just how ruthless Alexander could be. Once he had taken the city, there followed one of the most shocking events in Greek history. Alexander's army slaughtered all who fell into their hands, including women and children. Neither home nor temple saved the helpless Thebans. Greek allies, especially from the small Boeotian towns Thebes had oppressed, joined in the slaughter.[2] Ulrich Wilcken condemned the deed as "a terrible massacre of the population."[3] Alexander allowed his Greek allies to decide the fate of Thebes, knowing they had long-standing grudges against the city, going back to when Thebes joined the Persians almost 150 years earlier.[4] They voted to raze the city and sell the survivors into slavery. Alexander preserved the citadel to serve as a Macedonian garrison and saved a few people, such as the priests and priestesses and some friends of Macedonia, but he raised no objections to the harsh fate of Thebes.[5] According to Diodorus, six thousand Thebans died in the battle for the city. He numbered the prisoners at thirty thousand and said they brought a sum of 440 talents in the slave markets.[6] The fate of Thebes was especially stunning because the city had fallen quickly.

The Greek historian Polybius bitterly placed the responsibility squarely on Alexander: "But then every one pitied the Thebans for the cruel and unjust treatment they suffered, and no one attempted to justify this act of Alexander."[7] However, Polybius's judgment was not universally shared by Greeks. Some did indeed feel pain and indignation over the fate of the Thebans, but Thebes had many enemies and they justified the decision as a punishment for past misdeeds.[8]

Alexander was less interested in justice than teaching the Greeks a lesson they would not forget while he was far away in Asia. Also, 440 talents was a large sum, and Alexander's financial needs were great on the eve of his expedition to Asia.[9] He paid a price for his severity, however. He had shattered the Panhellenic illusions of the Corinthian League. Greeks and Macedonians treated each other with mutual distrust, and the Greek allies played a subordinate role in the war against Persia. As Fritz Schachermeyr asked, what good is an alliance whose leader could appear as "an angel of death?"[10]

Alexander's political purpose at Thebes suggests that his ruthlessness was calculated, and, indeed, he could be more generous if political considerations called for it. For example, when he wanted to win the loyalty of the Greek cities in Asia Minor, he followed a humane policy in his campaign against them. At Miletus, he came to terms with three hundred defenders who had fled to an island to make a last stand and whose bravery aroused his admiration: they could serve in his army as mercenaries. He spared the rest of the survivors and even granted them their freedom. There were some Persians in Miletus and those that they captured the Macedonians killed, leaving a clear message to the Greeks of Asia Minor that Alexander would favor them if they distanced themselves from the Persians.[11]

According to Pausanias, Alexander intended to sell the women and children of Lampsacus into slavery and burn its sanctuaries because he suspected the city of tilting toward the Persians. Only the last-minute intervention of the sophist Anaximenes, a favorite of Alexander, saved the people of Lampsacus.[12] After breaking into Halicarnassus, he ordered that those Halicarnassians who remained in their houses be spared.[13] Both Arrian and Diodorus said Alexander razed Halicarnassus, but this makes little sense after he had denied his soldiers a sack, and later evidence makes clear the city was not destroyed. More likely, the Macedonians cleared out an area for their siege of the citadels.[14]

Not all the Greek cities of Asia Minor escaped so lightly. Before Alexander arrived, the advance Macedonian force under the command of Parmenio took the Aeolian city of Grynium and sold its entire population into slavery.[15] Alexander's Ionian campaign was in no way a liberation. He treated the Greek cities of Asia Minor as conquered cities and disposed of them as he saw fit.[16]

Alexander's lenient treatment of Miletus and Halicarnassus were exceptional cases. Once the Macedonians had broken into a city, they were all too prone to slaughter its inhabitants as they did at Thebes. At Tyre, the siege had lasted seven exhausting months. During it the Tyrians had slaughtered some prisoners on the wall in sight of the Macedonians and then thrown the bodies into the sea. The Tyrians had also killed some Macedonian heralds Alexander had sent to them to demand surrender. The Macedonians were bitter and vengeful, and they killed indiscriminately.[17] The violation of the heralds, one of the most sacred customs of the ancient world, infuriated Alexander, and he ordered two thousand Tyrians crucified after the battle.[18] This was not the only time Alexander crucified prisoners. When he took the Rock of Sogdiana, he tortured and crucified the chief of the defenders, along with his relatives and nobles.[19]

Some Tyrians had reached the Temple of Heracles, including the king. These people Alexander granted a complete pardon, in contrast to Thebes where the temples had been violated. The women and children he sold into slavery along with the other survivors. In all, eight thousand Tyrians died, leaving only two thousand male survivors.[20] Arrian numbered those enslaved at thirty thousand, but the actual number was probably lower.[21] Diodorus said thirteen thousand noncombatants were captured. Some women and children may have fled to Carthage, but it appears that the official decision to send all of them to Carthage came too late to be carried out before Alexander unexpectedly seized control of the sea around Tyre.[22] One report claimed that Tyre's neighbor Sidon was able to save fifteen thousand of the Tyrian prisoners from slavery.[23] Fritz Schachermeyr succinctly summed up the siege of Tyre: "The siege lasted seven months as a triumph of modern technology. Now it ended as a triumph of brutality."[24]

At Gaza there were no male survivors. J. F. C. Fuller, a man who knew war, believed this meant the Macedonians gave no quarter.[25] Alexander sold the women and children into slavery and resettled the city with neighboring tribesmen.[26] Quintus Curtius tells us that Alexander lashed Batis to his chariot and dragged him to death around the city in imitation of Achilles' treatment of Hector. Curtius believed this cruelty was an early sign of foreign influence on Alexander.[27] Alexander's admirers reject the story because Homer, whom Alexander revered, denounced Achilles' deed.[28] But Alexander was wounded at Gaza, and this would have angered his army. In light of this, it is possible that his Thessalian cavalry may have suggested to him their custom of dragging a murderer behind their horses.[29]

Did the contrast between the savage treatment of Tyre and Gaza and the relatively moderate treatment of the Greek cities in Asia Minor reveal that Alexander distinguished between barbarians and Greeks in his treatment of cities? When the Athenian orator Isocrates had called for a Pan-Hellenic crusade against the Persians, he had repeatedly characterized them as inferior beings whom the Greeks could justly pillage, whose land they could take, and whom they could reduce to bondage.[30] Did Alexander's treatment of Tyre and Gaza reflect such an attitude? The fate of Thebes shows that we should be cautious in drawing such a conclusion. Although we cannot disregard the possibility that Alexander killed barbarians more readily than Greeks, other circumstances, such as the length of the siege, provocations during the siege, and the wounding of Alexander, weighed more heavily in the balance than the race of the victims. As we will see, Alexander was capable of moderating the treatment of barbarians when it was advantageous to do so.

Wounding Alexander especially aggravated the Macedonians. At Multan, where Alexander received his most serious wound and his troops thought he had been killed, they slaughtered every inhabitant, including the women and children.[31] Just how angry the Macedonian troops could become over even a small wound of Alexander was shown in the capture of a small fortified stronghold along the Indian frontier. The place was too small for Arrian to bother giving us its name. It fell quickly and easily, but Alexander received a slight wound in his shoulder. In revenge his troops killed everyone who fell into their hands. Alexander himself ordered the town razed.[32]

However, Quintus Curtius did not connect this massacre to Alexander's wound. He said that before Alexander was wounded he had already ordered that no quarter be given because he wanted to strike terror into a nation that had not yet experienced the arms of the Macedonians.[33] Indeed, a wound suffered by Alexander did not necessarily lead straight to a massacre. In the siege of Massaga, an Indian city, Alexander unwisely led the phalanx right up to the wall without proper support and was slightly wounded in the ankle by an arrow shot from the wall. The next day Macedonian rams quickly breached the walls, but the Indians resisted fiercely and over the next two days repulsed four efforts, supported by a siege tower, to storm the breach. Only when a bolt from a catapult killed the Indian commander did the Indians despair. Their defense was still intact enough for them to be able to open negotiations. In this case, the tenacious resistance of the Indians aroused Alexander's admiration, and he offered to incorporate them into his own army. The Indians agreed to this condition, and Alexander allowed them to march out of the city with their arms and camp next to the Macedonian army. However, the Indians had no intention of fighting against other Indians and planned to slip away during the night. Somehow Alexander learned of this, surrounded the Indian camp with troops, and slaughtered them all. He then took the undefended city and captured the mother and daughter of the king of Massaga.[34] In this case, Alexander's wound did not prevent him from attempting to make the Indians useful to him. It was their betrayal that doomed them.

Diodorus told a different story. He said that Alexander simply massacred the Indians of Massaga without provocation and paints a picture of the Indians heroically defending themselves in a bitter battle in which even the women fought. Not without difficulty, the Macedonians killed the Indians, women and all.[35] Diodorus's account more fits the usual Macedonian reaction to the wounding of Alexander.

Plutarch told yet another version of this massacre. In an apparent reference to Massaga, he tells us that the city had employed some very effective Indian mercenaries who specialized in defending cities. Alexander had learned to appreciate the skill of these troops in previous sieges and he was determined to eliminate them. When they agreed to a truce that granted them permission to leave the city, Alexander "fell upon them as they marched and slew them all."[36] Plutarch called this incident "a stain to his military career" but claimed that "in all other instances he waged war according to usage and like a king."[37] The violation of the truce, not the massacre itself, disturbed Plutarch. He knew well that the Macedonians frequently massacred the defenders of cities, especially in India. In other words, the massacre of a captured city's population was "according to usage."

As the fate of Massaga testifies, Alexander was particularly brutal in India. Sangala provides one example. The numbers of Indian dead Arrian reported were so high we must doubt them: 17,000 killed, 20,000 captured. Macedonian casualties were only 102 killed and 1,200 wounded. But if the proportion of Indian casualties to Macedonian casualties in Arrian's figures are even roughly accurate, what happened was more of a massacre than a battle. Alexander razed Sangala. When the inhabitants of a neighboring city fled upon hearing of Sangala's fate, the Macedonians murdered the five hundred people who had been too old or sick to travel.[38]

Alexander's campaign against the Mallians was especially destructive. He surprised them by crossing a barren desert, covering almost fifty miles in a day and a night. He caught the Mallians unarmed and slaughtered most of them. The survivors fled into a city. Alexander had only his cavalry with him, and so he surrounded the city with them, "using them like a palisade," in Arrian's words.[39] When the rest of the army caught up, Alexander sent a part of it to another Mallian city while he assaulted the walls of the first. The Mallians, weakened and demoralized, quickly abandoned their walls and sought refuge in the citadel. But the Macedonians, with Alexander playing an active role, took the citadel by assault and killed all its defenders, who numbered about two thousand. The defenders of the other city fled upon the approach of the Macedonians, but the Macedonian cavalry and light-armed troops pursued them and massacred all who fell into their hands. The only survivors were a few who escaped into a marsh.[40]

In 325 B.C. Alexander captured an Indian city named Harmatelia that had rebelled, and Arrian said that he killed all the Brahmans, who had led the revolt. However, Diodorus claimed Alexander spared the

Brahmans.[41] In the same year, in a campaign against the Indian ruler Musicanus, Alexander razed several cities and sold their inhabitants into slavery.[42]

Quintus Curtius reported an Indian campaign in which Alexander followed a more moderate policy. When Alexander approached a city that Quintus Curtius did not name but described as strong, the inhabitants of the city divided into those who wanted to resist and those who favored surrender. As a result of this division, they sent envoys to negotiate but at the same time prepared for a siege. In the end the defeatists acted on their own and opened the city to the Macedonians. Alexander pardoned all the inhabitants, although Quintus Curtius thought he could have justly punished those who wanted war. Instead Alexander contented himself with taking some hostages and moved on to the next city. This example of clemency caused all the cities in the region to surrender to him.[43] Another example of moderation came at the city of Artacana. Its citizens surrendered as soon as they saw Alexander's formidable siege towers. He not only spared them but also returned their property.[44]

More often, however, Alexander allowed his soldiers to sack cities. Having taken a city of the Indian ruler Oxicanus (also called Porticanus), Alexander handed over all the plunder to his army, keeping only the elephants for himself. Diodorus said that Alexander took two cities in this campaign and that his soldiers sacked them and then razed them.[45] In a campaign against the Kingdom of Sambus, he enslaved the populations of the cities and destroyed the cities. Diodorus claimed more than eighty thousand Indians died in this campaign.[46] In his famous speech to the Macedonians at Opis, Alexander referred to "the enormous sums you gained by pay and plunder, whenever a besieged place was plundered."[47]

One of the worst Macedonian sacks occurred at Persepolis in 330 B.C. Its Persian governor offered to turn the city over to Alexander, and at the least he surrendered the royal treasury.[48] But, apparently out of revenge for the Persian invasion of Greece, Alexander treated the city as if he had taken it by storm and unleashed his troops to plunder it. This was against the advice of Parmenio, who, with characteristic practicality, pointed out that Alexander was destroying what now belonged to him and that the sack would discourage other Asian cities from surrendering. Arrian, in a rare critical remark, said he did "not think that Alexander showed good sense in this action nor that he could punish Persians of a long past age."[49] Persepolis was opulently wealthy, and a wild sack followed. Crazed by greed, the Macedonian soldiers slaughtered all the men because holding them prisoners would have hindered the looting. So avaricious did the Macedonians become that they fought among themselves over the loot and not a few were killed. Some cut off the hands of rivals who were grabbing a disputed prize. Upper-class

Persian women fell victim to greed-crazed Macedonian soldiers who ripped off the women's fine clothes and jewelry and dragged them naked into slavery. The spectacle was so horrifying that entire Persian families killed themselves by jumping from the walls or setting fire to their houses. Curtius said that Alexander finally ordered his men to spare the women.[50] According to a popular story, at a drunken celebration of the sack of Persepolis, an Athenian courtesan named Thais suggested that courtesans and soldiers join Alexander in a victory march to set fire to the palaces. Alexander and his men marched out to the music of the courtesans. Thais and Alexander threw the first torches. When the fun was over, the entire palace area had burned to the ground.[51]

Quintus Curtius described another terrible act of vengeance during Alexander's campaign in Sogdiana. His army came across a small town inhabited by the Branchidae. These people had once lived near Miletus in Ionia and had sided with Xerxes in the Persian War. When the Greeks drove the Persians out of Ionia, the Branchidae had fled with them and settled in Sogdiana. Although 150 years in Sogdiana had partly dehellenized them, the Branchidae greeted Alexander with enthusiasm— a mistake, because the Greeks had not forgotten their Medizing and considered them especially odious because they had desecrated a temple on Xerxes' behalf. Alexander occupied the city with light-armed troops and surrounded it with the Macedonian phalanx. He then ordered the latter to sack the town and kill all its inhabitants. The Macedonians razed the town and even uprooted its sacred groves.[52]

W. W. Tarn calls this story "a clumsy fabrication," and, although we might question Tarn's objectivity because of his excessive admiration of Alexander, other modern scholars also have doubts. Neither Robin Lane Fox nor N. G. L. Hammond mentions the atrocity. Ulrich Wilcken dismisses it in a footnote as "an invention of the hostile tradition." George Grote, however, who was completely free of hero worship, found the report credible. We know enough about Macedonian treatment of captured cities not to reject it out of hand.[53] In any case, the campaign in Sogdiana was brutal. In two days the Macedonians captured five cities, killed all the adult males, and seized the women and children as booty. Alexander gave his troops free reign to sack these places.[54]

Results could be quite arbitrary. When Alexander drove the Uxii into their citadel, he spurned their pleas for mercy. But when Sisigambis, the mother of Darius who was in Alexander's custody, intervened on their behalf, Alexander reversed himself, pardoning not only the Uxii but also their Persian governor who had led the resistance. He left the city intact and free, contenting himself with the imposition of a tribute.[55] The fate of these people had hung on the whim of an unstable king who oscillated from vicious brutality to magnanimous clemency.

In a campaign against the Agalesseis in 325 B.C., Alexander besieged a number of cities and sold their inhabitants into slavery. One large city, packed with twenty thousand refugees, resisted fiercely, and its capture was costly. This angered Alexander, and he set fire to the city, burning most of its inhabitants alive. Another version reports that the Indians burned themselves to death to escape capture. Both stories agree that large numbers of women and children perished in the conflagration. Three thousand escaped to a citadel. When they surrendered, Alexander spared them.[56]

Another case in which Alexander used fire against a besieged force took place at the famous Kalat-i-Nadiri in Persia, a huge natural fortress that rises as high as two thousand feet on its highest side. The upper half was sheer vertical cliff. On top were grassy plains and many springs. It would later become the only fortress Tamerlane was unable to take, and it now baffled Alexander. In order to gain access to the top, the Macedonians had piled a large amount of wood against the fortress, and the wood accidentally caught fire. Alexander seized the opportunity and ordered his soldiers to pile on more wood until the fire reached the top. The flames then swept across the fortress, engulfing its defenders. The Macedonians forced all who tried to escape back into the fire. Only a few survived. Curtius said the defenders numbered about thirteen thousand soldiers "not fit for war." He did not mention the presence of women, but some may have been there. There is a medieval Persian epic that tells the story of a battle over the Kalat in which the hero's mother sets fire to the fortress and all the women commit suicide by jumping over the side. The legend may well be a distant echo of Alexander's siege.[57]

Certainly mass suicide was not unknown in siege warfare. As we have seen, entire Persian families leaped to their death during the Macedonian sack of Persepolis. An example of mass murder occurred during Alexander's siege of a stronghold in Caria. During a campaign to subdue the surrounding country after the fall of Halicarnassus, Alexander marched past a fortress height held by the Marmares. From this fortress the Marmares attacked Alexander's rear guard and inflicted heavy casualties on it. Angered by this, Alexander determined to take the fortress. After two days of assaults failed to capture it, he began to settle in for a siege. When it became apparent that Alexander intended to stay as long as was required to capture the fortress, the Marmares held a council of war. The elders advised surrender on whatever terms were possible. But when the young warriors refused to surrender, the elders decided that the strong young warriors should kill their wives, children, and elders and then break out of the fortress and flee into the mountains. After a last meal with their families, the Marmares warriors carried out this horrible decision. Some could not slay their families with their own hands, so they set fire to their houses and burned them alive. Under cover of the fire and darkness, the Marmares warriors then slipped through the Macedonian lines and escaped into the mountains.[58]

Our sources give short shrift to the Macedonian treatment of women, although, as we have seen, it is clear the Macedonians did not hesitate to kill them. Rape and enslavement were common. One peculiarity of Macedonian military operations was that the Macedonians did not allow large numbers of camp followers to trail along behind the army. Philip had banned women from traveling with his army in order to increase its mobility, a policy Alexander continued in his early campaigns. As his army marched farther from home, however, Alexander changed this policy to allow his men to take captive women along with them.[59] These women were expendable, however. In one curious case, Alexander made the incredible blunder of encamping along a dry riverbed during the rainy season in the Gedrosian desert. A flash flood carried away most of the women, children, and pack animals, but the soldiers were apparently on higher ground, because none of them were lost.[60] Donald Engels suggests that Alexander may have deliberately planned the loss of the women and children to lighten his army for the arduous desert march.[61]

Alexander showed more generosity in his treatment of women of aristocratic birth. Arrian tells us that the beautiful Roxane, daughter of King Oxyartes, was among the captives at Sogdiana and that Alexander fell in love with her. Arrian praised Alexander for treating her with respect and planning to marry her instead of raping her. Even the usually judicious modern biographer Wilcken romanticizes this episode. He believes that chivalry was "a typical trait of Alexander's character."[62] However, Quintus Curtius was scandalized that Alexander would marry "a woman who had been brought in among the entertainments of a banquet" and tells us that Alexander's men were ashamed of the liaison.[63] The truth seems to have been that Alexander humiliated Roxane by forcing her to perform before the Macedonians and then took her in marriage for political reasons.[64]

Roxane was second in beauty only to the wife of Darius, whom Alexander had also captured. Although Arrian wondered if there was some reason Alexander felt no passion for Darius's wife, he also praised him for not raping that prisoner either, pointing out that Alexander was young and powerful. He evidently regarded Alexander's self-restraint as unusual.[65] Quintus Curtius also praised Alexander for his treatment not only of Darius's wife, whom Quintus Curtius said Alexander protected from any shameful treatment, but also of the ladies-in-waiting, whom Alexander "treated with as much deference as if they had been born from the same mother as himself."[66] Arrian wrote that Darius could scarcely believe his ears when he heard that Alexander had not raped his wife.[67] Quintus Curtius told a similar story. When Darius heard that his wife had died in Alexander's custody, he was certain she had been raped to death and was only persuaded with difficulty that this was not so.[68] Plutarch also expressed surprise that Alexander respected the wife of Darius, even though she "was far the most comely of all royal women."[69] However, Plutarch's account is contradictory. Although he explicitly stated that Alexander kept his hands off of Darius's wife and kept her "as though guarded in sacred and inviolable virgins' chambers instead of in an enemy's camp, apart from the speech and sight of men,"[70] he revealed later that, after two years in captivity, she died in childbirth. Plutarch denied that she became pregnant by Alexander, but she must have gotten within the speech and sight of some man.[71] In another case, there is a report that the Indian queen of Massaga ended up pregnant by Alexander.[72] This may say more about Alexander's orientalizing tendencies in a society in which concubinage was common, but it is clear that he felt entitled to take whatever women he captured.

In fact, Quintus Curtius, perhaps with some exaggeration, reported that after the conquest of Persia Alexander quite gave himself up to drinking and whoring among the captive women.[73] He told the story that the Macedonians forced these women to perform before them but that one woman of unusual beauty and modesty caught the attention of Alexander. Upon inquiry he learned she was high born and set her free. He then made a general review of the prisoners and separated out the high born from the common.[74] The tale may be a romantic fancy of Quintus Curtius. We have already seen that Alexander forced Roxane to perform, and there could have been no doubt in his mind about her social status. There can be little doubt that Alexander distinguished between aristocrats and commoners in his treatment of women, but he treated all captured women as his property, which was, after all, the universal practice in the ancient world.

Chapter X

DEMETRIUS THE BESIEGER

In the Hellenistic age that Alexander the Great began, siege warfare became increasingly technical and the equipment more and more sophisticated. Demetrius the Besieger's siege of Rhodes in 305 B.C. exemplified this refined stage in the history of ancient siege warfare. At Rhodes Demetrius deployed an array of equipment astonishing in its variety and size. However, his failure to capture Rhodes revealed that elaborate siege machinery had not changed the fundamentals of ancient siege warfare.

Demetrius's siege of Rhodes was justly famous in ancient times because it pitted a man whose prowess in siege warfare was so great he earned the epithet "the Besieger" against a city that was the most powerful in Greece. Rhodes had come into existence scarcely one hundred years earlier when three small cities on the northern end of the island had united to form the city of Rhodes. Until the time of Alexander the Great the city had been under the sway first of Sparta, then Athens, then Carian princes, and finally the Persians. After the battle of Issus, Alexander imposed a Macedonian garrison on Rhodes, but the Rhodians expelled it after his death.

Finally independent, Rhodes experienced remarkable growth in the years after the death of Alexander. Its strategic location made it the crossroads for trade both east and west and north and south. Trade between Rhodes and Egypt especially flourished. Rhodes replaced Athens as the leading naval power in the Greek-speaking world. It was able to suppress piracy, and wealth flowed into the city.[1] The usual struggle between oligarchs and democrats had divided Rhodes in its early years, but there had apparently been a reconciliation during the Macedonian occupation when a moderate democracy emerged in which political tensions seem to have been low, an important asset for a city under siege.[2]

Wealth, power, and internal harmony enabled Rhodes to remain independent and aloof from the rivalries between the Diadochi. But this splendid isolation ended in 315 B.C. when Antigonus One-Eye of Macedonia began to seek Rhodes' aid in his growing rivalry with Ptolemy, the ruler of Egypt and Rhodes' main trading partner. Rhodes attempted to appease Antigonus without provoking Ptolemy. This delicate policy collapsed in 306 B.C. when Antigonus asked Rhodes to help in a campaign against Cyprus. Rhodes' refusal caused Antigonus to decide that Rhodes was necessary to his struggle with Ptolemy, and he sent his son Demetrius to conquer the proud city.[3]

Demetrius was just thirty-one years old but had already gained renown for his compelling personality. He was large, handsome, and haughty. In the Macedonian fashion he was a hard drinker and womanizer, but on campaign he was always completely sober.[4] As for his abilities in siege warfare, Diodorus wrote:

> For, being exceedingly ready in invention and devising many things beyond the art of the master builders, he was called Poliorcetes [the Besieger]; and he displayed such superiority and force in his attacks that it seemed that no wall was strong enough to furnish safety from him for the besieged.[5]

Such was the man who now sailed against Rhodes with a large and well-equipped army.

The prospect of a siege alarmed the Rhodians enough for them to accept an alliance with Antigonus. Demetrius, however, did not trust the Rhodians and demanded that the Rhodians open their harbor to his fleet and provide him one hundred hostages from the leading families of Rhodes. These demands convinced the Rhodians that Demetrius intended to bring them under Antigonid rule, and they resolved to resist. Rhodes was at the height of its wealth and power and the Rhodians could count on the support of Ptolemy, so the decision was far from desperate.[6]

However, the force that sailed against Rhodes was formidable. Demetrius's fleet consisted of 200 warships and 170 transports. They carried an army of forty thousand soldiers as well as cavalry. Pirates, who would benefit greatly from the defeat of Rhodes because the Rhodian fleet had relentlessly campaigned against piracy, also accompanied Demetrius. In addition, Demetrius brought with him a variety of catapults and all the equipment necessary for a siege. A fleet of almost a thousand privately owned merchant ships followed behind to make money from the voracious demand of Demetrius's large force for supplies and perhaps to profit from plunder should Rhodes fall.[7] The entire armada was so large that it filled the entire strait between the island of Rhodes and the mainland, causing great alarm among the Rhodians who watched the spectacle.[8]

Having landed his force on the island a little south of the city unopposed, Demetrius established his camp near the city, just out of reach of catapult range (see map 6). He sent out pirates and merchants (the distinction between the two seems to have been slight) to plunder the island by land and sea. His soldiers cut down trees and tore down farm buildings to get wood to fortify the camp. These fortifications were extraordinarily strong, consisting of a triple palisade surrounding the camp. Several large stockades reinforced the palisade. Turning his entire army as well as his ship crews to the task, Demetrius built a mole to create a safe harbor for his ships.[9]

These impressive preparations convinced the Rhodians that Demetrius was settling in for a determined siege, and they attempted to reopen negotiations with him. When Demetrius did not respond to their overtures, the Rhodians' most immediate problem was the demoralization caused by the prospect of a long siege, and they took immediate steps to prevent a collapse of morale or the outbreak of tensions in the city.

Most in doubt was the loyalty of noncitizens and slaves. Alien residents did not usually serve militarily, but the Rhodians asked them for volunteers for the duration of the siege. About a thousand alien residents came forward to join with the six thousand citizens who were fit for military service. The Rhodians sent the rest of the alien residents away from the city in order to reduce the number of mouths to feed during the siege and to eliminate a potential source of discontent. The city also gave slaves the opportunity to help defend the city, promising to buy from their masters those slaves who fought bravely and to emancipate them and make them citizens.[10]

As for the citizens, the city passed a decree stating

> that the bodies of those who fell in the war should be given public burial and ... that their parents and children should be maintained, receiving their support from the public treasury, that their unmarried daughters should be given dowries at public cost, and that their sons on reaching manhood should be crowned in the theater at the Dionysia and given a full suit of armor.[11]

These steps produced the intended effect. The city came together in the common defense, the rich contributing money and the craftsmen turning their skills to the production of arms. Rhodes employed some outstanding engineers, such as Dionysius of Alexandria, famous for his invention of a repeating catapult, Diognetus of Rhodes, and Callias of Aradus. The women of Rhodes sacrificed their hair to the torsion cata-

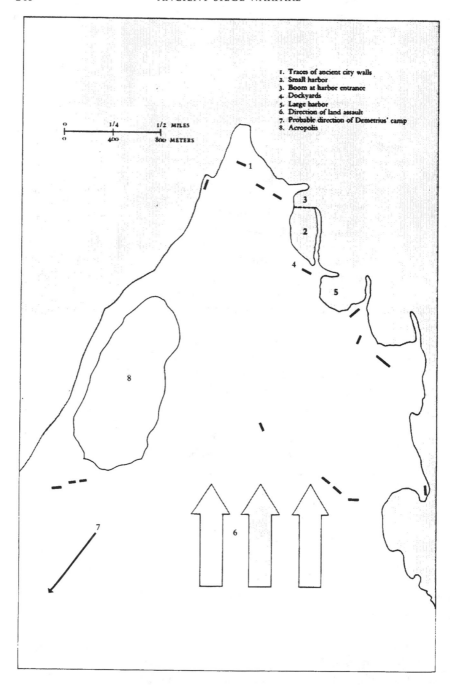

Map 6. Demetrius the Besieger's siege of Rhodes. *Used by permission of the publisher, Cornell University Press, from Richard M. Berthold,* Rhodes in the Hellenistic Age, *copyright 1984 by Cornell University Press.*

pults these men devised.[12] Construction workers and masons strengthened the wall and stacked stones near the wall for emergency repairs.
A surprise naval sortie by three of the swiftest Rhodian ships succeeded
in sinking a number of Demetrius's merchant vessels and capturing
others, undoubtedly boosting morale in the city.[13]

Demetrius's primary objective was to gain control of the harbor in
order to seal off Rhodes from the outside world. He mounted a stone-
throwing catapult on a cargo ship and an arrow-shooting catapult on
another cargo ship. Protective sheds covered both catapults. Ship-borne
siege towers four stories high supported the catapults. To prevent the
Rhodians from ramming or sinking the ships bearing the siege machinery, a spike-studded boom floated ahead of them. Demetrius fortified lighter ships with walls of planks. The plank walls protected long-
range arrow-shooting catapults and Cretan archers. Shooting through
embrasures that could be closed when not in use, the catapults and
archers killed the Rhodians who were trying to strengthen the harbor's
fortifications.[14]

The Rhodians fully realized the importance of keeping their harbor
open, and they took strong measures to defend it. They stationed catapults at two key points at the entrance of the harbor and covered the
catapults with protective sheds. They also built platforms on the cargo
ships in the harbor to make them capable of bearing catapults.[15]

Rough seas thwarted Demetrius's first effort to deploy his ship-
borne siege machinery against the harbor. When calm weather arrived,
Demetrius carried out a night attack that succeeded in capturing the
end of the mole at the mouth of Rhodes' largest harbor. He landed four
hundred soldiers there who fortified the place with a wall of stones and
planks behind which he stationed a variety of catapults. These catapults were only five hundred feet from the walls of Rhodes, well within
the range of Demetrius's powerful stone-throwers. During the next day,
these catapults disabled the Rhodian siege machinery in that area and
seriously damaged the city's walls, which were weak and low at that
point. But the Rhodians were able fire back at Demetrius's catapults,
inflicting sufficient damage to decide him to withdraw his machines at
the end of the day. The Rhodians pursued him with fire boats and managed to set fire to some of Demetrius's ships before being driven back
by the spiked boom and Demetrius's arrow-shooting catapults.[16]

The next day, Demetrius attacked again, but this time he not only
launched his siege ships against the harbor but at the same time attacked the city on land from several sides. He sustained these attacks
for eight days, during which his stone throwers smashed the Rhodian
siege machinery at the entrance to the harbor and damaged the harbor
fortifications. Some of Demetrius's soldiers actually captured a portion
of these fortifications. However, they were unable to hold their position
against a Rhodian counterattack.

Many of the ships that had landed these soldiers smashed on the rocky shoreline or ran aground. The Rhodians burned the grounded ships, leaving Demetrius's force in a desperate position. This action diverted the Rhodians, however, giving Demetrius the chance to land more soldiers, who attacked the city by escalade.

Fierce fighting over the walls inflicted heavy casualties on the attacking force, including some of Demetrius's most important officers. These losses forced Demetrius again to withdraw his forces. Seven days were required for him to repair his siege machinery and ships, giving the Rhodians time to repair their damaged walls.[17]

When Demetrius returned to the attack, he again concentrated entirely on the harbor. Once again he moved his siege machines forward, shooting flaming arrows at the Rhodian ships and throwing stones against the Rhodian wall. Demetrius pressed the attack with great determination, and the Rhodians were in considerable difficulty. However, they were able to extinguish the fire arrows, narrowly averting the loss of their fleet. Their situation was still precarious, however, and the democratic magistrates issued a desperate call to the Rhodian aristocracy to man ships to save the city.

Execestus, the chief admiral, took command of three of the best Rhodian ships, manned by the best rowers, and counterattacked Demetrius's siege ships in the hope of stopping the bombardment that was shaking the walls of Rhodes. Demetrius's catapults subjected the Rhodian ships to intense fire, but they advanced with such speed that they broke through the iron-studded beam that protected Demetrius's siege ships. Such was the skill of the crack Rhodian rowers that the Rhodian ships sunk two of the siege ships by repeatedly ramming them, a tactic for which Rhodian sailors were famous. In an age of grappling and boarding, the old art of ramming had declined, and it may have been new to the Macedonians.[18]

To save the third siege ship, Demetrius ordered it hauled back with ropes. The Rhodians, flushed with success, recklessly pursued. They soon found themselves surrounded by enemy ships, and one of the Rhodian ships, which carried Execestus, fell into Demetrius's hands. But the other two escaped, revealing the greatly superior skill of the Rhodian sailors.[19]

Undiscouraged, Demetrius built another sea-borne siege machine that was three times bigger than the ones the Rhodians had disabled. As he brought this monster forward, however, nature rescued the Rhodians. A violent storm arose that capsized the gigantic siege ships, whose size may have made them top-heavy. The Rhodians took advantage of the storm to make a sortie from the city against the men Demetrius had landed on the mole. Isolated by the ferocity of the storm, these men, who numbered four hundred, soon had to surrender. Thus the Rhodians retained control of their harbor. The importance of this was soon demonstrated when more than 650 soldiers from Knossos and Egypt sailed into the harbor to reinforce Rhodes. They undoubtedly brought abundant supplies as well.[20]

Having failed in two attacks on the harbor, Demetrius changed strategy. He now gave up hope of isolating Rhodes and directed his attacks against the city itself. He meant to break through Rhodes' defenses by the sheer weight of his attack. The centerpiece of the assault was to be the largest siege machine in the history of siege warfare, the helepolis, or "taker of cities." The helepolis was the work of Demetrius's leading engineer, the Athenian Epimachus.[21]

Diodorus and Plutarch provided detailed descriptions of this huge machine. The helepolis was built of timbers joined together with iron. The base measured forty-eight cubits square, or seventy-five feet square.[22] The tower rose nine stories to a height of 150 feet.[23] The walls of the tower inclined inward so that each floor was slightly smaller, the size of the tower diminishing from about sixty-five feet square on the first floor to thirty feet square on the top floor. Iron plates protected from fire the three sides that faced the enemy. This huge structure carried catapults of all sizes, including large stone-throwers.

Demetrius had built a smaller helepolis in Cyprus, and in that one the largest catapults fired from the middle floors and the lightest from the top floor. On the bottom floors were medium catapults capable of hurling stones weighing 180 pounds.[24] The catapults fired through ports. Shutters made of hides protected the catapults and their crews. The hides were stitched together and filled with wool to absorb the blows from enemy stone-throwing catapults. A mechanical device lifted the shutters when the catapult was ready to fire. Two staircases, one up and one down to avoid confusion, enabled the crews to move up and down in the helepolis.[25]

Remarkably, this large machine was mobile. It rested on eight large solid wheels with iron-plated rims. The rims were two or three feet wide. The wheels could turn on pivots. Men standing in rows at the bottom of the structure pushed on crossbars built into the base. Additional men standing behind the helepolis helped push it forward.[26] Demetrius selected the 3,400 strongest men from his army to push the helepolis. There was room for about 1,200 of them to push at one time.[27] According to Plutarch, when Demetrius later built another helepolis at Thebes, it was so heavy and slow that in two months it moved forward only two stades (about four hundred yards).[28] However, in siege warfare four hundred yards is a significant distance. In any case, there is no mention of such difficulties at Rhodes, perhaps because Demetrius mobilized thirty thousand sailors from his fleet to fill in the moat protecting Rhodes and to clear a pathway for the helepolis and other siege machines. Sheds and covered passageways protected the workers, who created a front seven towers and six curtains wide.[29] Demetrius paid a price for the redeployment of sailors. As events would show, the fleet sorely missed them.

The Rhodians could see the giant helepolis taking shape. Demetrius's catapults had already weakened their walls, and they knew they could not withstand another attack. So they built a second wall behind the weakened wall at the main point of attack. Stones were hard to come by, but the Rhodians tore down the outer wall of their theater and dismantled nearby houses to get stones. They even tore down some temples, promising the gods to build even finer ones after the siege.[30]

The Rhodians also took advantage of the absence of thirty thousand sailors from Demetrius's fleet by continuing to harass Demetrius's supply lines. A small force of nine ships, which then divided into three squadrons, made a series of daring attacks.

One squadron attacked a strategic island between Rhodes and Crete, sunk or burned a number of ships, impressed some of their crews, and captured a good many transports laden with grain. Another squadron, consisting only of light undecked ships, sailed across the Lycian sea to the mainland, where it surprised a fleet of transports. The warship convoying the transports was at anchor and its crew on shore when the Rhodian ships arrived. The Rhodians burned the warship and captured many of the transports. These light Rhodian ships even managed to capture a quadrireme that happened to be carrying royal clothing for Demetrius. The Rhodian navy consisted mostly of light ships to pursue pirates, and the Rhodians were skilled in their use.[31] Damophilus, the Rhodian commander, sent these royal clothes to Ptolemy, whom the Rhodians considered more worthy of them than Demetrius. All the sailors Damophilus had captured he sold into slavery.

The third squadron cruised around nearby islands and fell upon a squadron of freighters carrying materials for Demetrius's siege engines. Accompanying these materials were eleven of Demetrius's best engineers. A modern emendation of Diodorus suggests the captives' skill may have been shooting catapults rather than making them.[32] In any case, the Rhodians captured the entire kit and caboodle and brought them back to Rhodes.[33]

The skill of the Rhodian sailors was so famous that an old proverb said each Rhodian sailor was worth an entire ship.[34] That the Rhodians could be so successful with such small naval forces suggests not only that they were exceptionally skillful sailors but also that Demetrius had fatally weakened his naval forces by withdrawing thirty thousand sailors from the fleet.[35] However, the Rhodians had manpower problems of their own and, despite the success of these raids, were able to sail out against Demetrius's supply line only three times during the year-long siege.[36]

Despite these successes, the Rhodians felt sufficiently insecure to vote down a proposal to pull down statues of Antigonus One-Eye and

Demetrius that stood in the city. The Rhodian assembly thought it better that the statues remain standing as a reminder of past friendship in case the city fell to Demetrius.[37]

In the meantime, Demetrius had sent sappers to undermine the Rhodian wall. A deserter alerted the Rhodians to this threat just as the Macedonian sappers were about to bring down the wall. The Rhodians dug a deep moat behind the wall and by countermines stopped the progress of the Macedonian sappers. The Rhodians placed a Milesian named Athenagoras in command of the mining. Athenagoras commanded mercenaries sent to Rhodes by Ptolemy. Aineias Tacticus had warned about mercenaries and counseled besieged cities to place only trusted citizens in critical command positions. Not surprisingly, the Macedonians attempted to bribe Athenagoras. In this case the mercenary remained loyal. Athenagoras feigned betrayal, and when Demetrius sent one of his most trusted friends to parley with Athenagoras, the Rhodians captured him. The grateful Rhodians gave Athenagoras a golden crown and five talents for his loyalty.[38]

When Demetrius was ready to advance the helepolis, he placed it in the center, with four tortoises deployed on each side to protect the sappers. Two enormous iron-plated rams 120 cubits long moved forward under the protection of the helepolis. Huge protective sheds covered them. One hundred and twenty cubits is 180 feet, but such a long ram would be too unwieldy so it seems more likely the smaller Macedonian cubit obtained here, which would make the rams about 120 feet long.[39] These huge rams rolled on wheels. One thousand men were necessary to move them forward.[40]

At the same time he brought his machinery forward, Demetrius ordered his fleet to attack the harbor and his infantry to attack the city from all sides. Just as the bombardment was beginning, envoys from Cnidus intervened and persuaded Demetrius to break off the attack while they tried to persuade the Rhodians to come to terms. Diodorus said these negotiations continued "at great length" but finally failed. He did not say why, but the sticky issue was probably access to the harbor. In any case, Demetrius resumed the attack and soon knocked down the strongest Rhodian tower and a long stretch of curtain.[41]

At the same time he pressed his assault on Rhodes, Demetrius also sought to interrupt the Rhodian supply line. King Ptolemy of Egypt had dispatched a large supply fleet bearing 300,000 measures of wheat and legumes for Rhodes. (A measure was slightly larger than a bushel.) Demetrius sent warships to capture this fleet. However, a favorable wind enabled the supply ships to race into the Rhodian harbor, and Demetrius's ships returned empty-handed. Two other fleets of supply ships bearing a total of 90,000 measures of wheat and barley also reached Rhodes safely, further revealing Demetrius's weakness at sea.[42]

The arrival of such large amounts of food caused a surge in morale among the Rhodians, and they turned this new energy against Demetrius's siege engines. Catapults could be useful to defenders as well as besiegers, especially in the hands of people as resourceful as the Rhodians. They concentrated all of their stone-throwing and arrow-shooting catapults on the wall facing the helepolis and stockpiled a huge number of fire-bearing missiles. On a moonless night, the Rhodians opened fire under the protection of darkness. The intensity of the bombardment shook loose some of the iron plates protecting the helepolis, and the flaming arrows soon set it on fire. Concentrated fire from the arrow-shooting catapults shot down the Macedonian fire squads that rushed forward. Demetrius himself had to bring forward water to fight the fire, but the danger was so great he ordered the helepolis withdrawn beyond the range of the Rhodian catapults.[43]

Shaken by the force of the Rhodian counterattack, Demetrius paused to assess Rhodian strength. He ordered his soldiers' servants to gather and count the Rhodian missiles that littered the battlefield. The results were sobering. The Macedonian servants collected 800 fire missiles and more than 1,500 catapult bolts, making it all too clear that the Rhodians did not lack firepower. The damaged siege engines and the many dead Macedonians lying before the walls of Rhodes provided further stark evidence of the intensity of the Rhodian fire.[44]

Demetrius lost valuable time burying his dead and repairing his engines. The Rhodians took full advantage. They built a third crescent wall covering all the weakened portions of their wall. They dug a deep ditch behind the damaged section of their wall to prevent the Macedonians from easily assaulting the city.[45]

The Rhodians also continued to assert their superiority at sea. A Rhodian naval squadron captured the strongest of the pirate forces that Demetrius had enlisted in his service. It also seized several of Demetrius's grain ships and took them to Rhodes. Soon afterward, another grain convoy as large as the previous one arrived from Ptolemy. Fifteen hundred soldiers came along with it. Their presence was soon to prove crucial.[46] With Rhodian strength growing rather than decreasing, Demetrius was on a treadmill.

Nonetheless, he returned to the attack. The helepolis and other engines advanced again and drove the Rhodian defenders from their wall with intense fire. Demetrius's rams knocked down two curtains, but the tower between them remained in Rhodian hands after a fierce fight in which the Rhodian officer was killed. It was at this time that the second supply fleet from Ptolemy arrived at Rhodes. Demetrius decided to accept a truce urged upon him by envoys from Athens and other Greek cities. However, despite strenuous efforts by the mediators, Demetrius and the Rhodians could not agree on terms to end the siege.[47]

Having failed to gain terms satisfactory to him, Demetrius decided on a last attack. He planned to assault the breached wall at night with crack troops under the command of Alcimus and Mantias. Alcimus, the strongest man in Demetrius's army, usually wore a suit of armor that weighed a hundred pounds, double the normal weight. But he had replaced his armor with a suit specially made in Cyprus that combined great strength with lightness. Weighing only forty pounds, it could withstand a catapult bolt from twenty paces. Demetrius was the only other man in the army who possessed such armor.[48]

Demetrius hoped that the assault troops could penetrate into the city and draw the Rhodians away from the walls and harbors, which he planned to attack at dawn. The third wall and moat behind the breached wall must not have presented formidable obstacles, because the assault troops did break into the city and occupy the area around the theater.

But the Rhodian magistrates did not panic. They ordered all defenders to stay at their posts. They themselves led the 1,500 troops sent by Ptolemy against the Macedonians in the city. Desperate fighting ensued, and the chief Rhodian magistrate was killed. But the Rhodians were able to bring in reinforcements, who steadily wore down the Macedonians. Alcimus's armor did not save him; and after he and Mantias fell, resistance began to crumble. The Rhodians killed most of the Macedonian soldiers; only a few escaped from the city.[49]

The siege had now continued for one year, but Diodorus maintained that Demetrius was determined to press on. He was a stubborn man and did not flinch from high casualties. However, cooler heads prevailed. His father, Antigonus, advised coming to terms. Ptolemy also wanted an end to the siege. So when envoys from the Aetolian League arrived, both sides were ready to accept mediation. The terms favored Rhodes. It would remain independent, and there would be no Macedonian garrison in the city. The city would not pay a tribute. However, Rhodes would become an ally of Antigonus, except in case of war against Ptolemy, and Rhodes would surrender one hundred hostages selected by Demetrius.[50]

Demetrius had failed to take Rhodes because of his inability to control the seas around the island and because he was unable to overcome the formidable firepower of the Rhodian catapults. His failure showed that it was still impossible to take a well-defended city without isolating it from the outside world. The Rhodians demonstrated that catapults could be just as useful in defense as in offense. Demetrius tried to overcome these weaknesses by building ever-larger siege engines, but their giganticism proved vain in the end. Plutarch hinted that technology mesmerized Demetrius:

> Nay, he was actually thought to be a better general in preparing than in employing a force, for he wished everything to be at hand in abundance for his needs, and could never be satisfied with the largeness of his

undertakings in building ships and engines of war, or in gazing at them with great delight.[51]

Ancient siege warfare had reached a technological dead end that was not escaped until the introduction of gunpowder a millennium and a half later. Diplomacy, betrayal, and blockade remained fundamental to success in siege warfare.

PART FIVE

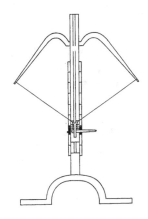

THE ROMANS

Chapter XI

EARLY SIEGES AND
THE PUNIC WARS

The Origins

Our knowledge of early Roman siege warfare is shrouded in legend. But if the legends reflect the reality, the first Romans were no more proficient in it than the early Hebrews and Greeks.

In one of the earliest stories of Roman history, the Sabines, having taken the Roman citadel by treachery, easily drove back the Roman effort to retake it. Only a prayer by Romulus to Jupiter sufficed to dislodge the Sabines, and Jupiter answered the prayer by moving the Sabines to come out from the citadel rather than having the Romans successfully assault it.[1] Later, when the Romans invaded the land of the Veientes in retaliation for a raid, the Veientes preferred to defend their fields in a battle rather than remaining behind their walls. However, when the Romans won the battle and drove the Veientes into their city, Romulus did not undertake a siege but contented himself with laying waste to their fields, a measure that Livy said forced the Veientes to sue for peace.[2] The avoidance of sieges and the preference for a decisive battle over the fields was an exact parallel to warfare in early Greece and was typical of a society of citizen soldiers.

Livy recognized how limited resources were in the early days. In a report of a campaign against the Volsci in 469 B.C., in which the Romans had driven the enemy into the city of Antium, he remarked that Antium was "a very opulent city for those days" and took it as a matter of course that the Romans did not besiege such a strong city.[3] It was only after a devastating field victory that Antium fell to the Romans. The Volsci were so demoralized that they surrendered after only a few days of

Roman blockade.[4] In a similar way, Livy showed no surprise that the citadel of Tusculum, which fell into enemy hands in a surprise night attack in 459 B.C., was impregnable to Roman assault. Only a blockade lasting for months starved the Aequi into surrender.[5]

Passive siege methods predominated in early Roman warfare. The Romans may not have even used siege machinery at this early date. Livy mentioned the use of siege machinery in his description of the siege of the Auruncan town of Pometia in the early fifth century, but his account is problematical because he gave several versions of the same siege. According to one version, the Romans employed "mantlets and other engines" against the Auruncan wall.[6] However, in other versions which Livy provided, apparently in a belief that differing accounts of the campaign in his different sources were describing separate actions, there was no mention of siege machinery.[7]

There is some evidence that the Romans created a corps of engineers early on. Livy reported that King Servius Tullius added two centuries of engineers in the sixth century B.C.[8] If true, their main task was probably to lay out and construct the famous Roman military camps. But the entrenchments and ramparts that protected these camps were the kind of works required for siege warfare, and we can see in this the beginning of the Roman army's great engineering skills and the famous capacity of the Roman soldier for digging.

The development of engineering skills is perhaps the reason that we hear of more aggressive tactics, such as the construction of siege ramps and undermining, beginning to develop in the late fifth century. In 436 B.C., the dictator Quintus Servilius showed considerable resourcefulness in the siege of Fidenae. Having invested the city with ramparts, Quintus sent troops with scaling ladders against the walls. The escalade failed, however. He then set about to tunnel under the wall into the city. Constant diversionary attacks distracted the defenders until the Romans had tunneled all the way into the citadel, a success that forced the city to surrender.[9]

Skill in escalade improved rapidly. In 417 B.C., the Romans succeeded in taking the town of Labici by escalade.[10] In 406 B.C., Numerius Fabius Ambustus took Anxur by capturing high ground above the city, from which he was able to launch harassing attacks against the walls. When these attacks diverted the defenders, Fabius's soldiers on the lower side of the city surmounted the walls by escalade.[11]

Despite the Romans' growing skill in escalade, their success shows that the walls of most Italian towns were not high. Really strong fortifications with high walls, such as citadels, remained impregnable to them. For example, as late as 405 B.C. the citadel of Artena, which was well provisioned and strongly defended, easily held the Romans at bay. Only the treachery of a slave allowed some Romans to sneak into the place, kill the sentinels, and sow enough panic to precipitate a surrender.[12]

It was not until the turn of the fourth century that the Romans began to mount major sieges. The siege of Veii, which began toward the end of the fifth century B.C.—about the same time the Carthaginians and Greeks were waging ferocious sieges in Sicily—was the first great Roman siege. Duration was the most striking characteristic of the siege of Veii. The tradition handed down by Livy and Diodorus said that Veii held out against the Roman siege for ten years, falling in 396 B.C. The figure arouses some suspicion because it is the same as the siege of Troy, but it is likely that the siege of Veii lasted at least seven years.[13] This was an extraordinarily long siege for farmer-soldiers such as the Romans to sustain. Plutarch realized this siege began a new era in Roman siege warfare:

> These [the Roman besiegers] had been accustomed to short campaigns abroad as the summer season opened, and to winters at home; but then for the first time they had been compelled by their [military] tribunes to build forts and fortify their camp and spend both summer and winter in the enemy's country.[14]

It was no accident that this siege took place shortly after the Romans had instituted pay for their soldiers. Even so, the decision to sustain the siege of Veii over the winter brought bitter complaints from the tribunes of the people, who pointed out the hardships this would impose on young men "who were no longer free, even in winter and the stormy season, to see to their homes and their affairs."[15] A forceful speech that Livy put into the mouth of Appius Claudius, one of the military tribunes, showed that Livy understood the implications of the siege. Appius argued that even if one discounted the strategic significance of Veii, the Romans needed to learn to maintain sieges:

> If it were of no moment to this war, it was yet, I assure you, of the utmost importance for military discipline that our soldiers become accustomed not only to pluck victory within their grasp, but if a campaign should be even more protracted, to put up with the tedium and await the outcome of their hopes, however long-deferred; and if a war be not finished in a summer, to stay for winter, nor, like birds of passage, cast about at once, on the approach of autumn, for shelter and covert.[16]

Long sieges were expensive, especially when the troops were drawing wages from the city. Thus the siege of Veii required an extraordinary tax. Again the tribunes of the people spoke against such a policy and when it was imposed blocked its collection. This brought the army at Veii to the edge of mutiny when its commanders could not make good on the pay.[17] Not surprisingly, the soldiers began to devote themselves more to pillaging the countryside than to pressing the siege.[18] When the siege dragged on, the logic of necessity became inescapable and the people had to pay the tax. Appius hoped to assuage them by allowing them to share in the booty.[19] He recognized that siege warfare required the support of the entire society.

In another speech that Livy gave to Appius, he expressed necessities of siege warfare that now began to guide Roman policy:

> Do you suppose there will be no great difference in men's opinions of us, whether our neighbors conclude the Roman People to be such, that if a city withstand the brunt of their first assault for a very brief time, it need thenceforward have no fears; or whether our name inspire such dread, that men believe that once a Roman army has sat down before a town, it will never budge, either from the weariness of a protracted siege or from the rigours of the winter, that it knows no other end of war but victory, and relies in its campaigns not more on swiftness than on perseverance? For perseverance, needful in every kind of warfare, is especially so in besieging cities, since fortifications and natural advantages make most of them impregnable, and time itself subdues them, as it shall capture Veii, unless the plebeian tribunes help our enemies, and the Veientes find in Rome those succours which they are seeking to no purpose in Etruria.[20]

With Appius's speech we enter into a new world in Roman warfare, a world in which war is no longer a matter of constant raids back and forth between neighboring peoples but instead becomes a matter of conquest and pacification. The ability to sustain long sieges was the key to this type of warfare.

Modern historians do not give much credence to Livy's detailed account of the siege of Veii.[21] But the main outline is by no means implausible from the perspective of siege warfare in its early stages. From the very beginning the Roman commanders at Veii positioned themselves for the long haul. They fortified their line of contravallation on both sides, against the city and against relief forces from Etruria.[22] Although Appius referred to "towers, mantlets, penthouses, and the rest of the equipment for storming towns," this machinery may be in Livy's imagination and, in any case, played a small role in the siege, which was essentially a long war of attrition.[23] Livy understood that the Romans were inexperienced in major sieges and in his account the Romans make mistakes. During the first summer of the siege, they aggressively pushed their siege works closer and closer to the wall, but they were careless about protecting their machinery. A night sortie caught the Romans by surprise, and the Veientes succeeded in burning most of the Roman siege works near the wall as well as killing quite a few Romans.[24]

Indeed, the Romans could never have captured a city as strong as

Veii with such rudimentary siege methods if Veii's Etruscan allies had come to its aid. The threat of Gauls from the north and petty jealousies among the Etruscans prevented that, and the Romans were able to wear down an isolated city.[25] Even so, Veii continued to stand undaunted by the Roman siege for a long time. Two developments broke the deadlock. First the Romans learned from the oracle at Delphi and from a captured Etruscan soothsayer how they could appease the gods and bring Veii to its foreordained fate.[26] Second, they appointed Marcus Furius Camillus dictator.[27] He became the first great Roman siege commander.

Camillus injected new energy into the siege, but most significantly he changed Roman tactics. He recognized that sieges were won more by work than by fighting and issued an order forbidding any man to fight except under a direct order. This put an end to the constant skirmishing and freed the Roman soldiers to dig. They strengthened the Roman siege works, but the main project was a mine into Veii reaching all the way to the citadel. The work on the mine went on twenty-four hours a day. Camillus divided the workers into six groups that rotated six-hour shifts until they had reached the citadel.[28] This was the beginning of the famous Roman capacity for labor, a capacity that enabled Livy to boast, "What soldier can match the Roman in entrenching? Who is better at enduring toil?"[29] Even if Livy's account is not strictly historical, there is no reason to doubt that the Romans learned much about how to organize the labor of siege at Veii.

When the workers had completed the mine, Camillus launched a major assault against the walls of Veii. The Veientes knew nothing of the mine and concentrated all their efforts on the defense of their walls. Picked Roman troops went through the mine and broke into the citadel. One group attacked the Veientine soldiers defending the walls and another group opened the gate. The completely unexpected presence of Roman soldiers within the walls spread panic among the Veientes. Keening women and children hurled tiles from the rooftops as the Romans poured into the city. To put a stop to this the Romans set fire to the buildings. The battle soon degenerated into a bloody massacre that Camillus halted only when he saw that the city was firmly in Roman hands.[30] Livy's story of the final battle may be largely imaginary, but he has given us a convincing account of Rome's first great siege.

According to Livy, Camillus took a keen interest in siege technology. Livy even claimed that Camillus wanted to use artillery in the siege of Antium in 386 B.C.[31] If this is true, it shows that the use of the catapult spread rapidly, because Dionysius's craftsmen had invented it in Sicily only about ten years before.[32] It is worth noting that Diodorus said the Romans learned the use of siege machinery from the Greeks.[33]

Roman logistics in these early years were primitive. In 320 B.C., when a Roman army laid siege to Luceria, a Roman ally that the Samnites had seized, the cavalry carried food to the army in leather pouches. But the Romans did not control the countryside, and Samnite foragers harassed the cavalry. The result was that the besieging Romans were as hungry as the besieged Samnites. Only the arrival of a second Roman army enabled the Romans to tighten the noose around Luceria and force the Samnites to surrender the town.[34]

By the beginning of the third century B.C. the Romans had developed the famous testudo. In 293 B.C. when the Samnite defenders of Aquilonia repulsed the first Roman assault on the walls, the Roman commander, Publius Scipio formed a small number of soldiers into a testudo, a formation in which the soldiers created a protective canopy by holding their shields over their heads. Polybius likened the testudo to a tiled roof.[35] The testudo charged the gate, which was not yet shut, and succeeded in bursting through and seizing the wall on both sides. This enabled the bulk of the Roman soldiers to enter the city just before night fell. During the night the Samnites abandoned Aquilonia.[36]

The Romans became very skilled with the testudo. Livy described a remarkable feat that employed the testudo during the siege of Heracleum, a Macedonian town near the Thessalian border, in 169 B.C. Young Roman soldiers were used to displaying their prowess with the testudo in games by marching into an arena and forming a slanting testudo in which the front rank stood erect and the back rank kneeled down and the middle ranks stooped just enough to form a slanting surface like the pitched roof of a house. They could hold this roof of shields so firmly that two soldiers could mount it and perform mock battles on it. At Heracleum such a testudo approached a lower section of wall and soldiers reached the top of the wall by climbing the slanted layers of shields covering the testudo. In this way they crossed the wall and took the city.[37]

Although our knowledge of early Roman history remains hazy, we can say that by the end of the fourth century B.C. the Romans had learned how to organize and sustain long sieges and had begun to develop skills and tactics, such as escalade, the use of elaborate entrenchments, and the deployment of the testudo, that would characterize Roman siege warfare until imperial times. They remained seriously deficient in the use of artillery and relied mostly on passive methods, but they had laid a firm foundation to develop the full siege capabilities that an imperial people would need.

The Punic Wars

When we reach the third century B.C., Roman siege warfare comes into clearer light because of the detailed accounts of Polybius, who was both an outstanding historian and an expert on siege warfare. His account of the Punic Wars made it evident that those wars were a Roman apprenticeship in siege warfare.

At the beginning of the First Punic War (264–241 B.C.), Roman siege methods remained simple. In the siege of Agrigentum (the old Greek city of Acragas) in 262 B.C., the ferocious assault methods the Carthaginians had employed in taking the same town 144 years earlier were absent. The Roman force of some forty thousand men relied entirely on blockade.[38] The siege was essentially a contest of logistics. It began just at harvest time (June in Sicily), and the first thing the Romans did was scatter to bring in the grain from the local fields. This left the foragers vulnerable to a Carthaginian attack, an opportunity the Carthaginian garrison at Agrigentum was quick to seize. Not only did the Carthaginians rout the Romans harvesting the fields, but they also mounted an all-out attack on the Roman camp. A sharp fight ensued, in which both sides suffered heavy casualties, but Roman tenacity carried the day and the Carthaginians were driven back into the city.[39]

Both sides then settled in for the long haul. The Romans blockaded Agrigentum with a double entrenchment, the inner one protecting against sallies from the city, the outer one protecting against relief forces and preventing anyone sneaking into the city with supplies. Two fortified camps, one west of the city and the other either south or east of it, anchored the entrenchments, and pickets patrolled the area between the trenches. The Romans established a supply base at the nearby town of Herbesus, which allied towns in the area kept supplied. In this way the Romans kept themselves well fed, enabling them to sustain the siege for the next five months.[40] The Carthaginian position was much more difficult. There were fifty thousand defenders locked up in Agrigentum, and after five months of siege they were terribly hungry. In the siege of Acragas 144 years earlier, the Greeks had supplied the city by sea from Syracuse. Like most Greek coastal cities, it was not directly on the sea, but rivers provided access to the sea. However, the Romans had invested the city so completely that even though they could not challenge Carthaginian naval superiority, they still prevented the Carthaginians from bringing food into Agrigentum.[41]

The Carthaginian commander, Hannibal, repeatedly sent urgent messages to Carthage pleading for assistance; the Carthaginians responded by sending a relief force of fifty thousand fresh troops, six thousand cavalry, and sixty elephants, placing it under the command of Hanno, who was already in Sicily.[42] Hanno immediately struck a hard blow against the Romans by capturing Herbesus. The disruption of the Roman supply system soon reduced the Romans to hunger, and they seriously contemplated raising the siege. Only the supreme efforts of the ruler of Syracuse, Hiero, who was a Roman ally, delivered just enough supplies to the Romans to enable them to keep Agrigentum besieged.[43]

The contest now was much more even, but the Carthaginians in Agrigentum were still hungry. Hanno successfully lured the Roman cavalry outside their camp and inflicted a defeat on them with his Numidian horsemen, but otherwise he simply tried to keep the pressure on the Romans by camping nearby. After two months, however, Hannibal had to notify Hanno that his mercenaries were beginning to desert and that he could not hold out much longer. A plague was weakening the Romans, so Hanno decided to risk an attack on the Roman camp. This was not unwelcome to the Romans, who, hungry and sick, were tired of siege.

In the battle that followed, the Romans were victorious. That night, however, the Romans, exhausted from the exertion of combat, relaxed their vigil, and Hannibal was able to escape with what was left of his army. The Carthaginians crossed the Roman trenches by filling them with baskets stuffed with chaff. When the Romans discovered what was happening, they proved more anxious to plunder Agrigentum than attack the fleeing Carthaginians, and Hannibal was able to reach safety.[44]

The siege had lasted seven months. The Romans relied entirely on blockade. If they had any siege machinery at Agrigentum, Polybius did not mention it. Agrigentum was strongly situated, and the terrain made it difficult to bring siege machinery against its walls. Although the Carthaginians had been much more aggressive when they besieged the city in 406 B.C., bringing towers forward and building siege ramps that made it possible to assault the walls in several places at once, they failed to breach the walls and the siege ended in the same way as the Roman siege: with the defenders fleeing at night because of famine. Blockade remained the most effective Roman siege method against a city as formidable as Agrigentum. The arduous siege had been costly. Diodorus claimed that thirty thousand Roman soldiers died at Agrigentum, but he numbered the size of the Roman army at 100,000, more than twice what it probably was. But his casualty rate of one-third dead is plausible enough. The siege left a sour taste in the mouths of the Romans. The Senate denied the honor of a triumphal march to the two Roman commanders at Agrigentum.[45]

Perhaps because of their difficulties before Agrigentum, the Romans began to employ siege machinery soon afterward. But the results were far from spectacular. Diodorus said that shortly after the fall of Agrigentum the Romans besieged the town of Mytistratus with siege engines for seven months but failed to take the town. Mytistratus fell three years later, in 259 B.C., after another long siege. Diodorus said it was the Romans' third try.[46] It appears the Romans were learning, because in the same campaign of 259 B.C. they also took the Sicilian town of Camarina by breaching its walls with battering rams.[47] Diodorus reported that the Romans got the rams from the Syracusans.[48] The Romans also used rams to knock down a tower in the siege of Panormus, an important city that soon fell.[49] These are the first reports we have of the Romans breaching a wall with siege machinery.

Toward the end of the First Punic War, the Romans began to employ more aggressive siege tactics, but not with complete success. By 250 B.C., the Romans had gained control of almost all of Sicily, and they determined to deliver the final blow to Carthaginian power there by besieging Lilybaeum, which was now Carthage's main entry point to Sicily. It was a formidable undertaking. The Carthaginians had strongly fortified Lilybaeum with walls and a moat ninety feet wide and sixty feet deep. Moreover, dangerous shoals in the entrance to the harbor made the seaward approach to the city difficult. Vergil immortalized these shoals in the *Aeneid* as the "pitiless shoals of Lilybaeum, with their hidden rocks."[50] The size of the Roman force sent to Lilybaeum is uncertain. Polybius said that the Roman fleet at Lilybaeum consisted of 200 ships. Diodorus said 240 warships and 60 light ships. One hundred and twenty is a more realistic estimate. Diodorus claimed the entire Roman force, sailors and soldiers, numbered 110,000, but his figure may be greatly exaggerated. In any case the Romans had to use men from the fleet to help build siege works and man the machinery.[51] Even though the Roman fleet succeeded in seizing the harbor, the loss of manpower prevented it from maintaining an effective blockade.

The Romans, against their usual practice, moved directly to assault Lilybaeum with siege works without first fortifying their own position. Indeed, Lilybaeum was the first thoroughgoing siege the Romans waged.[52] They built siege ramps, dug mines, and deployed siege towers, battering rams, and catapults. The mention of catapults is significant because it is the first we hear of the Romans using artillery. The Carthaginians waged an equally aggressive defense, building counterworks and digging countermines and constantly sending sallies, day and night, against the Roman siege works in an attempt to set them on fire. The Roman catapults must have been large stone throwers, because they breached the wall. The Carthaginians threw up a second wall behind the breach. The fighting over the siege works was bitter and Polybius said that "at times more men fell in these encounters than usually fall in a pitched battle." However, the Romans did succeed in destroying six towers.[53]

The Carthaginians employed some ten thousand mercenaries (seven thousand according to Diodorus), mostly Greeks and Celts, to help defend the town. Mercenaries always presented a security risk within a besieged city, as Aineias Tacticus had pointed out. As the Roman siege works pressed relentlessly forward, some of the mercenary commanders plotted to betray Lilybaeum to the Romans. But a Greek mercenary officer revealed the plan to Himilco, the Carthaginian commander at Lilybaeum, and he promised lavish bonuses to the mercenaries if they remained loyal, in this way barely averting a catastrophe.[54]

The Carthaginian leaders at home did everything in their power to relieve Lilybaeum. They sent a force of fifty ships carrying ten thousand

troops (four thousand according to Diodorus) to break through the
Roman blockade. Hannibal, son of Hamilcar, commanded this force,
and he proved a resourceful sailor. He anchored at some islands off Lilybaeum, and when a favorable wind blew up he boldly sailed straight
into the harbor through the treacherous shoals that guarded its entrance. The Romans were dumbfounded by his audacity and, fearing
the shoals and the wind, dared not attempt to block his way. At about
the same time Himilco was able to board his cavalry, uselessly cooped
up in the city, onto ships and send them to nearby Drepana, from
whence they could harass Roman foragers.[55]

This spectacular success boosted the morale in Lilybaeum tremen-
dously, and Himilco determined to take advantage of the opportunity
before the exhilaration waned under the stress of the siege. Accordingly,
at dawn he sent no less than twenty thousand men against the Roman
siege works. The Romans had anticipated the attack and had concentrated a larger number of troops to defend their works. The result was
a fierce battle in a highly constricted space that prevented any organized fighting. Desperate hand-to-hand combat by individual men made
for a chaotic battle. The Romans fought stoutly, but the confusion of
the battle made it impossible to prevent some of the torch-bearers from
getting to the siege works, and they were soon in flames. At this critical moment, Himilco's nerve failed him. Casualties were high, and he
feared the battle would leave his forces so weakened they would not be
able to defend the city. So he ordered a retreat just as it seemed the Romans might lose their siege works.[56]

A new opportunity soon presented itself to the Carthaginians. A
strong wind kicked up that blew so hard it extensively damaged the
Roman towers and penthouses that protected their machinery. Some
Greek mercenaries in Carthaginian employ recognized this as an ideal
chance to try again to fire the Roman siege works. This time smaller
forces struck at three different spots and succeeded in setting the Roman
works on fire. The wind, which was blowing into the face of the Romans, whipped up the flames. The smoke was so dense the Romans had
difficulty getting near enough to the flames to douse them, and the
Carthaginians, shooting with the wind at their backs, were able to drive
away those Romans who tried to approach the fires. The fires burned
until they damaged the Roman siege machinery beyond repair.[57]

The Romans, however, soon scored a success of their own. A daring
Carthaginian captain, Hannibal the Rhodian, had continued to flout
the Roman blockade by sailing in and out of Lilybaeum, a feat that gave
heart to the city's defenders and discouraged the Romans. Even when
the ten fastest Roman ships attempted to ambush Hannibal, he was able
to pass them "as if they were motionless," and when he offered battle,
the Roman ships backed down. In vain the Romans tried to block the
harbor by dumping all manner of stuff into its mouth. The depth was

too great and the current too rapid for this to work. Finally, though, the Romans were able to create an artificial bank in one part of the shoals and one of the speedy four-banked Carthaginian ships ran aground, enabling the Romans to capture it. Having manned this ship with a picked crew, the Romans were waiting for Hannibal the next time he sailed out of the harbor. With this ship, the Romans were able to catch Hannibal, board his ship, and capture both him and the ship. They now had two fine Carthaginian ships that finally enabled them to put an end to Hannibal's blockade running.[58]

Despite this success, the loss of their siege equipment decided the Romans to halt siege operations against the walls of Lilybaeum and instead to resort to circumvallation. They enclosed the city with a trench and a palisade, built a wall around their camp, and settled in to starve Lilybaeum into submission. But the Romans themselves were out of grain and were eating only meat, a food source that cannot have lasted long. Only a timely shipment of grain from Syracuse enabled the Romans to continue the siege. The Carthaginians on their part repaired their damaged walls. Apparently the city was well stocked, because Polybius reported they were confident of the outcome now that the Romans had ceased attacking the walls.[59] The blockade may not have been complete. Diodorus said that repeated Roman efforts to block the harbor with ships loaded with stones or with anchored logs all failed. The Romans sent a new fleet out to Lilybaeum to reinforce the siege, but an ill-advised effort to destroy the Carthaginian supply base at Drepana ended in disaster. The debacle left the Romans in a weakened condition, and the Carthaginians were able regularly to reprovision Lilybaeum.[60]

And so the situation remained until the end of war eight years later. With the end of offensive siege operations at Lilybaeum, the focus of the war shifted elsewhere, and it was the eventual Roman victory at sea, isolating the Carthaginian forces in Sicily, that forced Carthage to surrender. The Roman failure to reduce Lilybaeum contrasts with Dionysius's successful siege of Motya in 398 B.C. Lack of manpower hampered the Roman siege. Polybius reported that "the greater part" of the sailors who had to leave the fleet to work in the siege died, greatly weakening the Romans at sea.[61] Diodorus mentioned a plague.[62] Only when they suspended active siege operations were the Romans able to rebuild their naval forces. They were unable vigorously to pursue the siege and contend for naval supremacy at the same time. Without a naval blockade they could not starve Lilybaeum. Roman assault methods were rather plodding in comparison to the furious assaults of Dionysius. The Romans worked their way forward gradually, allowing the defenders time to take countermeasures. Even when Roman catapults damaged a section of wall, the Carthaginians were able to build a secondary wall that neutralized the breach.[63] Dionysius kept the Motyans off balance with enormous towers, catapults, and the sheer force of his attack.

The Romans had all this machinery at Lilybaeum, but they were not yet the equals of Dionysius in their use.

During the Second Punic War, the Roman siege warfare became more effective. This was evident from their siege of a city famous for past sieges, Syracuse. Both the Athenians and the Carthaginians had failed to take Syracuse and it remained a formidable city in 213 B.C., the year of the Roman siege. The wall guarding Epipolae, built by Dionysius over 150 years earlier, still stood. It ran along an escarpment that created a natural glacis, making access all but impossible except in a very few places. Nevertheless the Romans hoped they could capture the city by a simultaneous assault from land and sea.[64]

Marcus Claudius Marcellus, the commander of the Roman fleet, chose to launch his seaborne attack on Achradina, the section of the city formed by its expansion beyond the island of Ortygia onto the mainland. Ortygia and the mainland created a harbor known as the Little Harbor, and there the sea came all the way up to the base of the wall protecting Achradina on the seaward side.[65] The Roman force consisted of sixty quinqueremes, carrying archers, slingers, and javelin throwers to sweep the battlements of their defenders. Livy said that Marcellus also had battering rams mounted on his ships, but Polybius did not mention this.[66]

Under covering fire from his slingers and archers, Marcellus hoped to get his assault troops over the wall with a siege machine known as a sambuca, or harp. The sambuca was a large ladder carried by two ships lashed together. Protective covering mounted on the ladder created a sort of tunnel to shield the soldiers as they climbed the ladder. Ropes ran from the ladder to pulleys attached to the ships' masts. As the ships approached the wall, sailors pulled the sambuca up to the proper height to reach the top of the wall. A platform protected by wicker shields stood atop the ladder. Four men stood on the platform. Their task was to drive the defenders off the wall at the point where the sambuca came to rest. Once the sambuca was in place against the wall, these men removed the wicker shields and secured that section of the wall for their comrades who were rapidly climbing the ladder.[67]

But the Syracusans were well prepared with devices of their own. Archimedes, their famous engineer, had arrayed a battery of catapults of various sizes to prevent the Roman ships from approaching the wall. Some of the catapults were long-range stone throwers that began inflicting serious damage before the Romans had hardly begun their approach to the wall. As the Romans came closer, the Syracusans moved to slightly smaller catapults so that, whatever the range, they were able to subject the Romans to such a devastating artillery fire that the Romans were unable to reach the walls.[68]

The Syracusan catapults forced Marcellus to seek the cover of night. The darkness enabled the Romans ships to come near the wall, but

Syracusan archers, shooting crossbows, or scorpions, through loopholes that Archimedes had made in the walls, inflicted heavy casualties on the Roman marines. Archimedes had also mounted long swiveled beams on the wall with which the Syracusans could drop stones weighing over five hundred pounds on the sambuca or the ships, completely wrecking them. The most amazing of Archimedes' devices, though, were long arms mounted on the walls that could reach out and hook onto the prow of the Roman ships. Using Archimedes' famous principle of leverage, these arms were capable of lifting the prows out of the water and then dropping them from a considerable height. The unpleasant result for the Romans was that the ships either capsized or took on so much water they were swamped. The Romans had to give up their seaborne assault. Despite this setback, Marcellus retained his sense of humor, wittily remarking, "Archimedes uses my ships to ladle seawater into his wine cups, but my sambuca band is flogged out of the banquet in disgrace."[69]

At the same time the Syracusans were repulsing the Romans at sea, the other Roman commander, Appius Claudius Pulcher, was assaulting the walls on the landward side. Under his command were three legions, probably not numbering much more than twelve thousand men. He chose Hexapylus, a place along the northern edge of Epipolae, the high plateau over the city, where access to the wall was easiest, as the most likely point of attack.[70] Despite the narrow access to the wall, Appius was confident he could gain the advantage because of the large number of penthouses, arrows, and scaling ladders that he could throw into the assault.[71] But the intense fire of the Syracusans thwarted Appius as well. At long range the Romans incurred heavy casualties from Syracusan stone-throwing and arrow-shooting catapults. Those who did manage to get near the wall encountered withering volleys of arrows through the loopholes in the wall. Stones dropped by the Syracusan beams smashed the protective penthouses. The long crane arms that were hooking Roman ships at the other point of attack here hooked soldiers' armor, lifted them high above the ground, and dropped them back to earth. Against these measures the Romans were unable to mount an assault of sufficient weight to surmount the wall, and Claudius Pulcher called off the attack.[72]

Having failed on land and at sea, the Romans decided to give up assault and resort to blockade, just as they had done at Lilybaeum thirty-seven years earlier. During the next eight months they tried a number of stratagems but no more assaults, although reinforcements arrived to more than compensate for the heavy losses of the siege. They believed that famine was their best weapon against the Syracusan machinery and endeavored to blockade the city by land and sea.[73] The army established two camps to the north and south of the city to cut off the main coastal roads. The fleet operated in the Great Harbor, but it apparently did not establish itself there permanently and the effectiveness of the Roman blockade is not clear. Polybius reported that within eight months the Syracusans were very hungry and that this played an important role in the fall of the city.[74] Livy said that the blockade was ineffective and that Syracuse "was sustained by almost unhampered supplies from Carthage."[75] Although the ancient sources are silent on the subject, Brian Caven believes the Roman fleet was based at Messana rather than Syracuse and probably wintered at Lilybaeum at the other end of Sicily. Also manpower shortages hampered the Roman naval blockade, as they had at Lilybaeum in the First Punic War. Rome sent a fourth legion, but Carthage sent twenty-five thousand infantry and three thousand cavalry to Sicily to threaten the Roman siege. Thus the Roman blockade was never complete, and, depending on the comings and goings of the Carthaginian admiral Bomilcar, the Carthaginians even enjoyed periods of naval superiority during the siege.[76]

In view of the tenuous naval situation, the Romans now pinned their hopes on the exploitation of divisions among the Syracusans, many of whom retained a loyalty to Rome and some of whom had deserted to the Roman lines. These pro-Roman Syracusans in the Roman camp made contact with friends in the city, but the plot collapsed when an informer revealed it to the Syracusan authorities.[77]

Another possibility soon presented itself. The Romans had captured a Spartan who was on a mission from Syracuse to King Philip of Macedonia. The Syracusans were anxious to ransom him, and negotiations with the Romans followed. These negotiations took place near the northern wall of Epipolae, and while they were going on some Roman soldiers calculated the height of the wall by counting the number of courses, just as the Plataeans had done to calculate the height of the wall of circumvallation during the siege of Plataea at the beginning of the Peloponnesian War. The Romans discovered that the wall was not as high as they thought at that point, creating the possibility of an escalade. The Syracusans maintained a heavy guard in that sector of the wall, but when a deserter informed Marcus Marcellus, who had moved to the landward side after the failure of his seaborne attack, that the Syracusans were engaged in a festival of Artemis and had relaxed their guard, Marcellus saw his chance. Although little food was available for the festival, wine was still abundant. Marcellus knew that wine on empty stomachs made drunk and careless soldiers and decided to try an escalade. He massed one thousand men for the attack, which came against the same point where Appius Claudius's assault had failed. In complete contrast to their experience eight months before, the Romans met no opposition this time. Incredibly they found the wall undefended, the guards either still drinking at the festival or asleep in a drunken stupor. The mounting party crossed the wall under cover of darkness early in the night and opened the nearest postern, admitting

the rest of the Roman soldiers. The Romans quickly killed the few defenders on Epipolae. Most of the Syracusans were in main sectors of
the city and were not even aware that Epipolae had fallen into Roman
hands.[78]

The capture of Epipolae still left Achradina and the island (Nasus
or Ortygia) in Syracusan hands.[79] Marcellus sent his Syracusan allies to
promise mild terms if the defenders surrendered the rest of the city. But
deserters who had little hope of survival even if there were an amnesty
guarded the gates and walls and would allow no negotiations. Forced
to continue the siege from Epipolae, Marcellus then attempted to entice
the fortress at Euryalus, which dominated the western apex of Epipolae
and which threatened his rear, to surrender. The commander of Euryalus, however, hoped for Carthaginian relief and refused to capitulate.
Marcellus then encamped on the northern edge of Achradina, fortified
his camp with a wall, and loosed his soldiers to plunder the part of the
city under his control. However, he still hoped for some negotiated settlement, and so he maintained strict discipline and prevented his soldiers from shedding any blood. While the plundering was going on, the
fortress at Euryalus decided to surrender after all when the hoped-for
Carthaginian relief force failed to arrive. Marcellus was able to begin
siege operations against the walls of Achradina without that threat
from the rear.[80]

Despite the capture of Epipolae, the Roman position was far from
secure. Thirty-five Carthaginian ships were able to slip out of the Great
Harbor on a stormy night and made it to Carthage with the news of
the fall of Epipolae. The Carthaginians sent one hundred ships back to
Syracuse to reinforce the defenders. Also a land force of Greeks and
Carthaginians arrived at Syracuse, leaving the Roman besiegers under
siege themselves. The naval force succeeded in reaching the city during diversions created by a Syracusan sally from Achradina against
Marcellus's position on Epipolae and a simultaneous attack by the relief army on a Roman camp at Olympieum that guarded the area south
of the city along the Great Harbor. Both attacks were unsuccessful, but
they enabled the relief fleet to land.[81]

That old Syracusan ally, General Plague, also arrived. The season
was autumn and the heat was unbearable. Soldiers died faster than they
could be buried, and the stench of corpses filled the air. Both sides
suffered terribly, but none more so than the Carthaginians. Marcellus
moved his troops into shaded shelter on Epipolae, and the plague abated
among his soldiers. The Carthaginians, on the other hand, found themselves abandoned by their Sicilian allies, who scattered to their towns
and villages at the first sign of plague. The plague flourished among the
Carthaginians, who were in the marshy lowlands south of Syracuse and
who were unaccustomed to the climate. Livy reported that this plague
wiped out the entire Carthaginian force, including its generals.[82]

The loss of the relief army left Bomilcar, the Carthaginian commander in Syracuse, in a mood of desperation. Once again he escaped the Roman blockade and sailed to Carthage, where he convinced the Carthaginians to send a huge new relief force consisting of 130 warships and 700 transports full of supplies. This time, however, unfavorable winds prevented him from rounding Cape Pachynum at the southeastern point of Sicily and he had to put into the southern coast and wait. With unfavorable winds and with a Roman fleet stationed on the other side of Cape Pachynum, there was serious doubt that Bomilcar could reach Syracuse. The Syracusan commander, Epicydes, knew that the fate of Syracuse depended on this relief fleet reaching the city. He therefore left the city under the command of his mercenaries and slipped out by ship to stiffen Bomilcar's resolve. Under the entreaties of Epicydes and with the wind abating, Bomilcar put to sea to round Cape Pachynum. But his resolve remained weak, and when he saw the Roman fleet sailing against him, he made for the open sea and sailed to Italy. Epicydes knew Bomilcar's decision had doomed Syracuse; he did not return to the beleaguered city but fled to Agrigentum. Bomilcar's lack of determination corresponded to the Carthaginian army's failure to reach Euryalus; both were missed opportunities to embarrass the Roman besiegers.[83]

Abandoned by the Carthaginians and deserted by their leader, the remaining defenders of Syracuse killed the leaders of the pro-Carthaginian faction and opened negotiations with the Romans. The negotiations were interrupted when the Spanish mercenaries, fearful of their fate if the city surrendered, killed the new magistrates and took over the city. Further reflection, along with the encouragement of a Spanish emissary from Marcellus, convinced them that they were not in the same boat with the pro-Carthaginian Syracusans and they began to move toward the Roman side. Assured of the support of the Spanish mercenaries, Marcellus launched an assault against Achradina. The Syracusans on Nasus rushed to the defense of Achradina, allowing a Roman force to sail up to Nasus and capture the island. Once the Romans were in possession of Nasus and had penetrated into Achradina, the defense of Syracuse collapsed.

The Syracusans surrendered, asking only for their lives to be spared. This Marcellus was willing to do, but he turned over the city to his soldiers to sack. Roman guards protected the royal treasury and the houses of those Syracusans who had fled to the Romans at the beginning of the siege. There was no bloodbath, but there was bloodshed. Among those who fell to the rampaging Roman soldiers was Archimedes.[84] Livy's account is chronologically confusing, but the siege apparently had lasted two-and-a-half years.[85] By its end both sides were seriously short of food. Only the arrival of a Roman fleet laden with supplies it had cap-

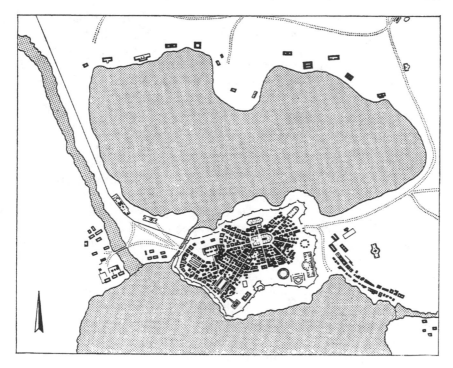

Map 7. New Carthage in Roman times. *From H. H. Scullard,* Scipio Africanus. *Used with the permission of the publisher, Thames and Hudson, Ltd.*

tured in a raid on North Africa prevented a famine among Syracusans and Romans alike.[86]

The key to the Roman victory at Syracuse was the ability to keep the city under siege for a long time despite the threat of a Carthaginian army and the lack of complete naval superiority. This was a considerable logistical feat that won the time necessary to exacerbate the factional divisions within the city to the point where the city was betrayed. Marcellus thus became the first siege commander to take the city of Syracuse, but Rome's gratitude was restrained. Because he had not completed the conquest of Sicily the Senate did not grant Marcellus a triumph and he had to settle for the lesser honor of an ovation in Rome, where he proudly paraded not only the riches of the city but also the catapults and other siege machinery and a representation of the great city that he had conquered.[87]

In contrast to the drawn-out siege of Syracuse was the sudden capture of New Carthage in Spain by Publius Cornelius Scipio in 209 B.C. (map 7). Still in his twenties, Scipio was a brilliant and charismatic commander. The capture of New Carthage was the first great achievement in a career that would culminate in the conquest of Hannibal, a feat for which he acquired his famous name: Scipio Africanus.

Excellent intelligence laid the groundwork for the rapid seizure of New Carthage. Scipio learned that the main Carthaginian forces were scattered about Spain, none closer that a ten-day march from New Carthage. Even though it was the main Carthaginian base in Spain, a garrison of only one thousand men defended the city because the Carthaginians believed their control of Spain was so secure that it was entirely safe. The population of the city was large, but it consisted mostly of tradesmen unaccustomed to war. Scipio had under his command about 25,000 infantry and 2,500 cavalry; such numbers could perhaps overwhelm the defenders if he could achieve surprise. Water surrounded most of New Carthage. It stood on a peninsula jutting out between a large natural harbor and a lagoon. The harbor surrounded the southern sides of the city; the lagoon protected the northern side. On the western side an artificial canal connected the lagoon to the sea; on the eastern side a narrow isthmus only about four hundred yards wide connected the city to the mainland. Careful interrogation of local fishermen yielded the information that the lagoon was shallow and that the water sometimes receded even more late in the afternoon.[88]

Because surprise was of the utmost importance to the success of his plans, Scipio shared them only with Gaius Laelius, the commander of the fleet.[89] When Scipio was ready, he commanded Laelius to sail to New Carthage with the fleet but not to arrive before the army. Scipio himself led the army across the Ebro to New Carthage. Because surprise was of the essence, he forced the pace and arrived at New Carthage seven days later, an incredible feat, because the distance was some three hundred miles; either the seven days listed by Polybius and Livy is corrupt or the start point was somewhere south of the Ebro.[90] In any case, there can be no doubt that the march was rapid. Scipio pitched his camp on the peninsula east of the city, blocking New Carthage's access to the interior. The sea protected the inner side of the camp; the Romans secured the outer side with a palisade and a double trench. However, Scipio left the section facing New Carthage open because he wanted his troops to be able to move rapidly against the town and because the terrain offered sufficient natural protection.[91]

Speed was essential to Scipio's plan. If he could not take New Carthage within a few days, his army would be terribly vulnerable to a Carthaginian relief army. The need for speed ruled out blockade or even the construction of siege works. His soldiers would have to take the city by assault. High morale was absolutely necessary. Scipio therefore spared no effort to inspire his troops. First he impressed upon the troops the strategic significance of New Carthage and convinced them of the soundness of his plan of attack. He offered the usual incentives, including mural gold crowns to those who were the first to mount the wall. (Livy reported that afterward the Roman marines and legion-naires almost came to blows over which of their champions had crossed the wall first and Scipio prevented a mutiny only by awarding mural gold crowns to them both.)[92] What most impressed his troops, however, was Scipio's revelation that Neptune had appeared to him in a dream and promised to help. The significance of this dream was not lost on soldiers contemplating an assault on a city almost entirely surrounded by water.[93]

The appearance before New Carthage of a Roman army twenty-five times larger than the garrison placed the Carthaginian commander, Mago, in a difficult position. With at least 2,300 yards and perhaps as much as 4,000 yards of wall to defend, he chose not to string out his troops along the wall.[94] Instead, he held them in reserve, placing half of them in the citadel and the other half on a hill in the eastern part of the city. In addition, he recruited two thousand of the most likely citizens, armed them as best he could, and stationed them at the gate facing the Roman camp. The defense of the walls he left to the other citizens. Rather than a line defense along the wall, Mago's disposition suggested an aggressive defense in depth with his best troops held in reserve to strike at the crucial moment. Events proved that Mago was too indecisive a commander to execute such a plan.[95]

Scipio began his assault about midmorning. On the seaward side Laelius attacked with his ships, which were equipped with a variety of catapults and with sambucas.[96] On the landward side, Scipio drew up his troops in front of the camp so they could advance across the isthmus against the city.

As soon as the bugle sounded the Roman attack, Mago sent his force of two thousand armed citizens through the gate to attack the Romans in front of their camp. This cannot have been unwelcome to Scipio. The Romans had the advantage of fighting near their camp, where Scipio had held them, probably in the hope he could draw the New Carthaginians out, while the New Carthaginians had to exit the city through a single gate and advance four hundred yards uphill before closing with the Romans. During the advance they had to shift from a column to a line, no easy task for untrained soldiers. Despite their inexperience, the New Carthaginians fought well, delivering a sharp attack on the Romans.

Modern scholars argue that Mago threw his Carthaginian garrison soldiers into the battle, because Polybius said that "both sides had picked out their best men" and because it is not clear why he would have kept them on the hills in the town. But Polybius's reference may have been to the picked men from the New Carthaginian civilians. It is not inconceivable that Mago may have withheld his own soldiers from what was almost a suicide mission.[97]

Scipio was able continually to throw reinforcements into the battle, and the terrain was favorable to the Romans. Finally sheer weight of numbers broke the New Carthaginian ranks. Many fell in the battle, but most were trampled to death in the stampede to get back through the gate. The New Carthaginians were just able to close the gate before the Romans could burst in.[98]

The townspeople on the wall watched this battle in horror; demoralized by the defeat, they abandoned the wall in panic. The Roman ladder bearers brought up the ladders and placed them in position against the walls without any opposition. But the walls of New Carthage were too high for an easy escalade. The Roman ladders had to be longer than optimal. Some broke under the weight of too many soldiers climbing them. All were shaky, and the soldiers near the top became dizzy from the height. These difficulties emboldened the New Carthaginians, and they returned to the walls, where they caused the Romans considerable trouble by knocking them off the shaky ladders or throwing beams and other heavy objects down on them from the battlements. The Romans continued the assault with great determination, but as the fighting dragged on into the afternoon, fatigue set in. Scipio recalled his troops to regroup.[99]

The Roman withdrawal caused jubilation in New Carthage, but Scipio now proved himself a master of psychology. Just when the New Carthaginians thought the battle was over, he ordered another attack against the gate. The defenders had exhausted most of their ammunition fighting off the first attack and their losses had been heavy. This second attack on the heels of the first demoralized them. Nevertheless they resisted bravely.

But now came Scipio's second psychological stroke. He sent five hundred soldiers carrying ladders to the shore of the lagoon with orders to wade across at an opportune time to launch a surprise attack. When a strong north wind kicked up, causing the water to become even more shallow, the soldiers remembered Scipio's dream of Neptune and moved across the lagoon confident that the gods were on their side.[100] The intense Roman attack on the gate had drawn the defenders to that sector, and when the escaladers crossed the lagoon and reached the wall they found it undefended. Because of the protection afforded by the lagoon, the Carthaginians had not built the wall so high in this sector and the Romans quickly mounted it and moved along it, sweeping the battlements of defenders. When they reached the gate, they descended and captured it from the inside just as the attackers from the outside were breaking through. Organized defense of the walls collapsed, and the Romans poured through the gate and over the wall. The fleet must also have reached the walls in the afternoon attack, because a dispute later arose over whether a marine or a legionnaire had crossed the wall first.[101]

The Carthaginian garrison put up little resistance. The Romans took the hill on which Mago had stationed half of the garrison without difficulty. Scipio led a thousand men against the citadel, but when Mago saw that the city had fallen and that the Romans were massacring all the adults, he meekly surrendered.[102]

Scipio had taken New Carthage in a single day. Except for scaling ladders and catapults, he had used no siege machinery.[103] No traitor betrayed the city to him. New Carthage fell because Scipio realized that he had sufficient forces to take the undermanned city by assault if he moved quickly enough. He proved a master of timing and psychology, demonstrating once again their supreme importance in warfare. Mago's position was essentially hopeless, given the relative strength of the forces. His failure to deploy his reserves meant that the entire defense of the city fell to untrained men pressed into emergency service. They fought much more bravely than Mago and his troops.

Two years after the capture of New Carthage, the Romans took another important Spanish town, Orongis. Orongis was a wealthy city in Andalusia, a rich agricultural and mining area. Scipio sent his brother Lucius with ten thousand troops and one thousand cavalry to capture it.

Lucius began by inviting the inhabitants of Orongis "to test the friendship of the Romans rather than their power," as Livy put it. Perhaps because there was a Carthaginian garrison in the town, Orongis rejected this diplomatic gambit. The Romans immediately dug a trench around the city, building earthworks on both sides of the trench. Lucius divided his army into three parts so that he could maintain constant assaults by rotating his troops.

But the city was heavily defended, and the first attack did not go well. The assault troops encountered heavy fire from the walls and had trouble approaching them. A few did reach the wall with scaling ladders, but the defenders pushed the ladders back with long forks or attempted to snare the Romans with grappling hooks. Lucius realized that one-third of his army was an inadequate attacking force for such a strongly defended city. He recalled the first wave but immediately threw the other two-thirds into the attack. This decisive action broke the spirit of the Spaniards, and they abandoned the walls.

The Carthaginian garrison had been helping to defend the walls, but the flight of the Spaniards left the Carthaginians terribly vulnerable both to the Roman attackers and to the Spaniards should the latter turn on them. So the Carthaginian soldiers also left the wall to draw up together for mutual protection.

Realizing their city's defense had collapsed and fearing that the Romans would make no distinction between Spaniards and Carthaginians in storming the city, the Spaniards tried to surrender. They opened a gate and rushed out, holding their shields in front of them to protect themselves from any missiles that might greet them but holding their right hands out to show that they were carrying no weapons and were trying to surrender. Either because they did not see the weaponless hands or did not trust the Spaniards, the Romans cut them down in front of the gate.

Lucius sent cavalry, supported by his most experienced troops, through the gate to occupy the center of the town. Breaking through the other gates with hatchets and pickaxes, Roman troops soon gained control of the entire city. Discipline was excellent; there was no plundering, and only those defenders who resisted were killed. Lucius placed under guard the Carthaginian garrison, along with about three hundred Spaniards who had been responsible for the rejection of his offer of alliance. But he still preferred an alliance to a ruined city, so he restored Orongis to the rest of the Spaniards. Livy reported that about two thousand of the defenders had died in the battle; the Romans lost about ninety men.[104]

The Romans had captured two of Spain's most important towns with sudden attacks that caught the enemy by surprise. Scipio's decisiveness, swiftness of action, and supreme sense of timing remind one of Alexander the Great. But Scipio's style of leadership represented a transition away from personal heroism toward a more managerial style. Heroic leadership was still very much in style in the Second Punic War. Hannibal, for example, manned the battering rams and suffered a serious javelin wound in his thigh under the wall of Saguntum. If the occasion called for it, Scipio himself was capable of risking his life. When an assault on the walls of the Spanish town of Iliturgi in 206 B.C. wavered in the face of fierce opposition, Scipio advanced close to the wall to berate the troops and even threatened to mount the wall on a scaling ladder himself. According to one report, Scipio suffered a wound in the neck. Consternation for his safety filled his troops, and they redoubled their efforts and took the city in four hours.[105]

But Livy believed Hannibal acted "somewhat incautiously" and Polybius called his leadership reckless, judgments that reflected a more sober view of the role of a commander.[106] Livy found the death of a Roman propraetor, Gaius Atinius, in 186 B.C. a cautionary tale. In assaulting a small Spanish town, which fell quickly and easily, Atinius was killed when he, "too carelessly" in Livy's judgment, approached the wall.[107] The model of leadership that Livy and Polybius had in mind had been displayed by Scipio at New Carthage. Although Scipio took part in the battle and his presence did encourage his troops, his purpose was more to be in a position to manage the battle to better effect. Three men carrying large shields protected him, and Polybius said that "he consulted his safety as far as possible." Polybius summed up Scipio's purpose well:

For he could both see what was going on and being seen by all his men
he inspired the combatants with great spirit.[108]

Scipio was content to leave the mural gold crowns to others so that he
could maintain better command and control of the battle.

The spectacular capture of New Carthage and Orongis demonstrated that Roman siege warfare had reached a new level under Scipio.
Although siege machinery had played a minor role in the sieges of
New Carthage and Orongis, Scipio did not neglect his siege train. In
his North African campaign he mobilized an abundance of machines,
some of which he brought with him, others having been sent from an
arsenal in Sicily, and still more having been built in an arsenal Scipio
established in Africa. These arsenals were probably manned by craftsmen from New Carthage. But siege warfare remained a difficult enterprise.

In North Africa Scipio found the important towns too strongly defended for a sudden seizure to be possible. Nor was he able to sustain
formal sieges long enough to reduce these cities. He attempted to take
Utica, which would have been a valuable base for his operations against
Carthage, but after a forty-day siege accomplished nothing, and with a
Carthaginian army threatening him, he lifted the siege and went into
winter camp.[109] Appian said that he transferred his siege machinery to
Hippo, but when no progress was made there either, "he burned his engines as useless."[110] Appian must have been wrong, because Scipio resumed the siege of Utica the following spring and despite his frustra-
tions he cannot have regarded his machines as "useless."

Indeed, the Roman ships were loaded with siege machinery when
they returned to Utica. There was a very close call when the Carthaginians sent their fleet against the Roman fleet while it was carrying
Scipio's machines and caught it unprepared for a naval battle. Scipio
had divided his forces, and he was in Tunis, which he had just seized.
Luckily for Scipio, the Carthaginian fleet moved too slowly, allowing
the Romans to regroup. Although the Carthaginians captured some
transports, the Romans managed to prevent a disastrous defeat.[111]
Scipio never did take Utica.

Roman siege warfare had come a long way since the First Punic
War, but there were still limitations. Long sieges were simply not feasible when strong enemy forces were operating at sea and on land, and
it may be that Scipio was less skilled at the deployment of works and
machines than he was at sudden assaults in which surprise and dash
overwhelmed the enemy.

Although the Romans had significantly improved in siege warfare during the Punic Wars, at the beginning of the second century B.C.
they still did not match Hellenistic capabilities in artillery. Indeed, the

Romans never developed the large arsenals that formed so important a part of Hellenistic armament production. Most Roman artillery was requisitioned from nearby cities or built by engineers on the spot. These methods could not produce the lavish batteries of artillery characteristic of Hellenistic armies.[112]

Livy remarked on the contrast between Roman siege methods and Hellenistic methods as evidenced in the siege of Oreus in 200 B.C., a joint operation of the Romans and King Attalus of Pergamum. The Romans deployed mantlets, sheds, and battering rams. King Attalus used a variety of artillery, some capable of hurling large stones. His troops also undermined the wall. None of these methods brought a rapid fall of the city, but after the long siege wore down the defenders, the Roman rams eventually breached the wall. Roman troops penetrated the breach during the night, and at daybreak King Attalus's troops also entered the city, where they had collapsed the wall by undermining. The defenders fled to the citadel, but exhausted and hungry, they surrendered after only two days.[113]

Nine years later, in another joint operation between the Romans and King Philip of Macedonia, the Romans besieged Heraclea, a strategic city near Thermopylae, while King Philip besieged the nearby town of Lamia. There was competition between the Romans and Macedonians over who should take their town first. Again the Romans relied on siege works and machines while the Macedonians concentrated on undermining. However, the Macedonian sappers ran into flint, which was too hard for their iron tools. Heraclea fell first, allowing the Romans to embarrass King Philip by requesting that he withdraw from Lamia and leave the city to the Romans. They added insult to injury by not bothering to continue the siege.[114]

These examples indicate that the Macedonian superiority in artillery did not give them any great advantage over the Romans in siege warfare in the early second century B.C. Despite shortcomings in artillery, by this time the Romans could almost always breach a wall. But forcing a breach remained a difficult task. In the siege of Atrax in 198 B.C., Roman rams opened a breach, but crack Macedonian troops defended the city and the Romans could not cross the rubble against them. Even after the Romans had cleared the debris away, they could not force the breach. The Macedonians drew up in their famous phalanx bristling with sarissas, and in the narrow confines of the breach the phalanx was impenetrable. The Romans tried to increase their firepower by bringing up a siege tower, but its wheels sunk in the soft ground, causing it to lean to one side and frightening the soldiers into abandoning it. With winter coming on, the discouraged Romans lifted the siege.[115] In the same year a Macedonian garrison at Corinth also faced down the Romans after rams had breached the wall, although this time the desperate courage of Italian deserters in the city contributed to the successful defense.[116]

The difficulty of forcing a breach meant that attrition remained the main means by which the Romans captured cities. For example, Heraclea, which lay on a plain at the foot of Mount Oeta, offered the Romans good access to all sides except toward the mountain, where a powerful citadel overhung the city. Nearby forests provided abundant material for siege machinery, so the Romans had plenty, and they battered the walls around the clock. The Roman commander, Acilius Glabrio, divided his army into four parts and mounted simultaneous attacks from four directions. Competition between the four sectors added spirit to the Roman assault as the soldiers vied to be the first to break into the city.

The Aetolians mounted a spirited defense of their city, but exhaustion took its toll. As the days wore on and the Aetolians became weary, both the number and the force of the sallies declined. Lack of sleep tormented the Aetolians because there were not enough of them to rotate in the defense against the much more numerous Romans, who were working in relays.

After twenty-four days of continuous operations, Acilius ordered his army to suspend operations for several hours each night. The defenders could not help but take advantage of the respite to sleep. Once he had established this routine, Acilius launched a sudden assault in the middle of the night. The Aetolians were only half prepared, and in the confusion the Romans were able to get into the city through breaches and by escalade. The people of Heraclea fled to the citadel, but they so crowded it that defense was scarcely possible. Exhausted and demoralized, they quickly surrendered.[117]

By the second century B.C. the Romans were hiring mercenaries when they needed specialized troops for a siege. In the siege of Same on the island of Cephallania off the coast of Aetolia, they recruited one hundred skilled slingers from coastal towns in Achaea. The beaches along the Corinthian Gulf provided an abundance of stones for the training of slingers. Moreover, the Achaeans had developed a sling with greater range and accuracy than the Roman slings. This enabled the Achaean slingers to direct an accurate fire of stones against the Sameans from a relatively safe distance. Livy claimed that the Achaean slingers were so accurate they could hit the enemy not only in the head but on any part of the face at which they aimed. The deployment of this corps of slingers greatly diminished the effectiveness and the frequency of the Samean sallies.[118]

The siege of the Greek city of Ambracia in 189 B.C. demonstrated the versatility and the limitations of Roman siege warfare at this time. Ambracia was strong and offered little hope of an easy victory. But, influenced by the advice of local Greeks that Ambracia offered an advantageous situation to a besieger, the consul Marcus Fulvius made a careful personal survey of the city. What he saw was sobering. A rugged hill with a strong citadel protected one side and a river another side. Four miles of strong walls reinforced these natural barriers. However, Fulvius could also see advantages. On the side away from the hill and the river, a plain offered broad access to the walls. The river provided an abundant supply of water and the forests around Ambracia contained ample material for the construction of siege machinery. These favorable circumstances suggested to Fulvius that formal siege operations were feasible.[119]

As was the Roman custom, Fulvius established two camps on the plain before Ambracia. He also built a redoubt on high ground opposite the citadel. His troops dug a ditch and constructed earthworks connecting the camps and the redoubt and enclosing the city. Before they completed these siege works, one thousand Aetolian reinforcements made it into the city. They had hoped to attack one of the camps but abandoned the plan as too risky, allowing the Romans to complete the circumvallation of Ambracia.

After completing his siege works and building siege machinery, Fulvius assaulted the wall of Ambracia with battering rams. He deployed five of them, three advancing along parallel lines over level ground and the other two attacking in different sectors. At the same time, the Romans attempted to pull down the battlements with long sickle-shaped grapplers.

Although the energy of the Roman assault terrified the Ambraciots, they fought back with effective countermeasures. Cranes dropped leaden weights, stones, and tree stumps on the rams. They used iron hooks to catch the grapplers and pull them over the wall down into the city, thereby breaking the poles and capturing the sickles. Frequent sallies, at night and in daylight, kept the Romans manning the siege machinery off balance. Five hundred additional reinforcements were even able to cross the Roman siege works and enter the city. These troops carried out an attack on one of the camps that the Romans repelled only with difficulty and heavy losses. The Aetolians had planned a simultaneous attack on the camp from troops on the outside, but something went wrong and that attack did not take place. The Romans would have been in serious difficulty if it had.[120]

The failure of the Aetolian troops outside the city to carry out the attack on the Roman camp discouraged the defenders in Ambracia, and their sallies from the city ceased. Now they fought from the wall tops, but the Roman rams succeeded in knocking down portions of the wall. However, the Romans were unable to penetrate through the breaches because the Ambraciots had built new walls behind the breaches or fought so determinedly that the Romans could not fight their way across the rubble.[121]

Frustrated by the failure to exploit the breaches his rams had made, Fulvius turned to undermining. First he built a covered gallery, about one hundred yards long, which ran parallel to the Ambraciot wall. The gallery protected the miners and concealed their labor. They worked in relays around the clock and made rapid progress. At first the gallery succeeded in preventing the Ambraciots from knowing what was happening, but when the pile of dirt the Romans had taken out became too large for the gallery to conceal the Ambraciots realized the danger and began countermining. They dug a trench behind the threatened sector of their wall. Then, lining the side of the trench with thin plates of brass, they listened for reverberations to determine the location of the Roman mine. They then began digging a countermine against the Roman mine. The Romans had already extended their mine under the wall, so it did not take long before the Ambraciots broke into the Roman tunnel.

A brief hand-to-hand combat in the tunnel proved that neither side could break through the other's armor because of the limited space. The Ambraciots then employed an ingenious device to smoke the Romans out of their mine. The device consisted of an earthenware jar large enough to fill most of the tunnel opening. An iron tube extended through a hole into the bottom of the jar. The jar was filled with feathers and a few hot coals. An iron lid full of holes covered its mouth. The Ambraciots pressed this device into the mouth of the tunnel, filling in all around it except for two holes through which they could thrust their spears to prevent the Romans from approaching the smoke machine. Bellows attached to the iron tube served to fan the coals, which ignited the feathers, and blow the smoke into the tunnel. The feathers produced an acrid smoke that drove the Romans out of the mine.[122]

Despite this success, the Ambraciots were still shut in by the Romans and their long-range prospects were grim. But after two weeks of furious assaults without success, Fulvius faced the prospects of a long siege. Thus when mediators intervened to try to end the siege, both sides were willing to negotiate. The Romans held the advantage and the terms gave them control of the city.[123] But despite employing the full range of siege methods, Fulvius had failed to take Ambracia by assault.

The siege of Cassandrea in 169 B.C. provides a final perspective of Roman siege capabilities in the early second century B.C. Cassandrea was a Macedonian town located on the Pallene peninsula. It was the former Potidaea, the town that had endured a two-year siege before falling to the Athenians at the beginning of the Peloponnesian War. The Romans failed to take it.

A moat protected the city, and its garrison included two thousand Illyrians and eight hundred Agrianes, both formidable warriors. King

Eumenes of Pergamum joined the Romans with a naval force, and he attacked from land and sea on one side while the Romans, after cutting the city off from the mainland with entrenchments, operated on the other side. The Romans filled in the moat, a task they accomplished only with difficulty.

Their commander, Gaius Marcius, learned that where the Macedonians had constructed arches in their wall, perhaps to create sally ports, the walls were only one layer thick. His plan was to breach the walls in these locations while mounting a general escalade at the same time. The Roman rams easily broke through the thin walls, but Marcius had failed to ready an assault team to force the breach. Instead his troops were scattered along the wall in their attempts to scale it.

As Roman officers began to gather their troops for an assault on the breaches, the Macedonians launched a counterattack with their Illyrian and Agrianian troops. The disorganized Romans broke into flight, but the moat blocked their retreat. Six hundred Romans died, and almost all those who got back across the moat were wounded.

After this disaster, Marcius, in Livy's acid remark, was "less eager to form other plans" and he turned to the slow task of constructing siege works. King Eumenes was making little progress in his sector, so he, too, began to work his way toward the walls with siege works. Marcius and Eumenes counted on keeping Cassandrea isolated while they built their siege works. But Eumenes anchored his ships so far away from the shore line that ten Macedonian scout ships carrying Gallic troops were able to reach Cassandrea in the dead of night by hugging the coast. This success discouraged the Romans enough to cause them to lift the siege.[124] Roman siege capabilities could be formidable under a commander the caliber of Scipio, but under an incompetent commander like Marcius sieges could end in ignominious failure.

Conditions of Siege Warfare

In a speech that Livy placed in the mouth of King Eumenes of Pergamum, Eumenes called siege the "most wretched fate in war."[125] Well he might, and sieges weighed most heavily on the common people. For example, when Hannibal approached the Italian town of Nola in 216 B.C., the nobles wanted to remain loyal to Rome, but the plebeians favored submission because they feared "the many hardships and indignities they must suffer in case of a siege."[126] Livy said that there was a general pattern in Italy during the Second Punic War for the plebeians to want to surrender and the nobles to want to remain loyal to Rome.[127] A Roman army saved Nola, a feat that cost the defeatist plebeian leaders their heads.[128] Nearby Accerae refused to surrender or to undergo a siege. While Hannibal was building his siege works, the men sneaked out of the city during the night and fled to other Campanian towns.

Livy was vague about the women and children, but presumably they were with the men.[129]

Livy's descriptions of hunger under siege leave little doubt why the plebeians were reluctant to endure sieges. A garrison at Casilinum, for example, which consisted mostly of Praenestines, suffered extreme deprivation while under siege by Hannibal during the Second Punic War. The siege had continued through the winter, and by springtime the defenders were so hungry some were committing suicide by jumping off the wall and others were deliberately exposing themselves to the enemy's missiles. Some relief came when the Romans floated jars of spelt down a river that ran through the town. However, the Carthaginians soon discovered what was happening and stretched a net across the river to stop the jars. Still the defenders held out, eating boiled leather, rats, and roots. When they planted turnips, Hannibal's patience reached its end and he offered terms. The town surrendered, and Hannibal freed its defenders on payment of a heavy ransom. Almost half of the 570 defenders had died in the siege. The Romans rewarded the tenacity of the Praenestine defenders by granting them double pay and five years' exemption from military service. They also offered them Roman citizenship.[130]

Another ally of Rome that endured great suffering under siege that year was the city of Petelia in southern Italy. Having consumed all the grain, the people ate all the animals they could find, "the familiar and the unfamiliar" according to Livy, and then lived on boiled leather, roots, bark, leaves and grass. Only when they no longer had the strength to stand did they surrender the city to the Carthaginians.[131] Appian said that as the food ran out the Petelians pushed those who were unable to fight out of the city to their death at the hands of the Carthaginians.[132]

Sieges were ordeals for the besiegers as well as the besieged. Livy called siege warfare "toilsome and dangerous."[133] The relief of the common Roman soldiers is palpable in Livy's description of their mood when, in 190 B.C. during the war against Antiochus, they entered the well-stocked Hellespontine town of Lysimachia, which Antiochus had boneheadedly abandoned in panic after he had suffered a crushing naval defeat in the Battle of Myonnesus:

> This gave him [Scipio Africanus] much greater joy than the naval victory, especially after they arrived there, and the city, filled with supplies of all kinds stored up as if for the arrival of the army, received them, where they had pictured to themselves the prospect of extreme want and hardship in besieging the city.[134]

Siege operations in winter were especially arduous, and commanders were reluctant to undertake them. Livy provided a vivid portrait of King Antiochus in a state of great uncertainty over whether or not to sustain his siege of the Thessalian city of Larisa over the winter, a "time unsuited to all military operations and particularly to the siege and storming of cities." The approach of a Roman army decided the issue, and he lifted the siege, glad to have winter as a plausible excuse.[135]

Scipio Africanus tried a winter siege in Spain in 218 B.C. when he besieged a town near the Ebro River and learned the hard way how difficult it could be. Four-foot-deep snow covered the siege machinery. Livy said that the snow protected the machinery from incendiaries, but one can imagine the difficulty of operating siege machinery under such conditions. Scipio maintained the siege for thirty days and then accepted twenty talents to lift it.[136] Scipio Aemilianus conducted a twenty-two day siege of Nepheris in North Africa under winter conditions. Although the town fell, the Roman troops endured great hardship.[137]

Picked troops usually led assaults on walls. During the siege of Arpi in 213 B.C., for example, Quintus Fabius selected the bravest military tribunes and the best of the centurions to lead six hundred men on a nighttime escalade to capture the gate. With the help of a convenient downpour that drove the watchmen from the walls and muffled the sounds of the attack, they were entirely successful.[138] Specially trained troops were also available for scaling duties. These men were "light and nimble" and could scale cliffs by driving iron pitons into the rock for hand and footholds.[139] When Scipio departed for Africa in 204 B.C., he recruited his army from veterans of the siege of Syracuse because he was anxious to have troops experienced in siege warfare.[140]

The tedium and danger of sieges required high morale, something Roman commanders recognized early on. Prizes and plunder were the main incentives for Roman soldiers to assault fortified towns. The first man to mount a wall in an assault on a city received the mural crown of gold (*corona muralis*), the most valuable prize awarded for individual gallantry.[141] The hope of plunder remained the greatest incentive for siege troops.

Customarily war booty was put up for public auction, the proceeds going into the Roman treasury. But siege commanders often found it impossible to withhold any of the booty from their soldiers. In 415 B.C., a military tribune named Marcus Postumius Regillensis caused a mutiny by first exaggerating how much booty his men would find in the town of Bolae if they took it and then breaking his promise that the soldiers could have all of the booty. Postumius's clumsy and heavy-handed response to the mutiny exacerbated the situation, and his soldiers killed him.[142] About the same time, the consul Gaius Valerius refused to distribute the booty from the citadel of Carventum to the troops who had taken it by assault. Instead he held a public auction. As a result, the soldiers roundly booed Valerius when he entered Rome in an ovation decreed by the Senate.[143] In both cases, the failure to meet the troops' expectations for booty exacerbated civil tensions in Rome. In the latter case, there were serious military consequences as well, because the Romans lost the citadel of Carventum when the soldiers whom Valerius had deprived of their loot abandoned the citadel to plunder the countryside in search of what they thought was rightfully theirs. A long siege failed to recapture Carventum.[144]

In contrast, when the Roman commander Numerius Fabius Ambustus carefully allowed all of his troops an equal opportunity to plunder the rich town of Anxur, his solicitude for their welfare went a long way toward reconciling patricians and plebeians.[145]

Soldiers often took matters into their own hands. After the capture of Contenebra in 388 B.C., the military tribunes decided to turn over the booty to the state, but by the time they had issued the order the soldiers had already seized the booty for themselves and it "could not be taken away without offending them."[146] In another case, after the Romans captured the Volscian town of Satricum, the commander gave all the booty to the soldiers except captured slaves. These he sold for the benefit of the public treasury.[147]

Another example of how delicate the problem of booty could be occurred at a much later time, in 171 B.C., when a Roman army besieged two towns in Illyria. One of them surrendered, and the Roman commander rewarded it by allowing its inhabitants to retain all their possessions. He hoped this example of clemency would move the other town to surrender as well. When it held out against the Roman blockade, the Romans decided to abandon the siege. But fearful that his troops would be unhappy to have toiled in two sieges with no reward, the commander violated custom by allowing them to plunder the town that had surrendered.[148]

When Marcus Furius Camillus knew Veii was about to fall, he referred the question of booty to the Senate rather than decide himself how the riches of Veii should be distributed. His position was expressed in the Senate by his father, Publius Licinius: Rome should invite all the citizens to go to Veii (it was only nine miles away) and share in the plunder. Camillus had relied less on the promise of booty to motivate his troops than the threat of punishment. When he took command of the demoralized army at Veii, he had severely punished all soldiers who had fled to show that he would not tolerate undiscipline and cowardice.[149] Appius Claudius spoke against allowing the citizens to share in the booty, arguing that it would deprive the soldiers of their just deserts. He favored placing the booty in the public treasury but thought that if this were not done, at the very least it should be used to pay the soldiers. Such a step would reduce the war tax, thus benefiting all the citizens without depriving the soldiers. Licinius responded that such a policy would arouse the jealousy of the people. If the people shared directly in the booty, they would gain quicker relief from the economic hardship of the war and the psychological impact of victory would be greater.

Fearing the anger of the people, the Senate voted to allow them all to go to Veii to share the booty. Livy does not tell us how the soldiers reacted to this decision.[150] An awkward point arose when it came time to dedicate a tenth of the booty to Apollo, as Camillus had promised. The pontiffs ruled that the people were obligated to fulfill Camillus's vow but allowed the people to appraise the value of their booty themselves. The contribution also was made voluntary. Nevertheless, there was considerable resentment toward Camillus for his vow, and the festering controversy continued until he went into exile.[151]

The question of booty related not only to morale but to discipline as well. Looting and indiscipline went hand in hand. How could a commander satisfy his troops' desire for booty and yet maintain discipline? Plundering soldiers were terribly vulnerable to attack, and many a victory turned into defeat when the troops scattered to get their share of the booty. When Roman troops approached the town of the Privernates in 357 B.C., their commander, Gaius Marcius Rutulus, attempted to avoid the chaos of a sack by promising them the booty on the condition that they maintain discipline:

> I give you now for booty the camp and the city of our enemies, if you promise me that in battle you will play the part of men, and be not more ready to plunder than to fight.[152]

Just how dangerous greed could be to troops was reflected by an incident in the First Punic War. The Romans were operating against Therma, a town in Sicily, when they captured a gatekeeper who had left his post to relieve himself. The gatekeeper negotiated his freedom by promising to open the gate to the Romans during the night. One thousand troops approached the city at the appointed time and the gatekeeper, true to his word, admitted them. The officers, men whom Diodorus called "men of note," ordered the gatekeeper to lock the gate behind them so no other Roman soldiers could follow them in for a share of the loot. That proved to be a miscalculation of the first order, because the townspeople fell on the Roman troops and killed them all. The story is somewhat implausible, but Diodorus clearly thought it was instructive.[153]

Roman commanders often awarded all the booty to troops who conducted themselves with distinction. For example, in 324 B.C. Junius Brutus Scaeva awarded all the booty to his troops who had taken the Vestinian cities of Cutina and Cingilia by assault "because neither the gates nor the walls of the enemy had held them back."[154] Against Bovianum, the capital of the Pentrian Samnites and a rich city, Livy explicitly said that the Roman soldiers had no special animus toward the city but that "the hope of plunder spurred them on to capture it," even though it was strongly defended. Not surprisingly, the commander thought it wise to allot all of the booty to the soldiers despite its remarkable abundance, a decision Livy called generous.[155]

During a war in Samnium in 296 B.C., having driven the Samnite army into Etruria, the Romans began a campaign against the most strongly fortified Samnite towns. Livy said that the promise of spoils was so great that the Romans soldiers dispensed with siege operations and stormed the cities by escalade, taking two of them in the initial assault. A third city, Ferentinum, offered greater resistance, but "all obstacles were overcome by a soldiery grown used to plunder." Their commander, Publius Decius, brought in traders so the soldiers could sell the enormous booty.[156] With a glut of booty on the market, merchants were often able to reap huge profits by purchasing it at a fraction of its worth. During the African campaign at the end of the Second Punic War, Publius Scipio had to ban from the Roman camp some merchants who were taking undue advantage of the soldiers.[157]

By the Second Punic War, a standard procedure seems to have established itself. Polybius described how it worked after the fall of New Carthage. The military tribunes assigned men from each maniple to gather booty. (A legion consisted of thirty maniples; one maniple contained about 120 to 200 soldiers.) Never more than half of the men were detailed to collect booty; the rest remained on guard duty. Every man had taken an oath before the campaign to share the booty equally. When the pillagers gathered the booty together in a central location, the military tribunes divided it equally among all the soldiers, including those who had remained on guard duty, the sick and wounded, and those who were absent on special assignment. In this way the Romans attempted to gather and distribute the booty in an orderly and equitable way so that the army would not degenerate into an undisciplined mob during the sack of a city.[158]

Polybius greatly admired the Roman system and explained the perils that it avoided:

> For since most men endure hardship and risk their lives for the sake of gain, it is evident that whenever the chance presents itself it is not likely that those left in the protecting force or in the camp will refrain, since the general rule among us [Polybius means to contrast the Romans with the Greeks] is that any man keeps whatever comes into his hands. And even if any careful king or general orders the booty to be brought in to form a common fund, yet everyone regards as his own whatever he can conceal. So that as most of the men start pillaging, commanders cannot maintain any control and run the risk of disaster, and indeed many who have been successful in their object have, after capturing the enemy's camp or a town, not only been driven out but have met with complete

disaster simply for the above reason. Commanders should therefore exercise the utmost care and foresight about this matter, so that as far as is possible the hope of equal participation in the booty when such a chance presents itself may be common to all.[159]

Nevertheless, Roman troops continued to be difficult to control when opportunity of spoils presented itself. Tacitus captured their mentality perfectly in describing the thoughts of Roman soldiers besieging the city of Cremona in A.D. 69:

> If we wait for the light, we shall be met with entreaties for peace, and in return for our toil and our wounds shall receive only the empty satisfaction of clemency and praise, but the wealth of Cremona will go into the purses of the legates and the prefects. The soldiers have the plunder of a city that is stormed, the generals of one which capitulates.[160]

A final condition of siege warfare was the presence of women. Appian, in his account of the siege of Petelia, even claimed that the women joined with the men in making sallies against the Carthaginians and were "no less manly than the men."[161] If the story were true, we would have an extraordinary example of women's involvement in siege warfare. That neither Polybius nor Livy mentioned such an amazing event forces us to conclude that the story is not true.

Livy provided a more plausible example of the role of women in siege warfare in his account of the Roman effort to take the Illyrian town of Uscana in 170 B.C. The Roman commander Appius Claudius, having received word that traitors were ready to betray the city to him, approached it very carelessly. Finding the walls deserted, the Romans assumed there would be no resistance. But when they came within missile range, the men of Uscana burst out of two gates at once and attacked the unprepared Romans. The women added to the din of the attack by mounting the walls and howling and clashing bronze together. Slaves added their cries to the noise. The surprise and the noise unnerved the Romans, who turned and fled toward their camp. It was twelve miles distant, and pursuing Illyrians cut down many of the fleeing Romans. Appius was so frightened he did not wait to collect his scattered troops but abandoned his camp immediately. The women of Uscana had contributed to a rout that cost the lives of half the Roman force of four thousand men.[162]

When the Romans attacked the Gallic town of Gergovia during Caesar's campaign in Gaul, the women at first panicked. Terrified women swarmed on the walls throwing silver and clothes down and begging with bared breasts for the Romans to spare them. Some even lowered themselves from the wall to surrender to the Romans. But when Gallic soldiers who had been building fortifications on the other side of the city rushed up to attack the Romans, the women stopped appealing to the Romans and now urged on their men in the Gallic custom by disheveling their hair and holding forth their children. Thus encouraged, the Gauls routed the Romans in a battle in which forty-six centurions and almost seven hundred soldiers died.[163]

As we have seen many times, though, the most common role for women in siege warfare was that of laborers helping to repair walls and carry arms to the men. Despite the many examples of women providing indispensable help in a siege, their presence could disturb the men. When the Romans breached the wall of Jotapata, the Jewish women began a terrible wailing. Josephus, the Jewish commander, shut them up in their houses because he feared the keening would "unman" his soldiers. Clearly, up to the moment of the breach, the women had not been in their houses but had been playing an active role in the defense of the city.[164]

Chapter XII

THE AGE OF IMPERIALISM

The Siege of Carthage

A new era in Roman history began with the destruction of Carthage in 146 B.C. The elimination of Rome's old Mediterranean rival sealed the destiny of Rome as an imperial power. The three-year siege that was necessary to bring about the fall of Carthage reflected the truth that had existed since the first empires in Mesopotamia: ruling an empire required the ability to conquer powerfully fortified cities.

Carthage was no longer the power it had once been and made every effort to avoid war, including surrendering three hundred noble children as hostages and even agreeing to surrender all of its arms, including no less than 200,000 sets of armor and 2,000 catapults.[1] Only when these extreme measures failed to deter Rome from its determination to transport the Carthaginians to an inland site and raze the city did they resolve to resist. It was a desperate decision. Carthage had just lost a costly war with Numidia. It had surrendered all of its arms to the Romans. And its old ally Utica had defected to the Roman side, giving the Romans a secure base of operations only thirty miles from Carthage, an advantage that Scipio Africanus had not enjoyed in his African campaign at the end of the Second Punic War.

Despite these advantages, the old city proved a tough nut for the Romans to crack. The sea washed against a precipice on three sides of the city, which was located on a peninsula in the Gulf of Tunis; only an isthmus three miles wide connected it with the mainland (map 8). A single wall provided sufficient protection on the seaward sides. A triple wall across the isthmus protected the landward side. A low outer wall ran along the inside of a moat. A higher middle wall overlooked the

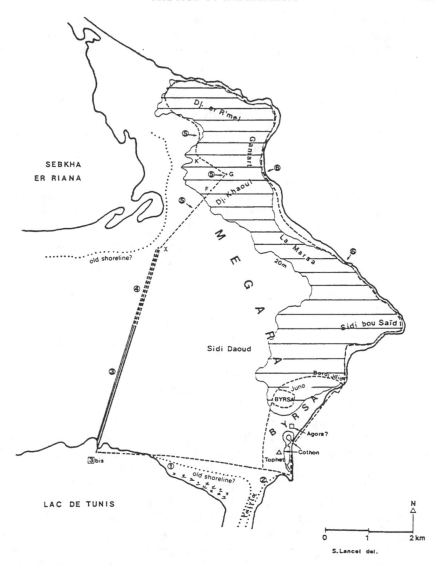

Map 8. Roman siege of Carthage. *From Serge Lancel,* Carthage, a History. *Copyright Basil Blackwell, Ltd., 1995. Used with the permission of the publisher.*

moat. The inner wall was massive, rising to a height of forty-five feet, not counting the battlements that lined the top. Every two hundred feet towers four stories high rose above the wall. The wall itself was thirty feet thick and contained stables and barracks on two levels capable of housing three hundred elephants, four thousand horses, twenty thousand infantry, and four thousand cavalry and storing their fodder and food. The city had no aqueduct, and so to keep its inhabitants supplied with water the Carthaginians had dug a series of giant cisterns to hold water from the winter rains.[2] This system worked better than an aqueduct in time of siege, because the Romans could have easily cut an aqueduct. The southern side of the city bordered on a lake and, as on the seaward sides, presented a single wall. A tongue of land, called Taenia, extended southward from the southeast corner of the city, dividing the lake from the sea. Appian said that this was the only place where the fortifications were weak, the wall being low and neglected. As in most cities of this era, the walls encompassed the suburbs; their circumference was about twenty-one miles.[3]

The Roman commanders, the two consuls of that year, Marcius Censorinus and Manius Manilius, were mediocre soldiers, and events soon proved them unequal to the task. Perhaps because of Carthage's strong fortifications, they moved slowly to attack, hoping that the Carthaginians would come to their senses and acquiesce to their fate. This delay gave the Carthaginians the opportunity to prepare for the siege. They freed their slaves to ensure their support and started a crash program to rearm the city. Carthage was a great mercantile city, and now not only its many artisans but women as well began working around the clock to produce arms. Every public place, including the temples, became workshops, and the men and women worked in relays. Their daily production reached one hundred shields, three hundred swords, one thousand catapult missiles, five hundred arrows and spears, and "as many catapults as they could [make]." Women cut their long hair to use for making the ropes that provided torsion for the catapults.[4]

The Roman delay also gave the Carthaginians time to organize logistical support from outside the city. It so happened that one of their more able commanders, Hasdrubal, was in the interior gathering an army because pro-Roman political enemies in Carthage had secured a death sentence against him. With the very existence of Carthage at stake, the Carthaginians appealed to him to turn his army against the Romans instead of his city, pleading that they had condemned him only under duress from the Romans. Being an inveterate enemy of Rome, Hasdrubal needed little persuasion and, with thirty thousand men under his command, was able to secure a supply line from the interior to Carthage. At the same time the presence of a large hostile enemy in the interior caused Rome considerable logistical difficulty.[5]

When Censorinus and Manilius realized that their logistical position was precarious, they moved to assault the city, believing they could relieve their situation by taking an unarmed city quickly and easily using only scaling ladders. Manilius advanced across the isthmus against the triple wall. He planned to fill in the moat, cross the lower walls, and attack the main wall by escalade. Censorinus attacked the soft underbelly of the city along the lake. His troops attempted to place scaling ladders against the single wall in this sector, resting them on ships

where the water came up to the wall and on land in places where there
was a narrow strip of land between the wall and the lake.

Manilius and Censorinus had completely underestimated the resourcefulness of the Carthaginians. Their newly built arms caught the
Romans by surprise and forced them to regroup. In a second attack,
Manilius succeeded in crossing the moat and breaching the outer wall,
but the strength of the Carthaginian fortifications soon convinced him
of the futility of attacking from the isthmus, and he withdrew. Censorinus had no better luck in his sector. Realizing that a full-scale siege
would be necessary and fearful of Hasdrubal's army operating in their
rear, the Roman commanders encamped their army. Manilius built a
fortified camp on the isthmus and Censorinus established his camp on
the shore of the lake.[6]

The Romans scarcely had room to operate along the shore of the
lake, and so Censorinus ordered his troops to fill in a part of the lake
to provide more space in which to deploy siege machinery. He also sent
a large number of troops across the lake to search for timber with which
to build the machinery. The danger that Hasdrubal's troops presented
became all too evident when a cavalry force under Himilco Phameas,
a very able Carthaginian officer, attacked these Roman troops as they
were felling trees. The Romans lost five hundred men, but those who
escaped managed to bring back enough timber to build two huge battering rams. The successful completion of the artificial shore enabled
the Romans to bring these rams forward. Six thousand infantry under
the military tribunes operated one of the rams and sailors under their
captains operated the other. The use of the sailors not only provided
manpower; it also helped motivate the troops by setting up a rivalry
between the army and the navy.[7]

The rams soon knocked down a part of the wall, and the Romans
could see into the city. But the Carthaginians stoutly defended the
breach and during the night began the construction of a wall to cover
the breach. They knew, though, that this hastily built wall whose mortar was scarcely dry could not withstand the Roman battering rams,
and so they also attacked the rams under cover of darkness. Although
the Romans beat the attack back, the Carthaginians damaged the rams
with fire badly enough to put them out of commission.

The situation that confronted the Romans the next morning was that
their rams were disabled but, as they could see, the Carthaginians had
not yet completed their new wall. They could also see that many of the
Carthaginians had no arms beyond the stones they had collected to
throw at the Romans.

With Carthaginian troops threatening foragers, a long siege promised to be a hungry one, and the Roman troops spontaneously surged
into the breach, desperate to seize what seemed like the last chance
to capture Carthage quickly. Only one man remained cool. This was

Scipio Aemilianus, one of the military tribunes. He recognized the dangers of an undisciplined charge into a city and kept the troops under his command out of the city, stationing them at intervals along the wall. When the Roman troops soon began pouring out of the city in even more disorder than they had gone in, it was Scipio's troops who fell upon the pursuing Carthaginians and prevented a major disaster.[8]

The bane of besieging armies, sickness, soon afflicted the Roman troops in Censorinus's camp on the shore of the lake, whose waters were shallow and stagnant under the hot summer sun. Hoping to escape the miasma of the lake, Censorinus moved his camp in July to the seaward side of Taenia, very close to the main Carthaginian harbor. The fleet also moved from the lake to the sea in order to support the new camp. This station left it vulnerable to burning ships, which the Carthaginians launched against the Roman fleet whenever there was a wind favorable to this purpose. According to Appian, the fire did great damage and "came a little short of destroying the whole fleet." By autumn the Romans had accomplished nothing, and Censorinus was probably glad enough that his consular duties called him to Rome to conduct elections.[9]

The departure of Censorinus and the weakening of the Roman fleet caused his camp to abandon active siege operations. That left the Carthaginians free to attack Manilius's camp on the isthmus. To achieve surprise they attacked in the middle of the night. Some of the Carthaginians carried planks rather than weapons in order to throw them across the ditch that protected the Roman camp. Manilius had only erected a palisade, and the Carthaginians were able to tear it down after they crossed the ditch. This attack threw the Romans into confusion, but once again Scipio, who had apparently transferred to the camp of Manilius, came to the rescue. He took his troops out of the west gate, circled the camp, and counterattacked the Carthaginians, driving them back to the city. A chastened Manilius now built a wall around his camp and constructed a fort on the north shore of the isthmus where his supply ships put in.[10]

With winter approaching Manilius began sending out foraging parties in force, deploying ten thousand troops and two thousand cavalry for the task. Despite such large forces, Phameas continued to harass the Roman foragers with considerable effect. Accordingly Manilius decided to attack Hasdrubal's base at Nepheris, which lay a little more than twenty miles southeast of Carthage. The road to Nepheris presented many opportunities for ambush, and the inexperienced Manilius soon fell into a Carthaginian trap from which only more heroics by the ubiquitous Scipio extricated him. Even so, three military tribunes died in this ill-fated expedition. The Carthaginians in the city also mounted an attack against Manilius's fort on the coast. It failed, but Manilius's precarious logistical position was all too evident.[11]

Eager to atone for the bungled march against Nepheris, Manilius moved against the town again, but again he had to return having accomplished nothing. The defection of the Carthaginian cavalry commander Phameas offered some encouragement, but at the end of the year the Roman position was no better than it had been at the beginning.[12]

The next spring (148 B.C.) a new consul, Gaius Calpurnius Piso, arrived to take command. In light of his predecessors' difficulties, he did not attack either Carthage or Hasdrubal in his stronghold at Nepheris. Instead he concentrated on capturing the coastal towns from which privateers were harassing Roman supply ships. He was mostly unsuccessful. Neapolis surrendered to him, but he razed it in violation of the terms of surrender. This piece of treachery determined the other towns, most notably Aspis (Clupea) and Hippagreta (Hippo), fiercely to resist him. A combined assault from land and sea at Aspis failed. Piso then moved to Hippo, but it withstood a summer-long siege and he went into winter camp at Utica having accomplished very little.[13] The Carthaginians remained in control of the countryside; they had now completed their rearmament; the Numidians who had promised to help the Romans when they expected a quick victory were now dragging their feet. Not surprisingly morale was high in Carthage.[14]

The lack of success at Carthage led to the election of Scipio Aemilianus as consul the next year (147 B.C.) and his assignment to the Carthage command, even though he was below the legal age for that office. He had gained his reputation rescuing incompetent Roman commanders, and upon his arrival back in Africa that necessity immediately confronted him again.

The very day that Scipio arrived at Utica, the Roman admiral of the fleet, Lucius Mancinus, launched an amphibious assault against Carthage. On patrol off the coast Mancinus had noticed that a section of the Carthaginian wall that ran along a particularly steep precipice was poorly guarded. He believed that men with scaling ladders might be able to ascend the precipice and get over the wall unnoticed by the Carthaginians. Picked sailors boldly climbed the ladders to the wall, but Carthaginian guards discovered them. Unwisely the few Carthaginian defenders at the point of the Roman landing impetuously attacked the Romans through a nearby gate, even though the Romans outnumbered them. The Romans routed the Carthaginian sally and pursued so hotly that they burst through the gate before the Carthaginians could close it. Mancinus, whom Appian called "rash and giddy by nature," was so excited by this success that he rushed forward with a disorganized force to exploit the opportunity. Most of the men with him were unarmed or poorly armed sailors. They established themselves in a precarious position next to the wall.

Only when night fell did it dawn on Mancinus how easily the

Carthaginians would be able to push the Romans off the precipice the next morning once reinforcements arrived. Only five hundred of the 3,500 men Mancinus had landed were armed. Even if they could maintain control of the gate, and that seemed doubtful, they had no food. Mancinus's desperate messages for help arrived in Utica at the same time as Scipio.

Scipio wasted no time setting to sea with a rescue force. He arrived early the next morning to find the Romans in a desperate situation. The Carthaginians were about to push them off the precipice, and Mancinus was wounded. Scipio's arrival gave the Romans one last burst of energy that enabled them to get down to the shore, where he took them aboard and carried them to safety. Mancinus was among the survivors. He returned to Rome in ignominy, perhaps wishing that Scipio had arrived one day earlier.[15]

Scipio immediately gave Roman strategy against Carthage more direction. The Romans had conspicuously failed to isolate Carthage from the interior. Scipio established a new camp on the isthmus, this one closer to the city than Manilius's in order to close off Carthage's land link with the outside world. This move forced the Carthaginians to establish a camp of their own less than a mile from their walls to keep that link open. The link was so crucial that the Carthaginians had to draw six thousand infantry and one thousand cavalry from the interior to reinforce their camp. Hasdrubal himself came to the camp and now assumed overall command of Carthage's defense.[16]

The Roman army over which Scipio took command was in shambles. Piso exerted no discipline, and the soldiers had grown accustomed to roaming the countryside for plunder or lounging idly in camp. The booty had attracted a multitude of camp followers who hoped to enrich themselves. Some of these men even joined the soldiers on plundering forays. The restoration of discipline was Scipio's first task.[17]

Scipio made plain to his soldiers what he would not tolerate:

> You are more like robbers than soldiers. You are runaways instead of guardians of the camp. Avarice has made you more like a set of holiday-makers than a besieging army. You are in quest of luxuries in the midst of war and before the victory is won. . . . I have come here not to rob, but to conquer, not to make money before victory, but to overcome the enemy first.[18]

He then banished from the camp all the hangers-on except for the bare minimum necessary to trade with the soldiers. He restricted this trade to specific times when the commanders could supervise it. He announced that he would enforce the regulation that prohibited a soldier from leaving camp without permission. Foragers could only bring back food, and this should be simple food befitting an army, not delicacies more appropriate for a banquet. By acting swiftly and decisively, Scipio overawed the soldiers and won their confidence.[19]

Having established discipline in his army, Scipio turned to the problem presented by the opposing Carthaginian camp. In order to complete his investment of Carthage he had to eliminate that camp. Scipio chose an indirect method. He gathered a force of four thousand men on a dark night and soundlessly approached the Carthaginian wall, probably on the northwest side near the sea, about three miles from the Roman camp, where there was just a single wall. The men carried scaling ladders, axes, and crowbars. Scipio divided his force to attack at two points. He planned to catch the Carthaginians by surprise and break into the suburb of Megara, which he hoped would panic the Carthaginians and draw them out of their camp in a rush to defend the city itself.[20]

The plan started badly when Carthaginian sentinels detected the Romans and sounded the alarm. The Romans tried to create confusion with loud shouting, but the attacks on the wall failed. Unfortunately for the Carthaginians, a deserted tower stood very near the wall. For the Carthaginians not to have torn down this tower was incredibly negligent. Perhaps the remoteness of the wall in this sector and its closeness to the sea made them complacent that no attack would be mounted in this area. In any case, elite Roman troops climbed the tower, threw down planks across to the wall, and fought their way onto the wall. They were able to open a gate and admit the rest of the Romans.

The Roman breakthrough into Megara threw the Carthaginians into confusion, and they retreated to the old city quarter of Byrsa in the southeastern part of Carthage near the harbor. The Carthaginians in the camp heard the commotion and, realizing that the Romans had broken into the city, rushed into Byrsa to prevent the fall of the city. They did not know that only four thousand Romans were in Megara and that Scipio had no intention of making a nighttime pursuit of the enemy through Megara, which was a bewildering maze of orchards, walls, irrigation ditches, and hedges. His purpose had been to draw Hasdrubal out of his camp, and with this accomplished, he withdrew his soldiers from Megara. The next day the Romans occupied the abandoned Carthaginian camp and burned it.[21]

Scipio now moved methodically to complete the investment of Carthage. He began to traverse the isthmus with a trench. His army worked in relays day and night. The trench ran very close to the Carthaginian wall, and the men had to work and fight at the same time along the entire three-mile front. After they completed this trench, the Romans dug a parallel trench not too far behind the first one and then connected the trenches with two more trenches, thus forming an entrenched quadrangle. The Romans filled the trenches with sharp stakes. They reinforced the trenches with palisades. Finally they built a wall twelve feet high and six feet thick along the entire frontage facing Carthage. Battlements crowned the wall and towers strengthened it at intervals. In the middle rose the highest tower upon which stood a wooden tower four stories high to serve as an observation post into the city. Scipio's soldiers completed this work in twenty days.[22]

In the wake of these Roman successes, dissension broke out in Carthage. The morning after the Roman attack on Megara, Hasdrubal brought some Roman prisoners to the top of the wall and, in sight of the Romans, tortured them by tearing out their eyes, tongues, tendons and genitals with iron hooks. He lacerated the soles of others or cut off their fingers. Some he flayed, a torture not reported in our sources since the days of the Assyrian Empire. Having inflicted these cruel tortures on them, Hasdrubal threw them from the wall alive. His intention was to dash the hopes of any Carthaginians who might be contemplating surrendering to the Romans in return for mild treatment. To avoid the terrible revenge that such crimes would provoke, he believed the Carthaginians would have no choice but to fight to the bitter end. Instead a dispiriting sense of dread seized the Carthaginian people, and they bitterly resented Hasdrubal's crimes. His political enemies in the Carthaginian senate denounced him. To secure his shaky authority, Hasdrubal had them arrested and executed. He now ruled as a tyrant through terror.[23]

Hunger soon added to the internal tension in Carthage. The completion of the Roman entrenchments cut Carthage's land link with the interior and prevented their farmers from reaching the fields near the city. Foreign merchants could no longer reach Carthage. Blockade runners brought in by sea the only food reaching Carthage, but only when the weather allowed it and even so it was a risky business. These meager supplies Hasdrubal distributed to his thirty thousand troops. The rest of the people went hungry, except for what they could glean from the gardens and orchards of Megara.[24]

Scipio next moved to tighten the naval blockade of Carthage even further by blocking the entrance to the harbor. The harbor was located in the southeastern corner of the city. Its mouth was formed by a wide quay on one side and the tongue of land called Taenia on the other. Scipio wanted to build a mole from Taenia to the quay, thus blocking the mouth of the harbor. This was a huge project. The mole had to be strong enough to withstand the waves breaking against it, so the Romans built it with large stones that they dropped in the sea. Almost one hundred feet wide at its base and twenty-four feet wide at the top, the mole required so many stones that the Carthaginians at first did not believe the Romans could complete it, or at worst, that it would take them a very long time. Once again, however, Scipio's ability to organize labor and motivate his men to work day and night in relays brought rapid progress.[25]

The prospect of losing the use of their harbor moved the Carthaginians to a stupendous effort. They began to excavate a new entrance to the harbor from the open sea where the depth and current would be too great for the Romans to build another mole. At the same time they scoured the city for old materials from which to build triremes and quinqueremes. Almost miraculously the Carthaginians were able to conceal these projects from the Romans. The women and children labored along with the men, despite their hunger and even malnourishment.

When all was ready the Carthaginians opened their new harbor entrance early one morning and sailed out with fifty triremes and a flotilla of smaller vessels. This fleet, which appeared to the Romans to materialize out of thin air, caught the Roman navy completely unprepared. The sailors and rowers were laboring on the mole, and if the Carthaginians had fallen upon the Roman ships they could have destroyed the fleet. But instead they simply showed off their new fleet and then sailed back into the safety of the harbor. The only explanation that Appian could provide for this lapse was that Carthage was destined to fall. Carthage had not fought a naval war for a long time, and it may be that the Carthaginian commander believed his untrained crews needed some practice before entering battle.[26]

By the time the Carthaginians sailed out again three days later to challenge the Romans, the Roman ships were fully prepared. An indecisive naval battle lasted until evening, when the Carthaginians decided to withdraw. But the numerous small boats which had supported the Carthaginian triremes, in their inexperience and haste clogged the entrance to the harbor. The triremes could not enter and had to dock along the commercial quay that extended to the south and formed the original entrance to the harbor. To protect themselves they anchored with their bows facing the sea. Here the Roman ships attacked, although they found it difficult to turn around once they had carried out the attack until some allied ships demonstrated a technique of anchoring a rope and then pulling themselves away after attacking, thereby eliminating the need for a hazardous turning maneuver. This tactic enabled the Romans to destroy a number of Carthaginian ships before the Carthaginians finally cleared the entrance to the harbor and retired.[27]

Scipio next decided to attack the commercial quay. It was wide and commanded both entrances to the Carthaginian harbor. At the beginning of the war the Carthaginians had built a low wall to protect the quay, but the Romans moved battering rams across the mole they had built to the quay and soon knocked down a part of the wall. The Carthaginians responded to this threat with a desperate nighttime sally against the Roman rams, which the Romans had apparently drawn back across the mole to Taenia. To reach the rams the Carthaginians had to wade or swim across the water, which meant that they could wear no armor. Naked and carrying unlit torches, they made their way to Taenia, where they lit the torches and set fire to the rams. Because they had no armor to protect them, the Carthaginians suffered heavy casualties, but they carried out the attack with such frenzy that the frightened Romans soon broke into flight. Only ruthless action by Roman officers stemmed the Roman stampede. Scipio himself cut down several fleeing Roman soldiers. The Romans took refuge in their camp, and the Carthaginians burned the Roman siege machinery and swam back home.[28]

The Carthaginians' triumph was short-lived. It gave them time to repair their damaged wall on the quay and reinforce it with several wooden towers, but the Romans built new rams and attacked again. They piled up mounds in front of the towers and threw firebrands on them. Other towers they set afire with burning ships. The fire drove the Carthaginians off the quay. Appian gruesomely reported that the Romans were unable to pursue because the quay was too slick with blood. This attack left the Romans in possession of the entire quay, which they fortified. At the north end next to the city wall they built a brick wall of equal height from which they were able to throw javelins and shoot arrows at the Carthaginian defenders. Scipio stationed four thousand men on the quay, which was now firmly in Roman hands.[29]

Winter was now at hand, but Scipio did not rest. He began a campaign in the countryside to bring it under Roman control. Nepheris was his main objective. A strongly fortified Carthaginian camp guarded Nepheris. Scipio established a Roman camp only a few hundred yards from the Carthaginian camp. Moving back and forth between Carthage and his camp, Scipio supervised both sieges. Having demolished the wall that surrounded the Carthaginian camp in two places, Scipio sent three thousand elite troops against the breaches. They attacked in waves, so that those in front could not retreat because of the press of men coming up behind them. Another force of one thousand men circled around to the other side of the Carthaginian camp; and while the Carthaginians were preoccupied with the main Roman assault, this force broke into the camp, setting off a panic among the Carthaginians and putting them to flight. Numidian cavalry pursued them remorselessly, and Appian claimed that seventy thousand Carthaginians died in this battle. Another ten thousand were captured and only about four thousand escaped the carnage. The capture of the Carthaginian camp opened the way to Nepheris, which the Romans took in a twenty-two-day siege during the dead of winter. The winter weather caused great suffering among the Roman troops.

The fall of Nepheris led to the collapse of resistance in the countryside. Roman siege works, the blockade, the loss of the interior, and winter weather left Carthage completely isolated. The twelfth-century Byzantine historian Zonaras wrote that famine now afflicted the city so severely that people were dying and there were even cases of cannibalism.[30]

The following spring (146 B.C.) Scipio began the final assault against Byrsa, the old city which was the most strongly fortified part of Carthage, and Cothon, the harbor area. The assault began from the quay that the Romans had captured the previous autumn. Cothon was a rectangular commercial port which led to a circular port where the warships docked. Hasdrubal expected that the Romans would assault the commercial port, and during the night he set fire to the wooden warehouses that lined the Carthaginian wall. In the confusion caused by the fire, a Roman force made its way around to battleship row. A double wall protected the battleships, but the Romans scaled the first wall and bridged the space to the second wall with planks, scaffolding, and even siege machinery. Hunger and demoralization weakened the Carthaginian resistance, and Cothon was soon in Roman hands.

From there Scipio immediately moved troops into Byrsa to occupy the marketplace. Darkness was falling, so he kept these troops in the marketplace during the night. The next morning he brought four thousand fresh troops into the marketplace to prepare for an assault on the citadel in Byrsa, where most of the Carthaginian population had taken refuge. A close-by temple of Apollo containing a statue of Apollo worth one thousand talents proved too tempting to these troops, who broke ranks and plundered the temple. Having temporarily satisfied their greed, they reassembled for the assault on the citadel.[31]

Three streets ascended from the marketplace to the citadel, into which more than fifty thousand people had jammed. Lining these streets were six-story apartment buildings, from which the defenders could rain missiles on the Romans as they climbed up the streets. Bitter house-to-house fighting was necessary in order to reach the citadel. After the Romans reached the top they set fire to the houses, keeping the streets clear of burning debris so that reinforcements could move up to the citadel. The fires created an inferno made all the more terrible by the cries of women and children, people jumping from the buildings, and the Romans frantically throwing aside the dead and the wounded alike in their effort to clear a way to the citadel.

Although lurid accounts of the fall of cities were a stock in trade of ancient Hellenistic historians, Appian may well have based his account on the lost description by Polybius, who was an eyewitness and who, the reader will remember, viewed with contempt formulaic methods of writing history.[32] Indeed, modern archeologists have found human bones beneath the debris of Byrsa, and they have also found the mass grave in which the Romans buried those who died in the battle.[33]

The fighting continued for six days. On the seventh day a few Carthaginians approached Scipio as suppliants offering to surrender if he would only spare their lives. To this Scipio agreed, and fifty thousand Carthaginians filed out of Byrsa under Roman guard.[34]

Hasdrubal, along with nine hundred Roman deserters who could expect no mercy, continued to resist from the Temple of Aesculapius. A stairway sixty steps high offered the only access to this temple, so the defenders easily kept the Romans at bay as long as their endurance held out. Finally hunger and exhaustion drove them into the inner sanctum of the temple. Hasdrubal abandoned the defenders and threw himself at the mercy of Scipio. Although Appian seemed doubtful of the story, he reported that Hasdrubal's wife berated him for cowardice and then killed their children and committed suicide.[35]

Scipio collected the gold, silver, and temple gifts and then allowed his soldiers to plunder Carthage for several days. He awarded many prizes for bravery, but those troops who had sacked the temple of Apollo received none. When officials from Rome arrived they brought orders to raze the city.[36]

Frontier Sieges

As the Roman Empire expanded the Romans found it necessary to conduct sieges in frontier areas, often under unfavorable circumstances. During the Jugurthine War at the end of the second century, the Romans successfully conducted two such sieges in North Africa.

In 108 B.C. the Roman consul Quintus Metellus determined to capture Thala, a remote desert city in which the Numidian rebel Jugurtha kept his family and his treasury. To reach Thala, Metellus faced the daunting task of marching fifty miles across a waterless desert. He prepared for the march by ordering that the pack animals carry nothing but water and a ten-day supply of grain. This enabled his army to cross the desert. He also arranged that all the Numidians in the area under Roman control near the river should meet him at a rendezvous point near Thala, bringing all the water they could carry.

When the Roman army arrived at its campsite, a heavy rain began falling, which the Roman historian Sallust said provided sufficient water for the army. Moreover, the Numidians were able to transport even more water than Metellus had anticipated. Even so, the Romans could not have conducted a protracted siege at Thala except for the existence of a few springs in the area. These made it possible for them to circumvallate the city with a ditch and a palisade and build a mound and other siege works.

Forty days of assaults with rams and other siege machinery were necessary to reduce the city. Metellus was fortunate in the help of the Numidians who brought water to his camp, the providential rain, and his access to spring water. Eliminate any one of these sources of water and he probably would not have been able to sustain the siege long enough to capture Thala. Even with the good luck it was a considerable logistical feat.[37]

The next year (107 B.C.) Metellus's successor, Gaius Marius, captured another Numidian desert city, Capsa. Marius had a guilty conscience because he had gained his election as consul by criticizing Metellus behind his back while serving under him in Numidia. Metellus's capture of Thala had embarrassed Marius, and when he replaced Metellus as commander in Numidia Marius was anxious to show he was the better soldier. Capsa offered him the opportunity.

Capsa was even more inaccessible than Thala. It was surrounded by a large, snake-infested desert. A single spring within the walls of the town was the only source of water and it could not satisfy the needs of the city's inhabitants, even though they were accustomed to living in a desert climate with little water. They had to rely on rain to supplement the spring. Sallust could scarcely believe Marius's rashness in deciding to march on Capsa, where he would be a three-day march from the nearest water. His army was even short of grain because of the predominantly pastoral economy of Numidia and because the Numidians had already brought their harvest in to fortified strongholds. He "must have put his trust in the gods," Sallust said, never a good policy for a military commander.[38]

Marius was a resourceful soldier, however, and he prepared for the march carefully. Marching his army along the Thais River to the closest point to Capsa, Marius fed his soldiers on cattle. This both alleviated the grain shortage and provided an abundance of leather for water bottles. When the army left the river, the soldiers and the pack animals carried nothing but water. They marched at night to escape the hot desert sun as well as detection by the Numidians. On the third night, the Roman army reached Capsa well before dawn, stopping in a concealed spot about two miles away.

The people of Capsa had no idea a Roman army had arrived in the vicinity, and at daybreak they went forth from the town on their ordinary business. Marius sent the cavalry and mobile infantry ahead to capture the gate, following as fast as possible with the main army. This lightning strike stunned the Numidians into surrendering.[39] We can admire the boldness that won Capsa for Marius, but he had been lucky. Sallust believed Marius had acted rashly and ill-advisedly.[40] If the people of Capsa had mounted a defense, his position would have been untenable.

Boldness and luck had brought Metellus and Marius spectacular successes in North Africa, but the best example of Roman effectiveness in frontier sieges was the Gallic campaign of Julius Caesar. We have an excellent source in the commentaries that Caesar himself wrote. By the time of Caesar, Roman siege capabilities had fully matured. Caesar was the first Roman general to make artillery a regular part of his siege train, although even he did not carry a great number of catapults with him.[41]

Siege machinery could intimidate people who were unfamiliar with it. The Gauls had no machinery. At the town of Noviodunum in northern Gaul the Gauls easily repulsed a Roman attempt to storm the place but quickly surrendered when the Romans filled in the ditch that surrounded the town and brought mantlets and towers forward. According to Caesar, "the Gauls were prevailed on by the size of the siege-works, which they had not seen nor heard of before."[42]

In a like manner a powerful stronghold of the Gallic tribe of Aduatuci surrendered when confronted with Roman siege machinery. This place sat on steep rocks, and the only approach to it was just two hundred feet wide. A very high double wall guarded this approach. When the Romans began constructing siege works and machinery, the Gauls laughed. They were especially contemptuous of the high mobile tower, which they believed the Romans would be unable to move. But when the tower began moving toward the wall, the Gauls, "alarmed at the novel and extraordinary sight," became convinced the Romans had divine help and obediently surrendered.[43]

Not all of the Gallic towns succumbed so easily. Gallic inexperience in siege warfare convinced the Gallic leader Vercingetorix to abandon the towns and wage a scorched-earth war against the Romans. One town, however, persuaded him to allow it to defend itself. This was Avaricum, which was confident that its natural strength made it impregnable. A river and marshlands surrounded Avaricum, making it impossible to invest the city. The single approach to the city was very narrow. Despite the Gauls' inexperience in siege warfare, Gallic walls were formidable. The Gauls built these walls by laying forty-foot-long timbers at right angles to the wall with the end of the timber forming the face of the wall. At the face, stones filled the space between the timbers; behind the face the Gauls piled dirt over the timbers, anchoring them very solidly. The result was a face of alternating stones and timber butts backed by a thick earthen rampart. Caesar admired these walls, for the stones made them resistant to fire and the long timbers anchored in the earthen rampart made them impervious to battering rams and grappling hooks.[44] Finally, Avaricum was well stocked with food.[45]

In contrast to the abundant supply of grain in Avaricum, Caesar faced severe logistical problems. It was winter, and Vercingetorix's scorched-earth policy deprived the Romans of a convenient grain supply. Gallic allies who had promised to bring grain dragged their feet. The severe grain shortage forced the Romans to bring in cattle from a considerable distance. The Gauls harassed every effort to forage. These problems resulted in a chronic food shortage. Despite the hunger, the morale of Caesar's army remained high. When he offered to raise the siege if the hardship were too great, his soldiers refused to hear of it.[46]

His offer to raise the siege was undoubtedly sincere, but it was also a shrewd psychological move. Voluntary hardship is easier to bear than involuntary suffering.

Hard work on an empty stomach is no easy matter, but hard work was the essence of siege warfare and Caesar put his men to work at Avaricum. They began both building a ramp and digging a mine, arduous labor. They also constructed two siege towers to move up the ramp as it rose in height. Mining was an important industry in Gaul, and the Gauls were good miners. They attempted to undermine the Roman ramp and countermined the Roman tunnel. When they broke into the Roman mine they blocked it with sharpened stakes and large stones.

The defenders on the wall kept the Romans away with stones and boiling pitch. They made frequent sallies, day and night, attempting to burn the Roman works and harass the Roman soldiers as they worked. The Gauls raised a wooden superstructure on their wall to maintain their height advantage as the Roman ramp rose toward their wall. Leather hides protected the superstructure from fire.[47]

Almost worse than the determined opposition of the Gauls was the cold rain that soaked the Romans as they labored on their siege works. It was a tribute to the toughness of the Roman soldier that Caesar's troops labored for twenty-five days under these conditions, building a ramp that was 330 feet wide and 80 feet high.[48] Caesar bivouacked next to the ramp to share the hardship and to keep the soldiers at their task.[49]

When the Roman ramp neared their wall, the Gauls made a last desperate effort to stop it. In the middle of the night they surfaced from a countermine to set fire to the ramp. At the same time they made sorties from two gates to burn the Roman towers. Men on the wall threw down brands and hot pitch on the ramp. The Romans were not unprepared, two legions always being on the alert in front of the Roman camp. Caesar acted quickly to organize the defense, sending some troops against the sorties and ordering the others to fight the fire. The fighting continued the rest of the night. The Gauls inflicted damage on the ramp and the towers but failed to destroy them.[50]

This failure demoralized the defenders of Avaricum. When Vercingetorix, who had been reluctant to defend the town in the first place, sent orders from his camp for the men of the town to flee through the swamp to live to fight another day, the men prepared to abandon the city. Even the pleas of the women not to desert them failed to dissuade the Gauls from their plan. Only when the women began to scream about the flight did the men abandon their plan out of fear that the Romans had heard and that their cavalry would be waiting for them on the other side of the marsh.[51]

The demoralization of the defenders set the stage for the fall of Avaricum. The night after the failed sortie against the Roman works, a hard rain fell. The exhausted and demoralized defenders failed to main-

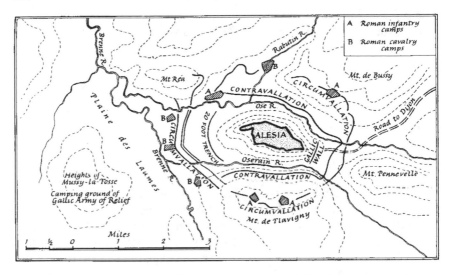

Map 9. Julius Caesar's siege of Alesia. *From J. F. C. Fuller,* Julius Caesar. *Rutgers University Press, 1965. Used with permission.*

tain a proper guard in the downpour, a failure Caesar was quick to notice. Under the cover of the rain he moved troops up behind mantlets. Offering prizes to those who first mounted the wall, Caesar ordered them to assault the town. The sudden assault caught the Gauls by surprise, and the Romans easily dispatched the few defenders on the wall. The Gauls drew up for a last stand in the marketplace, but the Romans did not follow them. Instead they deployed along the wall. When the Gauls saw what was happening, the threat of being trapped broke their spirit. Throwing down their arms, they fled for the gates.[52]

A fearful slaughter followed. The mood of the Roman soldiers was grim. They had endured almost a month of hunger and labor, and they were angry about the massacre of some Roman citizens at a Gallic town shortly before the siege of Avaricum. The lure of plunder did not distract them; they concentrated on killing. The infantry fell on the Gauls crowding at the gates; the cavalry cut down those who made it through. Out of the forty thousand Gauls in Avaricum, eight hundred survived.[53]

The second major siege in Caesar's Gallic campaign was at Alesia. Alesia was also virtually impregnable to assault (map 9). It sat on a high hill, two sides of which were protected by rivers. Except for a small plain on the eastern side, rugged hills surrounded the town. The difficult terrain meant that a line of contravallation had to have a circumference of eleven miles. Caesar's logistical position must have been strong, because he did not hesitate to begin to invest Alesia.[54]

Vercingetorix's army camped in front of Alesia, protecting the camp with a six-foot-high wall and a ditch. When the Romans began constructing siege works, Vercingetorix challenged them with his cavalry, but the Romans won such a decisive victory that he sent his cavalry away to mobilize a relief army, abandoned his camp in front of Alesia, and brought all of his forces into the city to await the relief forces. He estimated that he had enough food for about thirty days but could extend this time somewhat by putting his troops on short rations. The decision to send away his formidable cavalry force meant that Vercingetorix could not follow his usual tactic of harassing Roman foragers and was a godsend for Caesar. Perhaps the difficulty of provisioning a large number of horses under siege conditions forced such a drastic decision.[55]

Caesar's decision to invest Alesia carried several dangers. The need for foraging and the length of the contravallation stretched the defense of his perimeter thin. With a formidable army in the city ready to sally at any sign of weakness and a relief army rallying in the countryside, the Roman position could quickly become precarious. Accordingly Caesar constructed extraordinarily strong siege works. When the Romans had arrived at Alesia they had built twenty-three forts around the city. Now they began encircling the town with a ditch twenty feet wide. The walls of this ditch were perpendicular to make crossing it more difficult. Four hundred yards behind this ditch the Romans dug two more ditches, fifteen feet wide and fifteen feet deep. In the areas where the elevation was lower than the rivers, the Romans were able to divert water into the inner ditch.

The Romans used the earth from the ditches to construct a rampart behind them. A palisade twelve feet high with breastworks and battlements crowned the rampart. Towers reinforced the palisade at intervals of eighty feet. Sharpened stakes bristled from the ramp and palisade.

To make the terrain in front of the ditches and rampart more difficult to cross, the Romans constructed obstacles and booby traps. They felled trees, topped them, and sharpened the truncated branches. These they sunk deeply in the ground close together so that an attacker had to negotiate a thicket of sharp branches. The Roman soldiers punningly dubbed these obstacles *cippi*, which meant both boundary stones and tombstones.

In front of these obstacles the Romans dug eight rows of pits, each cluster shaped like the dots on the five side of a die. A flimsy cover of brushwood concealed these pits so that someone trying to cross them would fall through, impaling himself on the sharp stakes sunk in the bottom of the pits. Because of their pattern, the soldiers called these traps "lilies." The Romans also buried foot-long logs bristling with iron hooks in the ground in front of the lilies.

Not only did the Romans build these elaborate defenses facing the city (J. F. C. Fuller calls their scale "unprecedented in classical warfare"[56]); they duplicated them facing the other direction as well, creating a powerful set of fortifications fourteen miles in circumference, a distance too long for a relief army to invest. Even so, Caesar ordered his men to stockpile thirty days' worth of food in case a Gallic relief army made daily foraging impossible. When one remembers that Caesar's army contained seventy thousand men and imagines what a huge stockpile thirty days of food for that many men would be, one can begin to appreciate the logistical difficulties of siege warfare.[57]

Most of Gaul rallied to support Alesia, but mobilizing troops took time and food was running out in the city. Some in the city counseled surrender. Others believed a last desperate sortie should be mounted before they became too weak from lack of food. The only other option was to hold on grimly hoping for the arrival of the relief army. One leader even offered the advice, which Caesar called "remarkable and abominable cruelty," that the Gauls kill those "whose age showed themselves useless for war" and eat them to sustain resistance until the relief army should arrive. The Gauls decided to hold out; but, although they expressed their willingness to turn to cannibalism if necessary, they did not want to resort to so extreme a measure yet. Instead they expelled from the city its civilian inhabitants to reduce the number of mouths to feed. Thus a pitiful throng of people approached the Roman lines begging the Romans to enslave them and feed them. Caesar ordered his soldiers to turn them back. His commentaries did not tell the fate of these people caught in the no-man's land between the walls of Alesia and the Roman fortifications.[58] Presumably the Gauls allowed them back into the city.

At this critical point the relief army finally arrived. Its strength—250,000 infantry and 8,000 cavalry—revived the spirit of the besieged and put the Roman fortifications to the test. For the first time in a month, the defenders of Alesia ventured out of the city. They filled in the innermost Roman trench to prepare for an attack on the Roman lines. The Gallic army on the outside gathered in the plain. Around noon the Gauls attacked from both directions. Although the battle lasted until nightfall, the attack failed. The Romans drove the relief army back to its camp and Vercingetorix withdrew his discouraged men back into the city.[59]

The Gallic relief army spent the next day preparing for another attack by making hurdles with which to fill in the ditches and ladders and grappling hooks with which to attack the Roman rampart. This time they attacked at night. Raising a great shout to signal to the Gauls in Alesia to attack from the city, slingers and archers attempted to drive the Roman defenders from the rampart. The Romans replied with their own slingers and with one-pound catapults. The Gauls lost many men in the traps and obstacles in front of the Roman entrenchments, and the sally from the city took so long getting across the first ditch that it arrived too late to affect the battle. At daybreak the Gauls broke off the battle fearing a flank attack by the Romans.[60]

The Gauls made a final effort to break the siege. Because of difficult terrain, the Romans had been unable to enclose one of their camps within the fortifications. At this weak point the Gauls concentrated their last attack. Sixty thousand of the Gauls' best troops moved into position during the night. Keeping themselves concealed, they launched a surprise attack around noon. At the same time the rest of the army staged distracting attacks at other places in the Roman lines.

When Vercingetorix saw what was happening, he immediately led his men from the city to the attack, and this time they caused the Romans considerable embarrassment. Roman troops facing one direction became apprehensive that the enemy might break through from the other direction, and this fear drained their morale. Having run afoul of the Roman defenses before, this time the Gauls brought earth with them to spread over the traps and obstacles. The Gauls from the city directed their attack toward the more rugged areas and away from the strongest Roman fortifications.

Caesar staved off collapse by shifting reserves from place to place as needed. In the area around the camp the situation became so critical that Caesar personally took command, easily recognizable by his red cloak. His appearance inspired the Romans to one last supreme effort. The arrival of Roman cavalry and other reserves tipped the balance, and the Gauls fled. The rout was so complete that few of them returned to their camp.[61] The rout of the relief army ended all hope for the Gauls in Alesia. Vercingetorix valiantly offered to give himself up to the Romans, and the town surrendered.

In what Fuller calls "one of the most extraordinary operations recorded in military history," Caesar had overcome the worst nightmare of a siege commander: an attack by a powerful relief army. The thoroughness of his preparations, especially the powerful fortifications that he built, made it possible for his troops to carry on the siege while under siege themselves.[62] The Romans may not have been the most dashing besiegers, but they knew how to build.

The last siege the Romans conducted in Caesar's Gallic War was at Uxellodunum. This siege was described by Aulus Hirtius in his addendum to Caesar's *Commentaries*, commonly included as Book VIII. A Roman army under the command of Gaius Caninius Rebilus had driven a Gallic force led by Drappes and Lucterius into Uxellodunum. Uxellodunum was in a naturally strong position, protected by rocky precipices that were difficult to climb even without armed resistance (map 10). Caninius did not have enough troops to invest the city as thoroughly as Caesar had invested Alesia, but he believed he could prevent the Gauls from escaping the city. Toward that end he established three

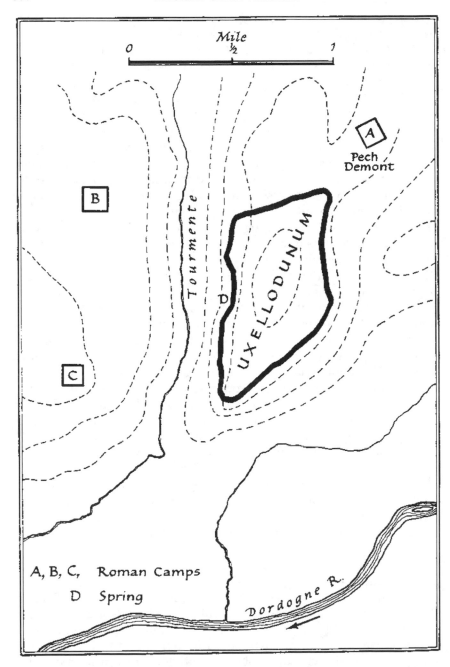

Map 10. Julius Caesar's siege of Uxellodunum. *From J. F. C. Fuller,* Julius Caesar. *Rutgers University Press, 1965. Used with permission.*

camps on high ground and began constructing ramparts around the city to the degree possible with the number of troops available.[63]

The memory of Alesia haunted the Gauls, and the sight of the Romans constructing ramparts deeply disturbed them. Lucterius had been in Alesia and knew firsthand the hunger pangs of siege warfare. Although a good supply of grain was on hand, he and Drappes decided to bring much more into the city. Leaving two thousand men to defend the town, they left with the rest of their troops to gather grain from among those Gauls who supported them or to requisition it from those who did not. At the same time their troops harassed the Romans, convincing Caninius still more that he did not have enough troops to defend a contravallation. He felt safer with his troops concentrated in the camps.

Having collected a huge store of grain, the Gauls established a camp about ten miles from Uxellodunum from which to convoy the grain into the city little by little. Leaving Drappes to guard the camp, Lucterius set out with the first convoy of food. Moving at night along narrow paths off the main roads, he hoped to reach the city undetected by the Romans. Roman sentries, however, heard the noises of the convoy, and scouts discovered what the Gauls were trying to do. Just before dawn the Romans attacked the convoy and killed most of the Gauls except Lucterius, who escaped on a fast horse with a few other men. However, he did not return to the camp to alert Drappes of the disaster.

When Caninius learned about the Gallic camp from some prisoners, he immediately sent out a force against it. Surprise was complete, and the Romans annihilated the Gallic camp, killing or capturing every man. Drappes was among the prisoners. The destruction of the bulk of the Gallic army enabled Caninius to resume the investment of Uxellodunum without fear of attack from the rear. The arrival of more Roman troops made the task easier.[64]

Caesar was in the last year of his command in Gaul, and he feared that if Uxellodunum withstood the Roman siege its success would encourage all the towns of Gaul to revolt after he left Gaul and ruin all that he had accomplished. He therefore placed the highest priority on capturing the city and hastened there to take personal command. By the time he arrived Caninius had completed the contravallation, but Caesar learned from deserters that the city still had a plentiful supply of grain.

A reconnoiter of the city showed him, however, that Uxellodunum drew its water supply from outside of the walls from the river that flowed by the city. This weak point became the main target for Caesar's attack. Because the river flowed through a deep gorge, it was impossible to divert it from its course. But the Gauls had to climb down a steep precipice to reach the river. Caesar stationed archers, slingers, and catapults in a position to kill anyone trying to reach the river from the city.

That forced the Gauls to draw all of their water from a spring just beneath the wall in a place that the Roman missiles could not reach. Nor was it possible for Roman soldiers to climb to the spring against the opposition of the Gauls. Caesar, though, was determined to deprive the Gauls of this spring. He ordered his troops to begin building a ramp up to it and to dig mines to the spring to drain its water away. When the ramp had risen to a height of sixty feet, the Romans built a ten-story tower that raised them higher than the spring and put them in a position to drive the Gauls from it.

The Gauls, knowing that they could not survive without water, fought back desperately. They rolled burning tubs full of grease, pitch, and shingles against the towers and mantlets and fought the Romans fiercely to prevent them from putting out the fire. Only when Caesar ordered diversionary attacks all around the walls did the Gauls break off the fight for the spring and return to defend the city. The Gauls continued to hazard the trip to the spring, but the commanding Roman position prevented them from getting enough water. Still they hung on grimly, even though people were dying of thirst. Finally the Roman mines reached the spring and completely drained it of water. Cut off entirely from water, the Gauls surrendered.[65]

The siege of Massilia (modern Marseilles) during the Civil War serves as an example of Roman resourcefulness in siege warfare in the first century B.C. Caesar was in Spain; he assigned his lieutenant Gaius Trebonius to conduct the siege.

Water surrounded Massilia on three sides, and the approach on the landward side was so difficult that blockade offered the best chance of reducing the city. In order to mobilize the labor necessary for the construction of siege works, Trebonius requisitioned draft animals and impressed laborers from the region around Massilia. These workers helped the Roman soldiers build a siege mound eighty feet high. It was tough going, however.

Massilia was no frontier town. It was well stocked with catapults, some powerful enough to shoot twelve-foot-long spiked beams that could penetrate the osier wickers that protected the Roman penthouses. The Romans had to build penthouses covered with thick timbers to withstand the strength of the Massilian artillery. As they pushed the mound forward, the Romans advanced behind a giant tortoise sixty feet high that also was built of heavy timbers. Hides and anything else the Romans could find protected the tortoise from firebrands. The Massilians made frequent sorties to try to burn these works. Caesar said that the Romans easily repulsed these sorties, but the impressed laborers can hardly have been happy with their duty.[66]

The Roman legionnaires who were responsible for protecting the workers from these sorties built a brick tower close to the city walls from which they could attack the Massilians. At first this tower was low and relatively small, about thirty feet square. The walls were five feet thick. The tower proved so useful, though, that the Romans decided to enlarge it into a powerful siege tower.

At first they raised the tower behind a protective shed, but when they reached the height of the shed they devised an ingenious method to provide independent protection. They constructed a roof out of beams and devised a means of leverage that could lift it one floor at a time. The beams of the roof extended enough so that they could hang protective screens from it. The Romans then built up the wall behind the protective screens. In this way they were able to raise the tower six stories without a single casualty. The beams of the floors in the tower did not extend beyond the wall so that they did not offer any wood as a target for fire missiles. Heavy mattresses protected the tower from stone throwing catapults. Caesar said that experience had taught the Romans that mattresses provided the best protection against artillery. Embrasures enabled the Romans to fire at the enemy from the tower.[67]

Having completed this powerful tower, the Romans built a mobile gallery sixty feet long. Parallel beams four feet apart formed the base of the gallery. On these beams the Romans affixed five-foot-high posts. Rafters were laid across the posts, and on these rafters the Romans built a sloping roof from strong wooden beams. Shingles nailed to the edge of the roof formed a lip to hold bricks that covered the roof. Finally the Romans roofed over the bricks with tiles and clay to protect it from fire and water. Leather hides and mattresses provided further protection. This gallery the Romans lifted onto rollers and, at an opportune moment, suddenly pushed it forward up to the main tower of Massilia.[68]

The gallery proved strong enough to withstand heavy stones that the Massilians dropped from the wall. When they realized that the stones could not crush the gallery, the Massilians tried to burn it by dropping barrels full of burning pitch and pine resin. This effort failed also. The barrels rolled off the roof and the Romans pushed them away from the gallery with long forked poles. Intense covering fire from the Roman tower hampered the Massilian effort to combat the gallery, and Roman soldiers working under its protective roof soon pried loose some stones at the base of the Massilian tower and collapsed a part of it. What was left of the tower listed badly. An effort to relieve the city by sea had failed, and the situation was desperate. Hoping to avoid the horror of a sack, the Massilians surrendered.[69]

Trebonius was glad to accept the surrender. Caesar had told him to avoid taking the city by storm because Caesar feared a massacre of the young men defending the city, something he wanted to avoid in a civil war. Being deprived of a sack made the Roman soldiers unhappy and may have contributed to a relaxation of discipline while they awaited the arrival of Caesar and final peace terms. The Massilians thought they saw an opportunity, and several days after surrendering they suddenly

launched a surprise attack on the Roman siege works. It was just after noon, and many of the Roman soldiers were napping. None were
on alert. The Massilians set fire to the siege works, and a strong wind
fanned the flames out of control. The mound, the sheds, the tortoise, the
gallery, the brick tower, and most of the machinery burned before
the Romans drove the Massilians back into the city. The next day the
Massilians attacked what was left of the Roman works, but the Romans
were prepared this time and the attack failed.[70]

The Massilian treachery had dealt a severe setback to the Romans.
There was nothing to do but begin again the laborious work of constructing siege works. They had stripped the area of all its timber in
building the first siege works, so it was necessary to build these works
from brick. The Romans devised a siege work that Caesar called "of a
novel kind that no one had heard of before." It consisted of two brick
walls, each about six feet thick. They roofed the walls with timber, reinforcing the roof with piles and cross-beams. They thickened the roof
with hurdles and covered it with clay to make it fireproof. They left
doors in the walls to allow sorties.

Working with a grim fury, the Romans progressed on their labor
much more rapidly than the first time. When they were finished they
had a work that protected them on both sides by thick brick walls. A
mantlet guarded the front, behind which the Romans were able to bring
up all their siege equipment in complete safety.

The speed with which the Romans worked dismayed the Massilians, who had imagined the destruction of the Romans works would
cause a long delay in the siege. The Roman work was so close to the
Massilian walls that it was within javelin range. The close range neutralized the catapults, the Massilians' most formidable weapons, because catapults were ineffective at such a short distance.

Shut in completely by the Roman works, the Massilians faced defeat
once again.[71] Their food stores were old and moldy, and a serious plague
had struck the town. Caesar's forces were in firm control of Gaul, Spain,
and the sea, so there was no hope of relief. This time Caesar insisted
that the Massilians turn over their arms and catapults and hand over
their treasury to him when they surrendered so there could be no repeat
of their treachery.[72]

Imperial Rome

The imperial Roman army had a highly developed siege capacity. Because of its practice of building fortified camps, a large number of
workmen always accompanied the army. A full array of construction
tools was part of the army's equipment.[73] These well-equipped laborers
could easily turn to the work of siege warfare. Pioneers were a regular
part of the Roman army. Their main task was to clear the way for the

army when it was on the march, but their skills could be just as useful to the army when it was at siege.[74]

By the time of Augustus the Romans had overcome the earlier shortcoming in armaments. A large arsenal existed in Rome. Each legion also contained engineers capable of designing and constructing siege machinery, including complicated artillery.[75] A siege train was part of the regular order of march, and by the time of Augustus it included artillery.[76] Mules pulled the towers and other siege machinery.[77] The crews that assembled and operated the siege machinery enjoyed a high status in the Roman army.[78]

The Romans attached great importance to their siege train. During a desperate retreat from Jerusalem after an unsuccessful siege in A.D. 66, the Roman commander Cestius abandoned the entire impedimenta of the army and killed the baggage animals. He spared only the mules that pulled the siege machinery, because this was the one part of the baggage train that Cestius considered too important to leave behind. Only in the final panic when the Jews utterly routed the Romans did they abandon the siege machinery.[79]

The Roman campaign in Judea that culminated in the destruction of Jerusalem in A.D. 70 offers perhaps the best view of the Roman imperial army at siege. The Jewish historian Josephus provided a detailed description of this campaign, especially for the sieges of Jotapata in A.D. 67 and Jerusalem in A.D. 70. He was present at both sieges and was the Jewish commander at Jotapata.

The siege of Jotapata, the strongest city in Galilee and therefore the key to the defense of northern Judea, was led by Vespasian, soon to be emperor. He began by erecting earthworks against the most accessible part of the city. Josephus admired the efficiency of the Romans as they stripped the surrounding countryside of timber and stones for the earthworks. Screens protected the Romans working on the earthworks. These screens were strong enough to withstand large boulders that the Jews dropped from their wall, but the noise was so deafening the Romans could scarcely tolerate it.

To force the Jews off of the wall, Vespasian deployed his catapults, 160 in all including both arrow shooting and stone throwers, as well as Arab archers and slingers (see figures 15 and 16). The fire of arrows and stones was so intense that it not only drove the Jews off the wall but also forced them to retreat from the area behind the wall. According to Josephus, a stone from a Roman catapult tore the head of a man standing next to him clean off. Josephus also remembered the infernal noise the stones made as they whizzed by.[80] When Vergil wanted to describe the terrible noise of Aeneas's flying spear, he recalled the loud whine of "rocks hurled from siege machine."[81]

Having been forced off their walls, the Jews turned to sallies against the Roman earthworks, a tactic at which they were very good because

Figure 15. Roman Catapult. *From J. F. C. Fuller,* Julius Caesar. *Rutgers University Press, 1965. Used with permission.*

of their experience in guerrilla warfare.[82] The Jewish sallies were so effective that Vespasian had to strengthen his earthworks by connecting them with a continuous rampart. When the Roman earthworks approached the height of the city wall, the Jews raised protective screens made of raw hides and brought masons up to increase the height of the wall. That, along with the constant harassing sallies, convinced Vespasian to resort to blockade.[83]

Although Jotapata was well stocked with food, there was a serious

Figure 16. Roman onager. *From J. F. C. Fuller,* Julius Caesar. *Rutgers University Press, 1965. Used with permission.*

water shortage, which Vespasian hoped would force the city to surrender. Jotapata had no springs and depended on cisterns, but the summer rainfall was insufficient to water the city. Josephus attempted to discourage the Romans by hanging soaking wet garments around the wall to make it seem as if there were plenty of water. He also was able to bring in extra water through a gully that created a gap in the Roman defense works. However, the Romans soon discovered what was happening and blocked the gully.[84]

When the earthworks were completed, Vespasian deployed battering rams against the walls of Jotapata. These rams employed a large beam, reinforced at the end by iron in the shape of a rams head. The beam balanced from a rope tied to a cross-beam held by posts driven into the ground. A large crew of men swung the ram back and then drove it forward against the wall. Protective covers shielded the crew and the ram from enemy fire. Catapults, archers, and slingers provided support fire.

The Romans brought their rams to the top of their earthworks so they could deploy them against the topmost part of the wall, which the Jews had just built to extend its height and which was weaker because of the hurried construction. At first the Jews parried the blows of the ram by lowering sacks filled with chaff to soften the ram's striking force, but the Romans cut the ropes holding the sacks with scythes attached to long poles. The Jews realized that the rams would make short work of these relatively weak walls and so they made a desperate sally against them. The sally caught the Romans by surprise, and the Jews succeeded in burning not only the ram but some of the Roman earthworks also.[85]

There followed a fierce struggle over the Roman siege works in which Vespasian received a slight wound in the foot from an arrow. The fighting continued through the night, but the Romans were able to reerect their ram and by morning had opened a breach in the wall.[86] Vespasian, still in command despite his wound, then deployed his troops to assault the breach. Interestingly, dismounted and heavily armored cavalry made up the shock troops. Crack infantry troops were to follow close behind. Archers, slingers, and artillery deployed in a semicircle to lend close support to the assault. The rest of the infantry prepared to attack the walls by escalade to divert the Jews from the breach. The remaining cavalry patrolled the perimeter to prevent anyone from escaping the city.[87]

Despite this formidable array, the assault on the breach failed. The Jews resisted fiercely, although the Romans exhausted them by attacking in relays. When Roman troops formed a testudo the Jews broke it up by pouring boiling oil on it. They also poured an extremely slippery substance, boiled fenugreek, on the planks the Romans had laid across the rubble in the breach. Toward the end of the day, Vespasian called off the attack. Josephus said that only six Jews died in the battle, although some three hundred were wounded. He did not number the Roman casualties but claimed they were heavy.[88]

After the failure of the assault, Vespasian ordered the construction of three siege towers fifty feet high. Iron plating protected these towers. The Romans erected the towers on their earthworks, and from that height slingers and archers, as well as light artillery, were able to keep the Jews off the walls. However, the Jews continued to defend their walls with sallies, and the siege dragged on.[89]

In the end the city fell through exhaustion and betrayal. A deserter informed Vespasian of the fatigued state of the defenders and advised him that exhausted sentries fell asleep toward morning. This information enabled the Romans to occupy the city by stealth one foggy night. Vespasian's son Titus accompanied the escalade party. Only a bloody mopping-up operation remained to secure the city. The siege had lasted forty-seven days.[90] Despite the overwhelming technological superiority of the Romans, they had been unable to take Jotapata until attrition and betrayal brought the city down.

The fall of Jotapata paved the way to the culmination of the Roman campaign in Palestine, the siege of Jerusalem (map 11). Josephus, in captivity after the fall of Jotapata, observed this siege from the Roman side and provided one of the longest accounts we have of an ancient siege. Extraordinarily strong fortifications guarded Jerusalem. Deep ravines made most of the city walls inaccessible. Interior walls presented multiple obstacles to besiegers. Exceptionally strong towers strengthened the walls. There was an upper city and a lower city, both on strongly fortified hills. Most strongly fortified of all was the famous temple, which stood on the highest point in the city.[91]

Despite the strength of its fortifications, Jerusalem was fatally weakened by internal divisions. Factional fighting raged in the city, and not even the Roman siege united the Jews in a common defense. So bitter was the enmity between the factions that they destroyed most of the vast stores of food in Jerusalem before the Romans even arrived.[92] There were gardens and fruit trees within the city and some scavenging for food outside the wall took place during the siege. Until the Romans circumvallated the city, some food was smuggled into the city from the outside. None of these sources sufficed for a city whose population was swollen with refugees. The Romans on their part enjoyed secure supply lines to Syria from whence they were able to bring in an abundant supply of grain.[93]

Water supplies favored the Jews. The exact sources of water in A.D. 70 are not entirely known, but it is clear that they were abundant. Huge cisterns capable of holding millions of gallons of water and an extensive system of underground tunnels, some of which may have led to springs outside the walls, provided all the water the Jews needed. The Romans, however, suffered from limited water supplies. Although a number of

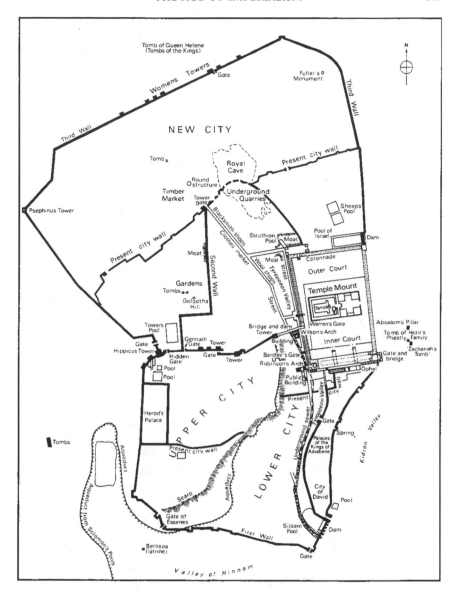

Map 11. Roman siege of Jerusalem. *From Jonathan J. Price,* Jerusalem under
Siege. *Copyright 1992 by E. J. Brill, Leiden, The Netherlands. Used with permission
of the publisher.*

water sources existed in the vicinity of Jerusalem, they could not meet the tremendous demands of a siege army numbering sixty thousand men in addition to many camp followers. Water shortages were the Romans' chief logistical problem, although the problem was not great enough to prevent the Romans from maintaining the siege for five months.[94]

In the early spring of A.D. 70 four Roman legions converged on Jerusalem.[95] Counting all the auxiliary troops, the total force numbered perhaps as many as sixty thousand troops. Vespasian had departed from Palestine, leaving his son Titus in command. Titus, a future emperor himself, had many good qualities, but he was only thirty years old and extensive military experience was not one of them. He lacked tactical imagination, but he was brave, as evidenced by his participation in the nighttime escalade at Jotapata. Although some regarded him as too reckless, his courage won his troops' complete loyalty.[96]

Titus had no wish to besiege such a strong city as Jerusalem, and so he sent an envoy, accompanied by Josephus, to negotiate a surrender. But the Jews wounded the envoy with an arrow when he approached the wall. Only then did Titus begin preparing for a siege.

He got off to a harrowing start when he ventured too close to the wall as he reconnoitered the city, and a Jewish sally almost captured him. He apparently had not learned at Jotapata how aggressive Jewish tactics were.[97] Emboldened, the Jews mounted a full-scale attack on one of the Roman legions, which had encamped east of the city on the Mount of Olives, and almost routed it before Titus arrived with reinforcements.[98] The Jews were not going to cave in easily.

Titus moved his main force forward to a camp four hundred yards from the northwest corner of the third wall, which protected a rather sparsely settled area in the northwestern part of the city called New City. This was the only area where Jerusalem was accessible to siege machinery.

The wall was not especially strong. Indeed, it had been incomplete until the Jewish rebels had hastily completed it after the war with the Romans had broken out.[99] Titus chose a point of attack somewhere along this wall, probably on its west side. The Roman troops tore down the suburbs to gain material for their siege works. Javelin-men and archers as well as arrow-shooting and stone-throwing catapults protected the soldiers who were building the works against Jewish sallies. The biggest of the catapults could throw a one-talent stone almost five hundred yards. The Jews soon learned to listen for the whiz of the catapulted stones and posted lookouts to warn of incoming fire. The Romans tried to frustrate the lookouts by blackening their stones to make them more difficult to see. The Jews also had catapults that they had captured from the Romans; but they were inexperienced with them, so their fire was not very effective.[100]

The Romans completed their siege works in about five days.[101] The siege works were very near the walls; the Romans measured the distance by tossing a leaded line to the wall and seeing that they were near enough for the battering rams to reach the wall they brought the rams forward. The Jews made repeated sallies against the rams, but Titus brought up reinforcements whenever the Jews threatened to be successful and they failed to put the rams out of action. The Romans built siege towers seventy-five feet high. Metal sheets protected them against fire, and their great weight protected them from being capsized. One of these towers collapsed during the night, causing a panic among the Romans. Probably the great weight of the metal sheets broke the wooden supports. These towers were so high the Romans were out of the range of Jewish fire, and they were able to clear away the Jewish defenders. After ten days of battering, the largest Roman ram finally opened a breach. The suburbs were of little use to the Jews, and they retreated to the second wall without defending the breach.[102]

The Romans then cleared the area between the third and second walls, and Titus moved his camp up to just out of bow range of the second wall, occupying the same site where Sennacherib had pitched his camp in the Assyrian siege of Jerusalem almost eight hundred years earlier.

The Romans attacked some spot along the second wall, which ran from the fortress of Antonia to the northwest corner of the upper city, probably in an area toward the center that was out of the range of Jewish artillery based at Antonia and at the Herod tower at the other end of the second wall.[103] After five more days of fierce fighting and battering they breached the second wall. The Jews retreated to the temple area and the first wall guarding the upper city, the most strongly fortified parts of Jerusalem.

Titus kept his troops under strict discipline, hoping that his generosity in preventing a sack would encourage the Jews in the temple area to surrender. Instead the Jews launched a fierce attack on the Romans. The Jewish attack caught the Romans in an unfamiliar tangle of narrow streets. Titus had carelessly failed to widen the breach, and the Roman troops were trapped. Only the archers that Titus brought up to cover the Roman retreat prevented a massacre, but the Jews succeeded in ejecting the Romans beyond the second wall. Three more days of hard fighting were necessary for the Romans to recapture the second wall. This time Titus razed the wall and securely garrisoned the area between the second and third walls.[104]

Titus now paused four days to pay his troops and allow the Jews to reflect on their prospects. The temple area was a powerful citadel, and he hoped that the Jews would surrender. The fall of two walls within nine days did indeed precipitate a wave of desertion from the city, but no surrender came. Indeed, the "desertions" may have been a calculated lowering of the population in the city to relieve the cramped conditions that would have obtained in the constricted area still in the possession of the Jews. The shortage of food provided another motive to reduce the numbers of the besieged. Some of the deserters sabotaged the Roman water supply, indicating that not all of them were demoralized refugees.[105]

After resting for four days the Romans began building earthworks against the final wall. Each of the four legions was responsible for building a ramp. Titus directed two of the ramps against Antonia, the key to the temple area, and two against the upper city.

Jewish resistance was tough. The Jews continued their usual sallies against the besiegers. Moreover, the Jews were now beginning to master the use of artillery. Josephus claimed they had three hundred arrow shooters (scorpions) and forty stone throwers, but this is an impossibly high number. Thirty is a possible emendation for three hundred, but the truth is that Josephus's figures are so unreliable that we cannot know how many catapults the Jews possessed.[106] Even so, we can believe Josephus's report that their fire seriously inconvenienced the Romans as they toiled on the earthworks.

Titus still hoped for a surrender and delegated to Josephus the task of persuading the Jews to give up. According to Josephus, Titus did not want his siege to end in the destruction of Jerusalem and its famous temple. Josephus, however, had to shout his pleas from beyond bow range as the Jews spurned his efforts.[107]

Although there was no surrender, a steady stream of deserters continued to escape from Jerusalem. Famine was the cause.[108] Although besiegers often prevented people from leaving a besieged city in order to keep maximum pressure on food supplies, Titus encouraged desertion by allowing the deserters to pass through the Roman lines. Most of those who deserted were apparently from the better-off classes; Josephus said that they swallowed gold coins, which were then recovered after passing through the bowels, to give them the wherewithal to reach safety. People from the lower classes, either through greater loyalty to the rebellion or from lack of resources, did not desert. They did, however, leave the city in search of food. But the Roman cavalry was vigilant against this, and Titus crucified all Jews who were caught foraging for food. Josephus provided a lurid description of degradation and violence as the Jews fought among themselves for food. Even if his account was influenced by his hatred of the rebel leaders, we can well imagine the grim conditions in the city.[109]

Reaching the temple area required huge earthworks, and it took the Romans seven days to complete them.[110] The Romans then brought up their battering rams to begin demolishing the wall. But the Jews succeeded in collapsing part of the earthworks at Antonia by undermining. A daring sally against the other earthworks by three Jewish soldiers set fire to the rams; and as a fierce fight over the rams ensued, the Jews poured from the city, drove the Romans back, and burned down the earthworks. So strong was the attack that the Jews drove the Romans back to their camp. Only the fortitude of the Roman guards at the camp, who faced capital punishment if they abandoned their posts, prevented the Jews from breaking into the camp. A flank attack by crack troops under Titus drove the Jews back into the city. In one hour the Romans had lost the labor of seven days.[111]

In the wake of this catastrophe Titus held a council of war to discuss the Roman options. Some of his advisers favored an all-out assault, arguing that the Jews were in such a weakened state that they would be unable to resist a mass assault. More cautious advisors believed the Romans should continue the same strategy and rebuild their earthworks. Others thought that in view of the famine in Jerusalem the city could be most easily brought to surrender by a blockade.

Titus rejected all of these plans. He believed that a costly assault was not wise against a city as divided and famine-stricken as Jerusalem. The Romans had scoured the area for materials for their earlier earthworks, so there was now a dearth of material for new works. A blockade would be difficult to enforce in the rugged terrain around Jerusalem, especially because the Jews were so familiar with it. Titus also did not like the long period of relative inactivity a blockade would impose on his army. In view of these considerations, he decided the surest and quickest way to bring the siege to a close was to circumvallate Jerusalem. The construction of the wall would keep his men active, and only a wall of circumvallation would seal the city off entirely so that famine could do its work. If hunger did not soon bring the Jews to surrender, it would surely leave them in such a weakened condition that the Romans could resume building siege works against only feeble opposition.[112]

The Roman troops, undoubtedly relieved to be working outside the range of Jewish weapons, set to the task of circumvallation with what Josephus called "preternatural enthusiasm." The construction was organized by legion and company, and they competed with one another to see who could build fastest. The wall was about five miles long. Thirteen forts, connected to the wall on its outer side, added another 2,200 yards to its length. Josephus said that the Romans completed the wall and forts in an incredible three days. If true, a Roman legion was capable of building over one-third of a mile of wall per day, demonstrating the Romans' unrivaled ability to organize labor.[113]

Josephus described great suffering in Jerusalem. A measure of grain sold for a talent, and people were reduced to eating cow dung and offal.[114] Children bloated by starvation wandered the streets. Corpses littered the city because people were too weak to bury them. The intolerable stench eventually forced the Jews to throw rotting corpses from the walls down into the ravines. When Titus saw the bodies piled in the ravines with a "thick matter oozing from under the clammy carcasses," he was visibly upset. A deathly silence hung over Jerusalem.[115]

So enfeebled were the Jews that Titus began the construction of new earthworks. This time he built them only against the Antonia fortress, the stronghold that dominated the temple area. The works were much larger than formerly. The most difficult task was procuring materials, which had to be brought in from over nine miles away, a four-and-a-half-hour trip.[116] The Romans so thoroughly stripped the suburbs of Jerusalem and the surrounding area of timber that the once-beautiful countryside looked like a desert. Josephus could hardly bear to see the utter desolation around Jerusalem.

The Jews attempted one last sally against the earthworks. But they were too weak to carry it off with the organization and dash that had so discomfited the Romans, and the sally fizzled out in the face of artillery fire and stout resistance from the Romans. The scarcity of timber slowed the progress of the Roman earthworks, and twenty-one days were required to complete them.[117]

When they had finished their earthworks the Romans began battering Antonia. It was now midsummer. The Jews hampered the battering by dropping boulders on the rams, but the weight of the stones caused the mine that the Jews had dug under the Romans' first earthwork to collapse, weakening a part of the wall. One group of Roman soldiers approached the wall in a tortoise formation and managed to pry four stones out with crowbars. At nightfall the wall still stood; but it was in a weakened condition, and during the night it collapsed.[118]

But the Jews had hurriedly built a second wall behind the point of the Roman attack, and when the Roman soldiers saw this they were dismayed. The new wall was undoubtedly weak, but the thought of assaulting it discouraged troops who were fatigued by a long and arduous siege. An exhortation by Titus was necessary to arouse the Roman soldiers to attack.[119] His speech seems to have had little effect. Josephus related a somewhat improbable tale that it inspired one man to assault the wall singlehandedly. Only eleven other men followed his example, and the attack failed.[120]

In the end the Romans took the wall by stealth. Two days after the failed assault, some Roman soldiers sneaked up to the wall during the night, scaled it, killed the guards, and sounded a trumpet to alert the Roman army that they had gained possession of the wall. Titus led his elite troops forward immediately, and the Romans pushed the Jews back to their last point of defense, the temple. A fierce struggle ensued, but the Romans failed to break into the temple.[121]

The Romans retreated to Antonia and began razing its foundations to clear a way for an attack on the heavily fortified temple. Ominously for the Jews, the temple sacrifices, which they had maintained throughout the famine, now ceased because of lack of sheep.[122]

Josephus said that Titus was most reluctant to fight a battle over the temple and made every effort to persuade the Jews to save their temple by surrendering. When these efforts failed, Titus ordered a night attack, hoping to catch the Jews by surprise. He gathered the thirty best men from each century for the attack. Josephus said that Titus wanted to lead the attack himself but that his subordinates persuaded him it was too dangerous. He decided to watch the attack from Antonia.

The attack failed. The Jewish guards were alert, so the element of surprise was lost. A bitter eight-hour battle followed in the confined spaces around the temple. Roman discipline served them well in the confusion of a night battle, but they failed to break into the temple by sunrise. After several more hours of fighting, Titus called off the battle.[123]

The Romans then returned to the slow work of clearing the area before the temple. They took seven days to complete the leveling of Antonia. They then began building four embankments leading up to the north and west sides of the temple. The task was not easy because of the scarcity of timber, which the Romans had to bring in from a distance of twelve miles (a five-hour trip),[124] and the constant sallies of the Jews, including a major assault on the Roman camp.

As the Roman embankments moved inexorably toward the temple, the Jews attempted to sever the temple's connections with Antonia by burning the northwest porticoes.[125] In burning the western portico, the Jews lured the Romans into a disaster. Having filled the portico with highly inflammable materials, the Jews retreated from it. Some of the Romans unwisely climbed into the portico in pursuit, and when the Jews set it afire the Romans were trapped.[126]

As the Romans toiled on their earthworks, they pounded the western wall of the temple with stone-throwing catapults. Six days of bombardment accomplished nothing. The Romans also attempted to collapse the northern gate by prying out stones from its foundation. With great effort they succeeded in dislodging several stones in the front of the foundation, but the inner stones continued to support the gate.

When the siege ramps were completed, the Romans deployed their rams against the temple wall. But Titus did not wait for a breach. Instead he ordered an escalade of the porticoes.[127] By this time the Jews had no food. Josephus reported that they ate leather and grass. One woman killed her baby and ate him.[128] Despite their extreme hunger the Jews repulsed the escalade.[129]

Titus next ordered his troops to burn the temple gates. The fire soon spread to the remaining porticoes, burning all day and through the night. The next day Roman troops cleared a road through to the gates and extinguished the fire. The Jews continued to resist, and as the Romans fought into the inner court one of the Roman soldiers threw a burning brand into the temple. Josephus claimed that Titus tried to restrain his troops from burning the temple but lost control of them so furious were they to wreak revenge on the Jews. However, Josephus was so sycophantic toward Titus that his report may not be reliable. Most modern scholars believe that Titus ordered the destruction of the temple, as Sulpicius Severus reported. Even though Severus wrote three hundred years later, he probably drew his information from Tacitus.[130] In any case, the temple was soon in flames. Before the flames had spread, Titus managed to violate the innermost chamber of the temple, revealing his lack of respect for Jewish beliefs.[131]

The burning of the temple led to the general destruction of Jerusalem as Titus unleashed his men to sack and loot the city. There was a fearful slaughter. Some of the Jews escaped into underground tunnels. Others took refuge in the upper city for a last-ditch stand. Another eighteen days of labor were necessary for the Romans to build mounds to the upper city. Hunger had so sapped the strength of the Jews that their resistance was feeble. As soon as the Romans, probably in early September, opened a breach with rams, the defense of the upper city collapsed. Mopping-up operations continued for several days amidst great carnage.[132]

Jerusalem had held out for five months. The defenders were badly divided and indeed had destroyed most of their food supply fighting each other before the Romans arrived. Hunger gripped the city for most of the siege. The Roman army was superbly equipped and trained for siege warfare. The ability to organize labor in the construction of siege ramps and a wall of circumvallation was more decisive than Roman artillery. Jerusalem was an unusually tough nut to crack and the Jews proved to be tenacious fighters, but the five months the city was able to hold out against the might of imperial Rome was another testimony to the limitations of siege warfare in the ancient world.

Chapter XIII

TREATMENT OF
CAPTURED CITIES

The Laws of War and Clementia

Two principles governed the Roman treatment of captured cities: the laws of war and *clementia*. The laws of war were the customary rights of those who took a city by assault to sack the city, kill or enslave the men, and rape the women and enslave them and their children.

These laws went back at least to the time of Xenophon in the fourth century B.C.[1] In the *Cyropaedia* Xenophon had the virtuous Cyrus say that "it is a law established for all time among all men that when a city is taken in war, the persons and the property of the inhabitants thereof belong to the captors."[2] Polybius expressed the same law, adding that guilt or innocence of crime was irrelevant to the laws of war as they applied to the rights of conquerors over a captured city.[3] Livy commented that because "there are certain laws of war which are legitimately to be experienced as well as practiced, it is sad rather than unjust to the sufferer, that crops be burned, homes be destroyed, men and animals driven off as booty."[4]

Clementia was a sense of clemency that moved one to forgo these rights of the conqueror. Dionysius of Halicarnassus captured the ideal of *clementia* in the words of Titus Larcius, who was urging a lenient policy toward the Latins after the Romans had defeated them:

> And finally he [Larcius] asked them to take as examples the best actions of their ancestors for which they had won praise, recounting the many instances in which after capturing cities by storm, they had not razed them nor put all the male population to the sword nor enslaved them . . .[5]

Vergil, too, expressed the ideal of *clementia* in Anchises' famous words to his son Aeneas foretelling the destiny of Rome:

> ... Make peace man's way of life;
> spare the humble but strike the braggart down.[6]

War is a stern schoolmaster, however, and Roman practice frequently violated these ideals.

For one thing, the laws of war could not restrain crazed troops. In 190 B.C. the Roman commander Aemilius Regillus had negotiated terms with the people of Phocaea, whose city he was anxious to take intact because of its excellent harbor. He promised not to treat them as enemies if they opened their gates to him. But when his troops marched into Phocaea, they became enraged when they learned of these terms and broke ranks to plunder the city. In vain Aemilius tried to restrain them. But his argument that it was a violation of custom to sack a surrendered city, as opposed to a captured one, and that, in any case, soldiers could plunder only with the authority of their commander, did nothing to restrain his rampaging soldiers. The best Aemilius was able to do was to order the Phocaeans to gather in the marketplace where he could protect them. After the sack, he restored the city to the Phocaeans, but the Roman soldiers kept their loot. Apparently their rank insubordination went unpunished.[7]

The laws of war more often justified violence than restrained it. In 212 B.C. the Romans stripped Syracuse of all its wealth, including its artworks, and carried it back to Rome. The plundering of such a famous Greek city caused considerable indignation. Polybius thought the Syracusan booty began the corruption of Rome and judged that the Romans would have done better "to add to the glory of their native city by adorning it not with paintings and reliefs but with dignity and magnanimity."[8] Some senators, most notably Titus Manlius Torquatus, believed that in view of the great service of Syracuse to Rome in the past under King Hiero and out of self-interest for the value of such a powerful ally in the future, the Romans should have spared the city from plunder and restored its freedom. Marcellus, the Roman commander at Syracuse, had a clean conscience, telling the Senate that he had acted "by the law of war." The Senate decided that the looting had been legal, and it voted to confirm the actions of Marcellus. But it had qualms about the justice of the act and somewhat lamely voted also to restore the property to the Syracusans "so far as could be done without loss to the republic."[9]

As for *clementia*, the Romans measured it drop by drop. The famous dictator Marcus Furius Camillus enjoyed some reputation for clemency. In a speech to the Faliscans, he claimed that the Romans "bear no weapons against those tender years which find mercy even in the storming of a city."[10] His phrase "even in the storming of a city," made clear that he believed if ever circumstances justified the killing of children, it was in city warfare. Indeed, when the Romans broke into Veii in 396 B.C. they massacred many of its inhabitants before Camillus halted the carnage. He sold the survivors into slavery.[11]

When the city of Capua surrendered to the Romans after having betrayed Rome in the Second Punic War, its citizens appealed to Roman clemency, "known to them frequently in many wars." The Romans arrested fifty-three Capuan senators who had supported the desertion of Rome. The two commanders at Capua, Quintus Fulvius Flaccus and Appius Claudius, disagreed on what to do with these prisoners. Claudius was willing to entertain pleas for pardon, which he wanted to refer to the Roman Senate. Fulvius Flaccus took a hard line. He believed only two Capuans deserved mercy, a woman named Vestia Oppia who had sacrificed every day for the victory of the Romans, and one harlot named Pacula Cluvia who had sneaked food to some captives in Capua.[12] The rest of the Capuans, he argued, deserved summary punishment. Claudius thought they had agreed to await instructions from the Senate, but Fulvius hurried to the two neighboring towns where they had imprisoned the fifty-eight Capuan rebels and, after torturing them, executed them by decapitation. At the second town, a letter from the Senate reached Fulvius, but he put it in his pocket without reading it and proceeded with the executions. The letter, as he had anticipated, reserved judgment on the rebels for the Senate.[13]

The Romans confiscated all the gold and silver in Capua, of which there was a great deal, but they did not sack or destroy the city.[14] Although they were tempted, Campania was too valuable to Rome for the Romans to want to leave such a bitter legacy of the siege. So the city survived intact.[15]

As for its people, the Senate followed a deliberate policy. The nobility received the harshest treatment. The Senate dealt with each noble Capuan family separately after careful investigation. Some suffered confiscation of their property and enslavement of the entire family, except for daughters who had married men from other cities. Others were imprisoned with the possibility of further action against them later. The property-owning classes below the nobility retained some of their property and their freedom. The Romans returned cattle and slaves, except for adult males, to their owners. Despite this, however, a hard fate awaited them. The Romans transported them to various places in Italy, determining the distance from Capua by the extent of their disloyalty to Rome.[16] The city of Capua was left to the lower classes—tradesmen and artisans, resident aliens, and freedmen. There would be no self-government; a Roman magistrate would come out every year to hold trials.[17]

Livy took pride in this example of Roman justice and clemency:

The matters concerning Capua were settled according to a plan that was in every respect praiseworthy. Stern and prompt was the punishment of the most guilty; the mass of citizens were scattered with no hope of a return; no rage was vented upon innocent buildings and city-walls by burning and demolition. And along with profit they [the Romans] sought a reputation among the allies as well for clemency, by saving a very important and very rich city, over whose ruins all Campania, all the neighboring peoples on every side of Campania, would have mourned.[18]

A Macedonian ambassador took a different view:

Capua indeed, tomb and monument of the Campanian race, survives, its people buried, exiled, driven away, a city despoiled, without senate, without people, without magistrates, a monstrosity, more cruelly left habitable than if it had been destroyed.[19]

For him, the Romans had made Capua a mausoleum and called it clemency.

The customs of war distinguished between the storming of cities and the conquest of the countryside. Polybius revealed the distinction between siege warfare and conventional warfare when he reported that Hannibal, as he marched through Italy, put to death all the adult Romans whom he encountered, "as at the capture of cities by assault."[20] In other words, Polybius considered massacre standard in cities but not in the countryside. He could only explain Hannibal's order as a manifestation of an inordinate hatred of the Romans. When Hannibal killed all the inhabitants of Geronium after a brief siege, Polybius did not think any explanation was necessary.[21]

Livy also thought the laws of war sanctioned massacre and enslavement in siege warfare. In a speech he put in the mouth of Alorcus, a Spaniard in the service of Hannibal who was trying to persuade the Saguntines to accept Hannibal's terms of surrender (surrender of the city and all its wealth; the people allowed to leave with only the clothes on their backs), Livy had Alorcus say:

but even this you ought, I think, rather to endure than to suffer yourselves to be massacred and your wives and children to be forcibly dragged away into captivity before your eyes, in accordance with the laws of war.[22]

With massacre and enslavement so common that Livy can have Alorcus justify them under the "laws of war," it is clear that no one considered it wrong to take such actions toward a city that had refused to surrender.

There is no doubt, however, that Livy considered it wrong to massacre the inhabitants of a town that surrendered. When Hannibal ordered the sack of the Italian town of Victumulae, which had opened its gates to him and accepted a Carthaginian garrison, Livy's outrage was apparent:

> Being commanded to give up their weapons they [the people of Victumulae] complied: whereupon a signal was suddenly given to sack the town, as if they had taken it by storm. Nor was any cruelty omitted which historians generally deem worth noting on such an occasion; but every species of lust and outrage and inhuman insolence was visited upon the wretched inhabitants.[23]

Polybius was also capable of outrage over atrocities in siege warfare. When the people of the Cretan city of Cydonia betrayed the neighboring city of Apollonia, with whom they had taken solemn oaths of friendship, by seizing the city, killing all the adult males and dividing the women and children among themselves, Polybius denounced these actions as "a shocking act of treachery universally condemned."[24]

The laws of war often took a back seat to practical considerations. As we have seen, a Roman commander (Quintus Mucius) on campaign in Illyria in 171 B.C. allowed his soldiers to sack a city that had surrendered and that he had agreed not to plunder. His reason was that he had unsuccessfully besieged a nearby town and he feared a mutiny if his soldiers carried out two sieges without taking any booty.[25]

Necessity could override the laws of war. When the Numidian city of Capsa surrendered without resistance to Gaius Marius during the Jugurthine War at the end of the second century B.C., the Romans nevertheless killed all the adult Numidians and razed the city. Sallust, no admirer of Marius, justified the violation of custom by the necessity of war:

> The consul [Marius] was guilty of this violation of the laws of war, not because of avarice or cruelty, but because the place was of advantage to Jugurtha and difficult of access for us, while the people were fickle and untrustworthy and had previously shown themselves amenable neither to kindness nor to fear.[26]

The laws of war could be ambiguous. After the Greek city of Ambracia fell to the Romans in 189 B.C., a dispute arose over whether or not the city had fallen by force. After a two-week siege, the Ambraciots, after considerable pleading by mediators, had decided to open their gates to the Romans. The Romans agreed to allow an Aetolian garrison in the city to leave safely, and the Ambraciots agreed to pay the Romans 150 talents in gold. When the Romans entered the city under these terms, they carried away its statues and paintings, but Livy said that "nothing else was touched or harmed."[27] Nevertheless the theft of artworks from a city that had voluntarily opened its gates to the besiegers was a violation of custom.

Later the political enemies of Marcus Fulvius, the Roman commander, attacked him in the Senate on this score. Ambraciot citizens whom they brought in to lodge complaints even charged that Fulvius had sold their women and children into slavery, even though Livy had explicitly said that nothing but Ambraciot artworks had been seized.[28] Fulvius's enemies had enough leverage to push through a senatorial decree stating that Ambracia had not fallen by force, thus implicitly condemning his actions.[29] Fulvius insisted that he had taken Ambracia by force, pointing out that his troops had fought over the walls for two weeks and that over three thousand Ambraciots had died in the battle. He conveniently did not mention that his soldiers had been unable to force their way into Ambracia and that the Romans ended up entering the city under negotiated terms.[30] Fulvius got his way, and the Senate voted him a triumph in which he proudly displayed the siege machinery that had failed to reduce Ambracia.[31]

Another violation of the laws of war took place at the end of the Third Macedonian War in which the Romans defeated the Macedonian King Perseus. The Roman victor, Lucius Aemilius Paullus, ordered two Macedonian towns and a Thessalian one sacked in 167 B.C. before commissioners from Rome arrived to implement the terms of peace, which included the provision that the Macedonians would keep their own cities. Paullus had grievances against each town: one had switched back to the Macedonian side after having surrendered to Rome; one had abused some Roman soldiers who had fallen into their hands after the war because they did not know the war had ended; one had resisted the Romans more tenaciously during the war than the surrounding towns. But the laws of war could not justify the sacking of cites in peacetime. Paullus simply wanted to give his troops one last chance for plunder.[32] They were ungrateful. Later they opposed his triumph in Rome because they believed he had allotted insufficient booty to them.[33]

Not all complaints to the Senate about violations of the laws of war were unsuccessful. For a time in the second century B.C. the Senate exercised the right to overturn decisions made by commanders in the field.[34] An egregious case of Roman misconduct took place at the Greek town of Locri in southern Italy late in the Second Punic War. Scipio had captured the place in 205 B.C. and then returned to Sicily, where he was preparing for the invasion of Africa. He had left Locri under the command of his lieutenant Quintus Pleminius. Even though Locri had been held under a Carthaginian garrison and had been betrayed to the Romans by the Locrians, Pleminius maintained no discipline at all and allowed his troops to rape and steal. They even looted temples.

When the military tribunes attempted to stop the outrages, a virtual civil war broke out among the Romans. Scipio had to interrupt his work in Sicily to sail to Locri and restore order. Because Pleminius represented Scipio's authority, Scipio sided with him. He ordered the tribunes to be sent to Rome for trial.

Pleminius, who had lost his ears and nose in the fighting, was not satisfied with this relatively mild measure. After Scipio returned to Sicily, he tortured the tribunes to death and left their bodies unburied. Those Locrians who had complained to Scipio about the stealing and raping received the same treatment.

Livy strongly disapproved of Pleminius, remarking that he "brought infamy and odium not only upon himself but also upon his general."[35] Indeed, the Locrians brought their case to the Roman Senate, where they received a sympathetic hearing. The violation of the temples especially outraged the Senate, and it ordered the arrest of Pleminius. In addition, the Senate restored the stolen goods to the Locrians and organized rites of expiation for the violation of the temples. Pleminius died in prison before his case could come to trial. The scandal damaged Scipio, but efforts by his political enemies to deprive him of his command failed.[36]

Another case of senatorial intervention came in 170 B.C. when the Greek city of Abdera, which was an ally of Rome, brought charges against the Roman praetor Lucius Hortensius. Hortensius had demanded 100,000 denarii and 50,000 pecks of wheat from Abdera. The Abderites had requested a delay while they sent envoys to the consul to discuss this command, as was their right as an ally. Apparently Hortensius agreed to this, but after the envoys left he treacherously stormed the town, beheaded its leaders, and sold everybody else into slavery. The Senate found Hortensius's action a "disgraceful occurrence" and sent orders to Hortensius that

the Senate had resolved that an improper war had been undertaken against the people of Abdera, and that it was just that all who were enslaved should be sought out and restored to freedom.[37]

The Senate had rendered a similar decision about the Boeotian town of Coronea, whose inhabitants had been enslaved by Licinius Crassus in 171 B.C. Not only did Rome free the Coroneans; it also imposed a fine on Crassus.[38]

Julius Caesar interpreted the laws of war precisely. If the town surrendered before the Roman battering rams touched the walls, Caesar accepted terms that required the towns to give up arms and hostages but left the towns free. He granted these terms at the Gallic towns of Noviodunum, Bratuspantium, and Vellaunodunum.[39] If there was any trickery, however, Caesar was ruthless. After he had granted these terms to an unnamed stronghold and even ordered his troops to leave the town so there would be no random stealing or raping, some of the

Gauls attacked the Romans that night with hidden weapons. The attack failed, and Caesar sold all fifty-three thousand inhabitants of the town into slavery.[40]

Caesar dealt harshly with cities that resisted to the end. As we have seen, there was a terrible massacre at Avaricum in which the Romans indiscriminately killed men, women, and children. Only eight hundred Gauls out of forty thousand survived. The Roman troops were full of vengeance for a recent massacre of Romans at the town of Cenabum and had endured a long siege, so Caesar was indulgent toward them:

> In such fashion the troops, maddened by the massacre at Cenabum and the toil of the siege-work, spared not aged men, nor women, nor children.[41]

At Alesia Caesar demanded the surrender of arms and the arrest of the Gallic chiefs. He spared the men from two of the Gallic tribes with which he wanted to maintain good relations. The rest of the Gauls he enslaved, one slave for each Roman soldier as booty for the siege.[42] Since he commanded some forty thousand men, that meant he enslaved at least forty thousand people. This is a very large number, and because ancient numbers frequently are unreliable some have doubted it. But given the large scale of slavery in the Roman world, just as many have found the number plausible.[43]

Caesar's most savage treatment of a captured city came at Uxellodunum. He was determined to make an example of this city to the rest of Gaul. He therefore ordered his soldiers to cut off the right hand of every Gallic warrior captured at Uxellodunum. This punishment so mortified Drappes, the Gallic chieftain in command at Uxellodunum, that he starved himself to death.[44] In short, if a town resisted, the laws of war permitted enslavement, massacre, and mutilation.

Aulus Hirtius said that Caesar had already established a reputation for clemency in Gaul so that he did not have to fear that atrocity would drive the Gauls to last-ditch resistance. The modern historian J. F. C. Fuller takes the opposite view, calling Caesar's treatment of the Gauls "barbarities" and suggesting that had Caesar acted more humanely the conquest of Gaul might have been easier. Fuller cites Aineias Tacticus's advice to avoid savage treatment of captured cities in order not to arouse desperate resistance in future sieges as words Caesar should have heeded.[45]

Caesar's reputation for clemency was not entirely undeserved. Especially during the Civil War he was anxious not to split Roman society irrevocably. The treatment of Massilia was his greatest act of clemency. Massilia had treacherously violated a truce that followed the city's surrender to Caesar's army. Caesar's soldiers had had to rebuild their siege works and start the siege all over again. This was usually a situation that led to massacre. But upon Massilia's second surrender, Caesar satisfied himself with demanding the surrender of arms, ships, and artillery as well as Massilia's public treasury. His soldiers were undoubtedly dissatisfied, but Caesar sent most of them back to Italy, leaving a strong garrison in the defeated city. Caesar said that he spared the Massilians "more on account of the name and antiquity of their state than for anything they had deserved." They were very lucky indeed.[46]

Regard for the laws of war could lead only to hypocrisy. In A.D. 49 the Armenian town of Uspe tried to surrender to a besieging Roman army, offering ten thousand slaves in return for mercy for the freeborn. The Romans rejected the offer because they did not think they could guard such a large number of people and because they believed it would be inhumane to kill the people of a surrendered city. Instead they carried on with the assault on the city, which was weakly fortified, and massacred its people. In this way they avoided violating the laws of war.[47]

Massacre, Torture, and Enslavement

The legends of early Roman siege warfare preserved terrible memories, reflecting how the Romans themselves viewed the nature of siege warfare.[48] Massacre, torture, and enslavement were all common.

Although it is difficult to sort out the different legends of the siege of Suessa Pometia at the end of the sixth century B.C., all the versions agree that the town suffered horribly. According to Dionysius of Halicarnassus, the Roman commander Lucius Tarquinius killed all the men of military age and turned over the rest of the population, including women and children, to the soldiers as booty.[49] Livy said that Lucius's own share of the sack was so great that he was able to sell it for forty talents, a sum sufficient to inspire in him a plan to build a grandiose temple for Jupiter.[50] Pometia apparently recovered rapidly, because a few years later the Romans took it again. This time they cut off the heads of the chief magistrates of the city and sold the rest of the population into slavery. The Romans destroyed the town utterly and sold off its land.[51]

Livy's characteristically uncritical use of his sources apparently led him to relate different versions of the same war against Pometia as separate actions. One of the versions agreed that the Romans slaughtered their prisoners and stated that they killed three hundred hostages as well.[52] Apparently these hostages were all children.[53] Another version simply said that the Romans sacked the city.[54] However, the different versions were not mutually exclusive and, under the circumstances, there is little reason to doubt that the Romans dealt with Pometia harshly. To be sure, such early stories of Roman siege warfare existed more in legend than in history, but they were not implausible by the standards of ancient siege warfare.

Massacre was a consistent theme in the stories of early Roman siege warfare. In a war against the Volsci, the Romans routed the enemy in a field battle and then burst into the city of Velitrae on the heels of the fleeing soldiers. Livy says that "more blood was shed there [in the city], in the promiscuous slaughter of all sorts of people, than had been in the battle itself."[55] In more direct words, the Romans indiscriminately killed women and children in Velitrae. Those who survived lost their land, and the Romans colonized the place.[56]

Another early example of Roman brutality took place at Tusculum in 459 B.C. The Aequi had captured the citadel, and it had taken the Tusculans and Romans months to starve them out. After the Aequi finally surrendered, the Tusculans stripped them and forced them to march under the yoke. The Romans were apparently not satisfied with this humiliation, for they overtook the Aequi as they were making their way home and massacred them all.[57] When the Romans drove the Etruscans from the city of Sutrium in 386 B.C., heavy infantry massacred those caught in the city while light infantry pursued those who escaped, killing all who fell into their hands. Only nightfall stopped the slaughter.[58]

The Romans could make distinctions in their killing, as they did at Nepete, a town they captured in 386 B.C. The Nepesini were Roman allies, but a faction within the town had betrayed it to the Etruscans. The Romans had asked the Nepesini to break with the Etruscans and surrender their town to the Romans. But the Nepesini pleaded that they were powerless to do so, and the Romans had to resort to a siege. When the town fell, the Romans killed only the Nepesini who had surrendered the town to the Etruscans. The others regained their property and continued to live in their town, although under a Roman garrison. As for the Etruscans, the Romans killed them all.[59]

Such brutality was matched by Rome's enemies. When the Gauls captured Rome in 390 B.C., furious at the continued Roman resistance from the capitol, they killed everyone, old and young, women and children.[60] In 406 B.C., the Volsci massacred a Roman garrison at Verrugo.[61] The Romans avenged the Volscian torture of Roman captives when they took the Volscian city of Anxur and gave no quarter to the defenders. In this case, however, the Roman commander Numerius Fabius Ambustus ordered a halt to the massacre and about 2,500 Volsci were able to surrender.[62] The Volsci seem to have been particularly ferocious. Livy said that when they captured the town of Satricum, which was a Roman colony, in 382 B.C., they "abused their victory by cruel treatment of the captives."[63] He gave no details, but he must have meant the Volsci tortured the captives.

One of the most horrible cases of cruelty by Rome's enemies came in the Samnite war of 320 B.C. The Samnites were besieging the Roman colony of Fregellae and persuaded most of its inhabitants to surrender by promising safety to all who did so. But after the surrender, the Samnites cooped up the people of Fregellae in their city, set the town afire, and burned them all to death.[64]

Another case of Samnite cruelty occurred in 311 B.C. when the Samnites starved out a Roman garrison at Cluviae, a town in Samnium. Even though the Romans had surrendered, the Samnites tortured the prisoners and then killed them. The Romans immediately retook Cluviae and killed all the adult male Samnites.[65] After capturing the Roman colony of Sora, which had revolted in 314 B.C., the Romans sent 225 rebel leaders back to Rome, where they were publicly tortured and beheaded.[66] Soon afterward the Hernici captured the Roman garrison at Sora and treated the Roman prisoners with what Livy called "shameful rigour."[67] Clearly, torturing prisoners was not uncommon in Italy.

Torture and beheading were common Roman practices. After taking Sassula and other Tiburtine towns in a war against the Tarquinienses and Tiburtes in 353 B.C., the Romans treated the Tiburtes with clemency. But the Tarquinienses had recently killed 307 Roman prisoners of war in a sacrifice to their gods, a deed Livy called "an act of savage cruelty."[68] In revenge, after defeating the Tarquinienses, the Romans killed them all, except for 358 nobles. These the Romans took to Rome where they publicly tortured and then beheaded them.[69]

After the capture of Satricum, which had admitted a Samnite garrison, in 319 B.C., the Romans made a careful investigation to determine who was responsible for allowing the Samnites into the town. They tortured and beheaded the guilty.[70] Diodorus called this sort of treatment of rebels an "ancestral custom" of the Romans.[71]

Polybius confirmed that the Romans tortured and beheaded prisoners, a practice he called "according to their custom." In this case the victims were some three hundred Campanians who had served as a Roman garrison in Rhegium to protect the town from both Pyrrhus and the Carthaginians. But the Campanians betrayed the Romans and took Rhegium themselves (and cruelly abused its people), forcing the Romans to besiege the place in order to restore it to the people of Rhegium. Most of the Campanians fought to the death, but the Romans made an example of the three hundred survivors in order to restore Roman credibility with their allies.[72] That was in 271 B.C. In 141 B.C. the Roman commander Servilianus on campaign in Spain captured a number of towns and beheaded five hundred prisoners.[73]

Another massacre that occurred out of Roman vengeance took place at Luceria in 320 B.C. When the Romans broke into a Samnite camp at Luceria, they sought revenge for the humiliating Caudine Peace in which the Samnites had forced the entire Roman army to pass under the yoke; the troops "cut down without distinction those who resisted and those who fled, the armed and the unarmed, slaves and freemen, adults and children, men and beasts." Only the consuls' concern for the safety of Roman hostages in Luceria stopped this slaughter.[74] When Luceria revolted again in 314 B.C., there was no intervention. The Romans killed all the Samnites and Lucerians, presumably including the women and children. Serious consideration was given to razing the city, but in the end the Romans sent a colony to resettle it.[75]

A century later the Romans massacred another entire town in revenge. This happened in 206 B.C. at Ilurgia, a Spanish town that five years earlier had killed Roman soldiers seeking refuge there after a disastrous battle in which Scipio's uncle died. The people at Ilurgia knew the Romans were anxious to avenge this treachery and fought with a desperation born in fear. Women and children helped defend the town by carrying weapons to the soldiers and supporting them "beyond their powers of mind and body." The fierce resistance forced Scipio to take over personal leadership of the assault on the wall, and he may have been wounded.[76] All the causes of massacres thus came together at Ilurgia: vengeance, fierce resistance by the defenders, women and children involved in the fighting, the commander wounded.

When the Romans broke into the town, they did not turn to plunder. They concentrated on killing:

> They slaughtered the unarmed and the armed alike, women as well as men; cruel anger went even so far as to slay infants.[77]

Livy did not say so explicitly, but there were apparently no survivors at Ilurgia. The Romans obliterated the city. Appian said that the massacre was a spontaneous outburst by angry soldiers, not the result of a command.[78] It is much more likely that Scipio ordered the massacre as a lesson to other Spanish towns that were reluctant to succumb to the Romans.[79]

The nearby town of Castulo suffered quite a different fate. It had switched to the Carthaginian side at the time of the Roman defeat, but it had not killed any Roman refugees. Moreover, alarmed by the fate of Ilurgia, the people of Castulo betrayed the Carthaginian garrison in the town and opened the gates to the Romans. There was no massacre at Castulo.[80]

Another example of massacre and torture serving to intimidate other towns took place earlier in the Second Punic War. A disaffected Carthaginian commander betrayed the Greek Sicilian town of Agrigentum to the Romans by secretly opening a gate for them. When the Carthaginians and Agrigentines realized the Romans were in the city, they desperately tried to flee, but the Romans closed the gates and slaughtered the wretched crowds clamoring at the gates. Of the survivors, the Romans tortured and executed the leaders and sold the rest of the population into slavery. This savagery had the desired effect; in the next few weeks, sixty-six Sicilian towns submitted to the Romans, only six of which the Romans had to assault.[81]

Caesar sacked cities as an example to others. During his Balkan campaign against Pompey, Caesar entered Thessaly after he had disengaged from Pompey's army at Dyrrachium. The first city he came to was Gomphi, which resisted because it believed Pompey had the upper hand against Caesar. Even though the city had high walls, Caesar's soldiers took Gomphi by storm and Caesar allowed them to sack the town. When the people of Metropolis heard of the fate of Gomphi and opened their gates to Caesar, he kept his soldiers under strict discipline to protect the lives and possessions of the Metropolitans. The lesson was not lost on the rest of the towns of Thessaly, and almost all of them surrendered to Caesar.[82]

Discipline was a critical factor in the fate of cities captured by the Romans. In 314 B.C. the Romans captured three Auruncan towns by treachery when a group of young Auruncan nobles opened the gates to Roman soldiers. Thus the cities were taken quickly and easily, often a circumstance in which clemency prevailed. But the simultaneous seizures made it impossible for the Roman commanders to be present and "because the leaders were not present when the attack was made, there was no limit to the slaughter, and the Ausonian [Auruncan] nation was wiped out—though it was not quite clear that it was guilty of defection—exactly as if it had contended in an internecine war."[83] Livy's words implicitly condemned the slaughter, but the fate of these towns offered terrible proof of the frenzy of city fighting, a frenzy that only a strong-willed leader could halt.

Disagreements between commanders could have dire consequences for a conquered city. In 214 B.C. the Romans recaptured the town of Casilinum, which was held by Campanians friendly to Hannibal. Both consuls were present, and after an initial assault on the wall had failed, there was a disagreement between them. Quintus Fabius favored breaking off the siege and moving on to more pressing tasks. But the other consul, Marcus Marcellus, convinced Fabius that they should not raise a siege once started because it would encourage other towns to resist. The Campanians wanted to give up the town in return for a safe passage to Capua. Knowing that Fabius was eager to end the siege, they approached him to negotiate. They must have thought a truce was in effect, because they opened a gate. But Marcellus was in no mood to negotiate, and he seized the gate through which the Campanians were leaving to seek out Fabius. Marcellus's soldiers burst into the town and began slaughtering the Campanians. Once the Romans had captured Casilinum, however, the consuls stopped the slaughter. They sent the military captives to Rome in chains, except for fifty who had reached Fabius and apparently gained his promise to allow them to go to Capua.

The townspeople were sent to neighboring towns friendly to Rome to be kept under guard. They were lucky to have escaped enslavement.[84]

The need for money could influence Roman policy. After taking Panormus in 254 B.C. the Romans accepted terms under which the people could purchase their freedom for two minas. Fourteen thousand were able to produce this amount. The Romans sold the remaining thirteen thousand into slavery.[85]

By the Second Punic War, Roman brutality was systematic and calculated. Polybius explained what happened in the capture of New Carthage:

> When Scipio thought that a sufficient number of troops had entered he sent most of them, as is the Roman custom, against the inhabitants of the city with orders to kill all they encountered, sparing none, and not to start pillaging until the signal was given. They do this, I think, to inspire terror, so that when towns are taken by the Romans one may often see not only the corpses of human beings, but dogs cut in half, and the dismembered limbs of other animals, and on this occasion such scenes were very many owing to the numbers of those in the place.[86]

Scipio ordered an end to the killing only after the citadel had surrendered.[87] The massacre had taken place under completely disciplined conditions. The scene that Polybius described of human corpses lying with dismembered animals recalls similar scenes in the Biblical account of Jericho or Thucydides' description of the Thracians at Mycalessus, scenes that blurred the fundamental distinctions between humans and beasts.

Scipio was not always able to maintain discipline over his troops. Appian told the story that during the siege of Locha, a large town in North Africa, the Roman troops "suffered great hardships." When the fall of the city seemed imminent, the Lochaeans attempted to surrender under the terms that they would abandon their city. The Roman troops were already scaling the wall, and when Scipio ordered them to withdraw, they refused. Instead they continued into the city and slaughtered everyone they found, including women and children. There were some survivors, and Scipio sent them away into safety. He punished the troops by depriving them of the booty from the town, and he executed three officers, chosen by lot, from those who had either refused or been unable to maintain control over the rampaging troops.[88]

It was not unusual for commanders not to be able to maintain discipline after a siege. We have personal testimony of this problem from Julius Caesar. During the Civil War Caesar had shut up in a town some Roman troops loyal to Pompey. Toward evening the troops decided to surrender and join Caesar's forces. Caesar was most anxious to have these troops, and any delay would provide them an opportunity to change their minds. But he feared that if he occupied the city so late in the day, he would not be able to control his troops during the night and they would turn to plunder. The latter risk seemed greater, and he kept his troops outside the city while he continued to negotiate the delicate transfer of the besieged Romans to his army.[89]

Another controversial case involving a failure to maintain discipline took place in A.D. 69 at Cremona, an Italian town in northern Italy, during the civil war that brought Vespasian to the emperorship. Cremona had sided with Vespasian's rival Vitellius, and when it was taken by troops favorable to Vespasian, they were eager for revenge. The commander, Antonius, tried to restrain them and did manage to prevent the soldiers from massacring the population when they first broke into the city. But he then carelessly departed for the baths to wash away the grime of battle, and in his absence the troops went berserk. Forty thousand of them poured into the city and began raping and pillaging. An even larger number of sutlers and camp followers rushed in with the troops, revealing a complete breakdown of discipline. Tacitus believed such moral breakdown came naturally to the motley imperial army, with its "varieties of language and character," and in which "each man had a law of his own, and nothing was forbidden." The sack of Cremona continued for four days.

Cremona had been a loyal Roman ally in the war with Hannibal, and its fate aroused deep emotion in Italy. Tacitus said that Antonius was "ashamed of the atrocious deed" and attempted to mitigate it by ordering that no citizens of Cremona be sold into slavery. Indeed, the soldiers had been unable to sell their captives because of a general refusal by the people of Italy to buy them. When they realized they could not sell them, some of the soldiers began killing their captives, forcing friends and relatives to ransom the surviving ones. So the soldiers got their money anyway. Most of the blame fell on Antonius, but Tacitus indicted all of the generals for their failure to maintain discipline.[90]

The Romans enslaved people on a scale hardly matched since Assyrian times. Women and children were not the only victims. The Romans often sold men into slavery, although Livy did not consider it customary. For example, he related that after defeating the Volscian town of Satricum in 346 B.C., the Romans sold four thousand men into slavery for the benefit of the public treasury. But he expressed the opinion that those whom the Romans sold were more likely captured slaves, not prisoners of war.[91] However, Livy failed to explain what may have happened to the citizens of Satricum, so his retrospective opinion is not convincing.[92] In relating the capture of the Thessalian town of Mylae in 171 B.C. by the Macedonian King Perseus, Livy said that Perseus sold "even the free persons who survived the slaughter," the word *even* indicating that Livy did not consider this normal.[93]

However, there was no established practice and the Romans frequently did sell free men into slavery. When the Romans captured

Agrigentum after a seven-month siege in the First Punic War, Polybius said that they plundered the city, "possessing themselves of many slaves." Diodorus said they "carried off all the slaves." But the historian Zonaras understood this to mean that the Romans sold all twenty-five thousand inhabitants into slavery.[94] Zonaras's report is plausible, because Diodorus reported that the Romans were routinely selling the inhabitants of captured cities into slavery on this campaign, as they did, for example, at Mytistratus and Camarina.[95] One report said that the Romans killed the people of Mytistratus even though they had opened their gates to the Romans after the Carthaginian garrison fled during the night. The siege had lasted seven months, which undoubtedly put the Romans in a bad mood.[96]

In the desperate year of 215 B.C., the Romans sold all the inhabitants of three Italian towns that had deserted to the Carthaginians. They amounted to over five thousand people.[97] When the Romans went to Greece toward the end of the third century B.C., they took their practice of mass enslavement with them. The Greeks were capable of the same thing, but the arrival of the Romans brought a surge in the practice.[98]

Especially hard was the fate of Antipatrea, a city in Epirus that the Romans captured in 200 B.C. The Romans killed all the men of military age and placed the rest of the population in the hands of the soldiers to be sold into slavery. The city itself was completely destroyed.[99]

Although killing the males was an unusually harsh act, by this time the practice of enslavement had become a common means of pacification for the Romans.[100] Aemilius Paullus, in another punitive expedition in Epirus about 167 B.C., took seventy towns and enslaved 150,000 men.[101] Quintus Opimius sold all the inhabitants of Aegitna, a town in Gaul, into slavery in 154 B.C. after they had done violence to some Roman envoys.[102] When the Romans captured the Boeotian town of Haliartus in 171 B.C., again after a fearful slaughter in which they killed people of all ages as they stormed the city, they sold all 2,500 survivors at auction.[103] During the Jugurthine War in North Africa in 109 B.C., the consul Quintus Metellus waged a terror campaign in Numidia, sacking towns and killing all the adults. His purpose was to deprive the rebel King Jugurtha of his bases for the guerrilla warfare that he was waging.[104] Revenge may also have been a motive. After the city of Cirta had surrendered to Jugurtha, he had massacred the entire adult population, including many Italian traders who lived there.[105]

Self-interest could exercise a restraining influence on the Romans. After the terrible massacre at New Carthage in 212 B.C., less than ten thousand male survivors remained. They expected death or enslavement. But Scipio ordered them to return to their homes, and he returned what property had survived the siege to them. The artisans, who numbered about two thousand, were to serve as public slaves of the Romans, mostly manufacturing arms and siege equipment. The fittest of the young noncitizens and slaves he enrolled in the Roman navy, but he promised both the artisans and the impressed sailors that if they served loyally they could earn their freedom upon the defeat of Hannibal. Scipio's motive was to win goodwill for Rome, not only in New Carthage but in the rest of Spain as well. Apparently his policy worked, because Polybius said that

> by this treatment of the prisoners he produced in the citizens great affection and loyalty to himself and to the common cause, while the workmen were most zealous owing to their hope of being set free.[106]

But when he was angry, Scipio could be ruthless. After the Carthaginians had broken a treaty, he marched into the interior of Carthaginian territory in Africa, taking the towns without giving them an opportunity to surrender. He sold all the inhabitants into slavery.[107]

Encouraging surrender remained one of the most powerful inducements to grant generous terms. In 198 B.C. when the Roman consul Quinctius Flaminius broke into Elatia, its defenders withdrew to the citadel. Quinctius coaxed them out by offering to spare the lives of the Macedonian garrison and to leave the citizens of the town free.[108]

As Rome became a more self-conscious imperial power, it began to obliterate troublesome cities. In 146 B.C. both Carthage and Corinth met this fate. The Romans captured fifty thousand Carthaginians and razed the city. Most of the Carthaginians died in captivity; the Romans sold the rest into slavery. Carthage had resisted a three-year siege and could hardly expect mercy. But the Corinthians had abandoned their city after losing a field battle. Nevertheless, the Romans meted out the same treatment to them as they had to the Carthaginians, selling them into slavery and razing the city.[109]

The rebellious Jews suffered a great deal at the hands of the Romans. In 37 B.C. King Herod and his Roman allies required a five-month siege to bring Jerusalem to heel. When they finally broke into the city, both Herod's Jewish troops and the Romans went on a rampage in which they massacred men, women, and children. Herod tried to stop the slaughter but to no avail. Josephus said that the length of the siege had angered the Romans and Herod's troops wanted to exterminate their enemies, so "like madmen they wreaked their rage on all ages indiscriminately."[110]

A full-scale Jewish rebellion forced the Romans to mount a pacification campaign in A.D. 67. The future emperor Vespasian commanded this campaign. His first success was the capture of Gabara. Although the town was undefended and fell to the Romans easily on the first assault, they killed all the males, young and old. Vespasian razed not only Gabara but all the surrounding villages as well. He sold all captives into slavery.[111]

Another terrible massacre took place at Japha. When the Romans broke into the city after a siege, the Galilaean defenders continued to resist. Six hours of bitter street fighting followed. The women joined in, pelting the Romans with roof tiles. When the fighting was over, the Romans killed all the males except infants. In this massacre the Romans killed three thousand and captured 2,130. They sold the captives into slavery.[112]

When Jotapata fell after a difficult siege, there was another fearful slaughter. After the Romans took the citadel, the Jews were so hemmed in that they could not fight. What followed was more of a massacre than a battle. Many of the Jews avoided the massacre by committing suicide.[113] Even after the Romans had secured the town, the killing did not stop. During the next few days the Romans scoured the town, rooting out those who had hidden in underground vaults and caverns. Twelve hundred people were found. Again the Romans killed all males except infants and enslaved the women and babies. The city was razed. Josephus, who had commanded the defense of Jotapata, estimated the number of Jews who died defending the city at forty thousand.[114] In 1997 Israeli archeologists discovered a mass grave at Jotapata that they believe contained the remains of victims of the siege. The remains included many young children and teenagers.[115]

Even harsher was the fate of Gamala, taken by the Romans in the same campaign. In this case the Romans had suffered heavy casualties when they first broke into the city. Gamala perched precariously on a mountainside, and when one building collapsed in the fighting it started a domino effect in which much of the city fell into ruins, burying many Romans in the rubble. The catastrophe left Vespasian isolated with a few soldiers in the city, and only his coolness enabled him to escape. His soldiers were embarrassed by this fiasco, and when they captured Gamala they took out their frustration on its inhabitants. They gave no quarter, killing even the women and children, "for at that moment the rage of the Romans was such that they spared not even infants, but time after time snatched up numbers of them and slung them from the citadel."[116] In the face of such fury, many families leaped to their death into a deep ravine. Josephus reported that over five thousand people committed suicide in this way. Those killed by the Romans numbered four thousand.[117]

According to Josephus the slaughters at Jotapata and Gamala sickened Titus, the son of Vespasian. This is unlikely. When Vespasian sent him to subdue the small town of Gischala, Titus did indeed open negotiations, in Josephus's version because he knew that if his soldiers took the town by assault "a general massacre of the population would ensue." Josephus claimed that Titus "was already satiated with slaughter and pitied the masses doomed along with the guilty to indiscriminate destruction."[118] Therefore Titus approached the walls of Gischala and made a speech in which he offered to leave the people in possession of their property if they opened their city to him. He did not fail to point out the fate of those Galilaean cities that had turned down Roman clemency, revealing that his motives did not go beyond the common desire of any siege general to avoid a siege by offering generous terms.[119]

Gischala was under the command of a rebel leader named John, and he controlled the walls and the gates of the city. He did not allow the people to respond to Titus's offer, but he was willing to negotiate with Titus. John accepted Titus's offer but pleaded that it was the Sabbath, a day on which the Jews could not conduct business. He asked to delay the final agreement for a day. The real purpose of this request was to buy time for an escape. Titus was anxious enough for a peaceful conclusion to the negotiations that he agreed to the delay. He carelessly camped too far away from Gischala, and that enabled John to slip out of the city during the night, accompanied by his armed band of rebels. Many of the wives and children of the rebels went with them. John made for Jerusalem with all speed, and when the women and children could not keep up he ordered his men to abandon them.[120]

The next morning the people who had remained at Gischala opened the city to Titus. When he learned of John's flight, he sent his cavalry in pursuit. John made it to Jerusalem, but he had not been able to keep his force together and the Romans caught and killed some six thousand of his followers. They also rounded up about three thousand women and children, whom they brought back to Gischala. Despite the trickery of John, Titus treated Gischala leniently. He tore down a section of the wall as a token of victory and stationed a Roman garrison in the city. He did not harm the people or their property, not even attempting to identify and punish any rebels who remained in the city, fearing that an inquiry would bring a wave of false charges from people looking to settle personal grudges and lead to the punishment of innocent people.[121]

At Jerusalem, Titus showed that he by no means had had enough of bloodshed. One of the first things he did when he besieged the city was to crucify a prisoner. Josephus said that he did this "in the hope that the spectacle might lead the rest to surrender in dismay."[122] The failure of this crucifixion to bring about the desired result did not discourage Titus. Later in the siege he ordered mass crucifixions of Jews who had tried to escape from the city. Most of these Jews were noncombatants, poor people whom famine had driven from the city. Titus had them tortured before they were crucified. Josephus gave two reasons for Titus's cruelty: the risk of holding prisoners, along with the drain on Roman manpower it would require, and Titus's continuing hope that the sight of tortured and crucified fellow citizens would intimidate the Jews into surrendering.

Such contingencies of war frequently decided the fate of a city. In

A.D. 58 the people of the Armenian capital of Artaxata surrendered to a Roman army. This act saved their lives. But the Romans did not have a large enough force to garrison the city, so they completely destroyed it, leaving the people of Artaxata with no city.[123]

Josephus said the Roman soldiers at Jerusalem amused themselves by crucifying their victims in a variety of positions. The crucifixions continued at a rate of five hundred a day, so there was scarcely any space left for fresh crucifixions and the Romans had trouble keeping up the supply of crosses.[124] In this case the crucifixions failed to bring about a surrender. But we do have a successful example of this method. In the mop-up operations in Palestine after the fall of Jerusalem, the Romans besieged the fortress of Machaerus. This fortress sat atop a rocky prominence, a naturally strong position that made it virtually impregnable. The Romans, however, in the course of their siege operations were lucky enough to capture a young Jew from a prominent family who had distinguished himself greatly in the sallies against the Roman earthworks. The Romans brought this young man before the fortress and tortured him in sight of the Jews on the walls. That caused consternation among the Jews, and when the Romans brought a cross forward as though they were going to crucify their prisoner, the Jews surrendered. The difference was that Titus's victims were from the lower classes while this prisoner was from a family of great influence.[125]

In another effort to horrify the Jews into surrendering, Titus cut off the hands of several prisoners and sent them back to Jerusalem to plead with the leaders to save the city by giving it up to the Romans.[126]

If he treated prisoners harshly in the hope that his brutality would demoralize the Jews into surrendering, Titus also offered lenient terms in the same hope. When his troops broke through the second wall, Titus did not allow them to sack any of the houses or to kill any of the Jews who were trapped. Instead he offered to leave the Jews free and in possession of their property if they would surrender.[127] Also, if Titus was cruel in his efforts to persuade Jerusalem to surrender, he did not like gratuitous violence. When Arab and Syrian soldiers began disemboweling Jewish refugees from Jerusalem in search for gold coins that some of the refugees had hidden by swallowing, Titus reacted angrily and ordered an end to the gruesome atrocity, although such was the strength of his soldiers' greed that the practice continued in secret.[128]

Titus's control over his troops was tenuous. According to Josephus, when the Romans finally took the temple after a long and arduous struggle, Titus tried to save what was left of it. That seems unlikely, but the picture that Josephus drew of soldiers out of control, who could not hear orders and ignored signals, was typical of sacks. Many of the survivors of the siege had packed into the temple, and the Roman soldiers in their frenzy massacred them. Most of them were civilians, half dead with famine. As Josephus commented, "passion was the only leader."[129]

Josephus painted a lurid picture of the temple in flames and the soldiers massacring the people, both young and old. Josephus was probably an eyewitness, and he recalled an infernal noise that deafened the ears.[130] The Romans either burned or razed the temple, but only after Titus had violated and looted the inner sanctum. Six thousand women and children took refuge in a portico of the outer court. There the Romans burned them to death.[131]

Titus's treatment of the captives taken in the fall of Jerusalem varied with the circumstances. When the high priests surrendered after the destruction of the temple, Titus said that it was too late for mercy. He believed the priests should have either saved the temple by surrendering or died in the temple. He ordered their execution.[132] But when the remnants of the defenders retired to the upper city, an extraordinarily strong fortress, Titus offered to spare their lives if they surrendered.[133] The priests' surrender had done him no good, and so his actions were governed by his sense of appropriateness. His contempt for the rebels was unbounded, but their surrender would be very useful to him, so he offered them mercy.

When the rebel leaders spurned Titus's offer of mercy, he announced that from that point on the laws of war would govern his conduct. This was bad news for the Jews, because the laws of war authorized sacking a captured city and killing or enslaving the population.[134] However, expediency continued to govern Titus's actions. When the Idumaeans, who had lent crucial support to the rebels, offered to surrender, Titus agreed to spare them. The rebels thwarted the Idumaeans' plan to surrender, but nevertheless a steady stream of refugees came down from the upper city. These pathetic refugees Titus spared out of a sense of clemency, according to Josephus. The massacre at the temple had glutted the bloodlust of the soldiers, and they now looked more to profit than revenge. Accordingly, the Romans sold the women and children into slavery, although they did not get a good price because of the glut on the market and the lack of buyers. Titus held an inquiry to determine who among the male refugees deserved punishment. Most of them he allowed to wander away. One should remember that they had nothing and were suffering from famine.[135]

When the Romans finally took the upper city, they killed all who fell into their hands.[136] When Titus entered the upper city, the work of slaughter had begun to fatigue his soldiers and he issued orders to kill only armed men who were still fighting. This ended the indiscriminate killing, although the Roman soldiers continued to kill the old, presumably because they would be worthless in the slave market.

An inquiry sought out the rebels among the prisoners, and the Romans executed them. During the time the Romans spent in the inquiry, eleven thousand of the prisoners died of starvation.

Titus sent the tallest and best-looking young men to Rome for the triumph. The Romans sold all those who were under seventeen years of age into slavery. Of those over seventeen, the Romans sent some to forced labor in Egypt; the rest Titus distributed among the provinces to serve as sacrificial victims in the games.

There still remained several thousand Jews who had hidden in caverns under the city. The Romans killed all they could root out.

Josephus put the number of Jewish dead in the siege of Jerusalem at 1,100,000, an impossibly high number.[137] The Romans razed Jerusalem, except for the west wall, which served as shelter for the garrison, and three towers, which remained to remind posterity of the might of the city that had fallen to Roman prowess in siege warfare.[138]

Shame, Women, and Suicide

An aura of shame hung over Roman siege warfare. Livy, for one, considered fighting from behind walls shameful. He condemned the cowardice of Roman soldiers in one battle who "nowhere ventured to fight in the open field, but defended themselves by the position [on high ground] and their rampart, not by bravery and arms."[139] In a description of a battle between the Romans and Volsci and Aequi, Livy said that the latter "were ashamed that their victorious armies should depend for protection upon stockade, instead of valour and the sword."[140] In contrast Livy admired the defense by the Phocaeans of their city after the Romans had breached the walls:

> the townspeople resisted so stubbornly that it was easily apparent that they placed more reliance on arms and courage than in walls.[141]

Vergil echoed the same values when he had the Italian hero Turnus taunt the Trojans of Aeneas who were holed up in their fortified camp:

> The wall between us makes them bold;
> the ditch detains us and postpones their death—
> hence they are brave![142]

Turnus scorned the wiles of the Greeks:

> Who'll take the steel with me
> against their wall, and charge their coward's camp?
> I need no thousand ships, no Vulcan's armor
> against the sons of Troy. Let Tuscans all
> join them! I'll not sneak in by night and steal
> their 'Pallas the Less,' nor knife their temple guards—
> no fear! Nor horse shall hide us in its belly!
> In broad daylight I'll circle their walls with fire.
> We're no Greeks—no pale Pelasgian boys
> (they'll see), that Hector held off for ten years.[143]

Livy also contemned ruses in siege warfare, as we can see in his story of a ruse the tyrannical Lucius Tarquinius (sixth century B.C.) resorted to in the siege of Gabii. Tarquinius sent his son Sextus to Gabii to feign treason against his father and urge the Gabini to take up arms against the Romans. A series of convenient military successes established Sextus's credibility with the Gabii, whereupon he delivered them up to the Romans.[144] The reader will recognize a suspicious similarity between the ruse of Sextus and that which the Persian Zopyrus used to deliver Babylon to Darius, but what is interesting is that Livy considered such tricks "unlike a Roman."[145] He preferred the virtue that Marcus Furius Camillus displayed during the siege of Falerii. A traitorous schoolmaster led his pupils into Roman captivity, hoping to curry favor with the Romans. But Camillus was so outraged at such low treachery that he stripped the schoolmaster naked and had the children beat him back to the city with rods.[146]

The corrupting influence of siege warfare was reflected in Livy's story of Tarquinius's downfall. The end came for him when he was bogged down in the siege of Ardea and his army sunk into drinking and carousing. It was in this atmosphere of a demoralized army that Sextus raped Lucretia, an outrage that precipitated the overthrow of Tarquinius.[147]

Sallust also disliked siege warfare, as we can see by his account of the siege of Vaga, a Numidian city that had risen against the Romans and killed almost the entire Roman garrison in the town. The uprising caught the Roman soldiers by surprise, and they were not armed when the Numidians fell upon them. Women participated in the massacre, a circumstance Sallust found especially humiliating:

> Moreover, women and boys from the roofs of the houses were busily pelting them [the Romans] with stones and whatever else they could lay hands on. It was quite impossible to guard against the double danger and brave men were helpless before the feeblest of opponents. Side by side valiant and cowardly, strong and weak, fell without striking a blow.[148]

Contemptuous views of defenders who fought behind walls encouraged humiliating treatment of them. Enslavement was not only a hard fate but a humiliation as well. It was said that the Roman king Servius Tullius's mother had been captured by the Romans in the siege of Corniculum. But although King Tarquinius's wife, Tanaquil, had saved this woman from slavery because of her obvious nobility, his mother's brush with enslavement was enough that the humiliation clung to Servius.[149] Although the story is undoubtedly a legendary one, it reveals how humiliating the Romans believed it was for a woman to be captured in a siege.

The shame of rape was so great that men sometimes killed wives or daughters rather than see them carried away. Livy told the story of Verginius, who had killed his daughter rather than see her raped by the decemvir Appius Claudius. Verginius, pleading for the condemnation of Appius after the overthrow of the decemvirs, accused Appius of attempting to rip "a free maiden from her father's arms, as though she had been a captive taken in war."[150] Verginius meant for his listeners to pity his daughter rather than a captive of war, but his chilling analogy shows that he took it as a matter of course that captured women were not only enslaved but also often raped.

Not long after the fall of the decemvirs, the consul Valerius exhorted his men in battle by reminding them that they were fighting as free men for a free city. If they failed they would be worse off than they had been under the decemvirs, because then

> no one's chastity but Verginia's had been in danger . . . , no citizen but Appius had been possessed of a dangerous lust; but if the fortune of war turned against them, the children of all of them would be in danger from all those thousands of enemies. . . . "[151]

The Roman soldiers did not doubt that defeat meant the rape of their wives and daughters, a fear that drove them to victory.

Circumstances could mitigate the treatment of women. When Scipio captured New Carthage, over three hundred hostages from prominent Iberian families that the Carthaginians had been holding fell into his hands. He was anxious to make a good impression on the Iberian people and treated the hostages with great consideration, even giving them gifts. He was therefore puzzled when an old and dignified lady fell at his feet and begged him "to treat them with more proper consideration than the Carthaginians had done." It was only when he noticed the youth and beauty of some of the women hostages that he realized the Carthaginians had been raping them. Scipio assured the woman that "he would look after them as if they were his own sisters and children and would accordingly appoint trustworthy men to attend on them."[152]

Scipio's soldiers soon tested his restraint in a more direct way. They found a young woman of great beauty and, "being aware that Scipio was fond of women," presented her to him. Nothing could make clearer that the Romans considered captured women as property to be disposed of like the rest of the booty. Scipio, however, did not avail himself of the rights of the conqueror. He was much taken with the woman's beauty and told the soldiers that if he were in private life nothing would please him more than to take her, but because he was a general he did not think it was appropriate.

This restraint somewhat puzzled Polybius, who struggled to explain Scipio's motives:

giving them to understand, I suppose, by this answer that sometimes, during seasons of repose and leisure in our life, such things afford young men most delightful enjoyment and entertainment, but that in times of activity they are most prejudicial to the body and the mind alike of those who indulge in them.[153]

Scipio turned the woman over to her father, telling him to give her in marriage to whomever he pleased. Polybius said that Scipio's self-restraint made a good impression on the troops, but he did not say if they emulated his example.[154] We do not know what the young woman thought, but clearly Scipio was moved by a sense of austerity in public duty and a desire to gain the respect of his troops, rather than any regard for her feelings.

Livy put Scipio's restraint in a different context. He said that Scipio learned the woman's parents had betrothed her to an important Celtiberian and that he restored the woman to her fiancé in order to gain the gratitude of a man whose influence could help draw the Spaniards away from the Carthaginians and toward the Romans. Scipio's generosity had the desired results. The young man entered Roman service, bringing fourteen hundred crack cavalry with him.[155]

Circumstances were not so fortunate for most captured women, and for them death was often the end to be preferred over the humiliation of rape and enslavement. Suicide provided an honorable end to a lost siege. According to most sources, the fall of Saguntum to Hannibal in 219 B.C. ended in mass suicide. Although Polybius did not mention a mass suicide, Appian and Diodorus both said that the Saguntine men fought to the last man and the women killed their children and then committed suicide.[156] In Livy's account, Hannibal ordered his soldiers to kill not only the men but all the adult inhabitants of Saguntum, an order that Livy called "a cruel command." But Livy offered the extenuating circumstance that the Saguntines were committing mass suicide in any case.[157]

Polybius admired mass suicide by the inhabitants of a fallen city. In his account of the siege of Abydos by Philip V of Macedon in 200 B.C., he expressed great admiration for "the bravery and exceptional spirit displayed by the besieged, which rendered it especially worthy of being remembered and described to posterity."[158]

The Abydenes earned this praise by deciding to place their women in the temple of Artemis and their children in the gymnasium during the siege. The men swore an oath that if they saw the town was falling, they would kill the women and children. Polybius praised their resolution "to meet their fate and perish to a man together with their wives and children rather than to live under the apprehension that their families would fall into the power of their enemies."[159]

In the event, the men entrusted with this grim task failed to fulfill their oath. Polybius condemned them for having "sacrificed in hope of personal advantage all that was splendid and admirable in the resolution of the citizens by deciding to save the women and children alive. . . ."[160]

This betrayal did not stay the others from fulfilling their oaths, and they began killing their families and themselves by cutting their throats, hanging or burning them, and jumping from the walls. This grim scene so moved Philip that he called a three-day truce to allow the Abydenes to commit mass suicide:

> The Abydenes, maintaining the resolve they had originally formed concerning themselves and regarding themselves as almost traitors to those who had fought and died for their country, by no means consented to live except those of them whose hands had been stayed by fetters or such forcible means, all the rest of them rushing without hesitation in whole families to their death.[161]

The magnanimity of Philip had saved the honor of the Abydenes from those contemptible men who had tried to betray it. The story reflects the values of a society in which honor was so much more important than pity that Polybius was unable to imagine that pity had moved the men who were supposed to kill the women and children; he attributed their failure to carry out their grim duty to a desire for personal gain. The story also reveals the terror of siege warfare. Implicit in a preference for death instead of capture was the assumption that the fall of a city meant rape and enslavement for the women. The Abydenes were not the only ones in the world of ancient siege warfare who chose death over rape and enslavement.

Diodorus also admired suicide as an honorable escape from the horror of siege warfare. In his version of the end of Victumulae, he says that the men "with high courage" killed their families and then themselves, "considering a self-inflicted death preferable to death with outrage at the hands of their enemies."[162]

Unlike the Greek historians Polybius and Diodorus, Livy did not admire mass suicides. He called the actions of the Abydenes "madness."[163] He was even more outspoken about another mass suicide that took place at the Spanish town of Astapa in 206 B.C. This town had always hated the Romans. It was not strongly fortified, and its natural position was weak. When the Romans approached in the same campaign in which they had slaughtered the inhabitants of Ilurgia, the people of Astapa chose death rather than face conquest by the Romans. Women and children gathered in the marketplace. The men made a giant pyre out of their most valuable possessions. They proposed to charge out of the city to meet the Romans and die fighting. Fifty young warriors took on the duty of remaining behind to kill the women and children. In this case, the plan worked perfectly. The warriors fought to the last man, and when the Romans entered the town, they found all its people dead. Livy called this act "brutal and barbarous" and said that when the Romans discovered the terrible scene "they stood for a little while stunned with amazement."[164] According to Appian, however, the mass suicide struck the Roman commander with admiration and he refrained from razing the town out of respect for the courage of its people.[165]

Just before the Numidian city of Thala fell to the Romans during the Jugurthine War, its men ate a sumptuous last banquet, then burned all the treasure in the town and themselves rather than fall into Roman captivity. Sallust did not bother to report the fate of the women and children nor the Roman reaction to the mass suicide. He only implied that the Roman soldiers were greatly vexed that the destruction of the treasure deprived them of their booty.[166]

Another fearful slaughter took place in 177 B.C. at the town of Nesattium on the Histrian peninsula at the head of the Adriatic Sea. The Romans diverted the flow of a river that provided the Histrians with water, making it impossible to defend the town. Rather than surrender the Histrians took their women and children to the top of the wall and cut them down before the eyes of the Romans. They threw the bodies down from the wall. While this atrocity was going on, the Romans scaled the walls and entered the town. The Histrian chief committed suicide, but the men who had killed their families did not. The Romans killed many of them, but many were also captured, apparently preferring the slavery into which they were soon sold to the death they had deemed the only honorable end for their wives and children.[167]

When Capua, which the Romans had strangled with a long circumvallation, was about to fall in 211 B.C., some of its leaders chose suicide to the humiliation of defeat. Capua controlled Campania, the richest agricultural district in Italy, and its defection to Hannibal in Rome's darkest hour was a betrayal that the Capuans knew was unforgivable.

A Capuan senator, Vibius Virrius, who had led the revolt against Rome, had no illusions about what lay in store for him and Capua, and he did not want to live to see it:

> I shall not see Appius Claudius and Quintus Fulvius, emboldened by their insolent victory, nor shall I be dragged in chains through the city of Rome as a spectacle in a triumph, so that I may then breathe my last in the prison, or else, bound to a stake, with my back mangled by rods, may submit my neck to the Roman axe. Nor shall I see my native city destroyed and burned, nor Capuan matrons and maidens and free-born boys carried off to be dishonoured.[168]

Virrius intended to prepare a last feast and end it with a draught of poison, and he invited the other Capuans to join him. As Livy drily remarked, "This speech of Virrius more men heard with approval than had the courage to carry out what they commended." Twenty-seven Capuan senators, however, joined Virrius's feast and drank the poison.[169]

When the other senators failed to take such a severe view and surrendered the city, a Capuan citizen named Taurea Vibellius took it upon himself to uphold Capuan honor. He killed his wife and children "that they may suffer no indignity" and then strode up to the Romans as they were leaving the executions of the Capuan rebels and ran himself through with his sword.[170]

A final example of a besieged people choosing suicide as an honorable end comes to us in Josephus's account of the siege of the Jewish fortress at Masada in A.D. 73. When it became apparent that defeat was inevitable, the Zealot leader Eleazar called on the Jews to commit suicide as the only way to prevent the rape of their wives and to end their lives in freedom. Although Josephus despised the Zealots, he admired Eleazar's courage. He put two long speeches in Eleazar's mouth to dramatize the decision.

In the first speech Eleazar exhorted:

> Let our wives thus die undishonoured, our children unacquainted with slavery; and when they are gone, let us render a generous service to each other, preserving our liberty as a noble winding-sheet.[171]

Eleazar's eloquence did not convince all of the Jews to heed his counsel. Josephus attributed their hesitation to a soft-hearted compassion for their wives and children and to their fear of death. He reported that Eleazar "feared that their whimpers and tears might unman even those who had listened to his speech with fortitude."[172] Thus the necessity of a second speech.

In this speech Eleazar claimed that Jewish law enjoined them to commit mass suicide rather than succumb to rape of their wives.[173] There was no such law, but perhaps Josephus inferred such a requirement from the laws of purity that would be impossible to maintain in captivity. And there were examples in the Jewish scriptures of suicide providing an appropriate way out of shame and dishonor.[174]

This second speech sufficiently fortified the Jews to carry out the terrible deed. The men killed their wives and children; then ten of the men killed the others; then one of the men killed the other nine; and then the last committed suicide. However, two women and five children survived by hiding in a cavern. Josephus did not denounce them; he commended one of the women, a relative of Eleazar, as "superior in sagacity and training to most of her sex." Nine hundred and sixty Jews had died in what Josephus called a pathos.[175] As for the Romans, Josephus said that they were "incredulous of such amazing fortitude" and "admired the nobility" of such undaunted courage.[176]

Despite Josephus's pride in the courage of the Jews at Masada, his account betrayed some ambiguity. He understood their deaths as an appropriate punishment for the rebellion, the fulfillment of God's will. He himself had evaded suicide at Jotapata, an evasion he did not see as shameful, because the Romans were offering mercy. In the course of a speech in which he tried to dissuade his soldiers from committing suicide, Josephus called suicide "repugnant to that nature which all creatures share, and an act of impiety towards God who created us."[177] He also alluded to a law that required the body of someone who had committed suicide to lay unburied until sunset because suicide was "hateful to God."[178] Although there was no direct prohibition of suicide in the Torah, suicide had become sufficiently problematic in the Judaism of Josephus's day that he could only justify it as an act of divine punishment.[179]

EPILOGUE

If my eyes roll and I mutter,
if my arms are gloved in blood right up to the elbow,
if I clutch at my heart and scream in horror
like a third-rate actress chewing up a mad scene,
I do it in private and nobody sees
but the bathroom mirror.
                    —Margaret Atwood,
                    "The Loneliness of the Military Historian"[1]

The reader, like the military historian, may by now yearn for privacy to scream in horror. Fortunately we are at the end of our story. The siege of Jerusalem represented the high point of Roman siegecraft. As the empire waned, offensive siege warfare became secondary to defense.

Especially under Diocletian (emperor A.D. 284–305) the Romans developed a defense-in-depth based on a system of walled towns and fortifications. Roman frontier fortifications became much stronger and more sophisticated than they had been in the early empire. These fortifications and their defense sustained the empire for a long time, because the Germans lacked adequate siege capabilities; but the age of expansion was over.[2] Indeed, the increasing number of Germans in the Roman army reduced the number of soldiers experienced in the use of siege machinery.[3]

By the fourth century ordinary legions no longer possessed artillery because of a lack of artificers and men with the technical ability to build and deploy them. Special artillery legions tried to fill the gap, but lack of skilled personnel limited their effectiveness. E. W. Marsden speaks of "the almost total lack of fighting men who really knew what engines of war could do for them and who knew how to make full use of them."[4]

Despite these limitations, siege warfare continued in the late empire and in Byzantium; it was too essential to ancient warfare to be neglected. Although medieval sieges were more often conducted against castles rather than towns, the methods and conduct of sieges remained much the same.[5]

Siege warfare continued to present the face of total war. Mutilation and massacre remained common all through the Middle Ages. In his book on medieval siege warfare, Jim Bradbury speaks of "the normal consequences of being stormed: death, slavery, exile, loss of property, rape, torture and almost any horror one could envisage."[6] In the late 1040s William the Conqueror cut off the hands and feet of the men defending Alençon, who, it seems, had insulted his parentage. Liudprand of Cremona said that Tedbald, the marquess of Camerino and Spoleto, campaigning in the tenth century, castrated most of the garrison of a place he captured.[7] When the Christian crusaders took Jerusalem, the streets ran with blood, just as they had when the Romans took the city in A.D. 70. During the sack of Limoges in 1370 the English cut the throats of over three thousand prisoners.[8]

Henry V found the passage in the Book of Deuteronomy, which counseled killing all the male defenders of a city that resisted and enslaving the rest, entirely adequate for his purposes.[9] Shakespeare caught the nature of siege warfare in a speech he put into the mouth of Henry V demanding the surrender of Harfleur:

> If I begin the battery once again,
> I will not leave the half-achieved Harfleur
> Till in her ashes she lie buried.
> The gates of mercy shall be all shut up;
> And the flesh'd soldier, rough and hard of heart,
> In liberty of bloody hand shall range
> With conscience wide as hell; mowing like grass
> Your fresh-fair virgins and your flowering infants.
> What is't then to me if impious war,
> Array'd in flames, like to the prince of fiends,
> Do, with his smirch'd complexion, all fell feats
> Enlink'd to waste and desolation?
> What is't to me, when you yourselves are cause,
> If your pure maidens fall into the hand
> Of hot and forcing violation?
> What rein can hold licentious wickedness
> When down the hill he holds his fierce career?
> We may as bootless spend our vain command
> Upon th' enraged soldiers in their spoil
> As send precepts to the leviathan

To come ashore. Therefore, you men of Harfleur,
Take pity on your town and of your people,
Whiles yet the cool and temperate wind of grace
O'er blows the filthy and contagious clouds
Of heady murder, spoil, and villainy.
If not, why, in a moment look to see
The blind and bloody soldier with foul hand
Defile the locks of your shrill-shrieking daughters;
Your fathers taken by the silver beards,
And their most reverend heads dash'd to the walls;
Your naked infants spitted upon pikes,
Whiles the mad mothers with their howls confus'd
Do break the clouds, as did the wives of Jewry
At Herod's bloody-hunting slaughtermen.[10]

Henry's speech includes all the familiar themes of siege warfare: the terror of a sack; the impossibility of controlling rampaging troops; rape; the killing of old men, women, and children; smashed heads and spitted babies. Shakespeare thus shows us that the conventions of siege warfare had changed little since ancient times.

The continuity of practice is quite remarkable. In the siege of Rouen in 1419 by Henry V, the defenders expelled the excess population. Henry forced them into a ditch, where he left them to die.[11] Bradbury calls Henry's action "part of the normal siege code."[12] On September 18, 1941, German Field Marshal Wilhelm von Leep ordered German troops to turn back any Russian civilians attempting to leave the besieged city of Leningrad. There were three million civilians trapped in Leningrad; one million of them died during the siege. After the war the victorious allies brought Leep to trial at Nuremberg for war crimes. Leep claimed that international law sanctioned his action under the principle of military necessity. When the judges consulted an American book on international law, they discovered the following statement:

It is said that if the commander of a besieged place expels the non-combatants, in order to lessen the number of those who consume his stock of provisions, it is lawful, though an extreme measure, to drive them back so as to hasten the surrender.[13]

The Nuremberg tribunal acquitted Leep.

Modern war in general presents a nature all too familiar to the student of ancient siege warfare. A recent report on events in East Timor stated that among various Indonesian atrocities, "children's heads were smashed against rocks."[14] Raping is commonplace in modern war, for example, in the Soviet campaign of rape in Germany in 1945 and the Serbian campaign against Bosnia in 1991–1992.

The study of siege warfare suggests that restraints on the violence of war only succeed when expediency and restraint coincide and that often not even expediency is strong enough to limit violence. Social and moral conventions that give structure and order to human life are fragile, and when they break down, anything is possible. As Josephus struggled to come to terms with the destruction of Jerusalem and especially the burning of the temple, he concluded that God had abandoned the city and was now on the side of the Romans.[15] His opinion reflected the ancient idea that the gods forsook cities unhappy enough to fall in siege warfare. The Romans encouraged the desertions of the gods from the cities they besieged through the rite of *evocatio*, in which they promised to establish a cult in Rome as good as the cult in the doomed city.[16] When we ponder the profound moral and social disorder in siege warfare, we might conclude also that the gods abandoned cities when the walls came tumbling down.

NOTES

INTRODUCTION

1. Joshua 6:20–21. Biblical quotations are from the Revised Standard Version.

2. Thucydides 7.29. All translations from Greek and Roman works are from the Loeb Classical Library editions unless otherwise noted.

3. Iliad 22.58–71. Translated by Richmond Lattimore (Chicago: University of Chicago Press, 1951).

4. Robert O'Connell, *Of Arms and Men: A History of War, Weapons, and Aggression* (New York: Oxford University Press, 1989), 43.

5. *Ancient Near Eastern Texts Relating to the Old Testament*, ed. James B. Pritchard (Princeton: Princeton University Press, 1969), 459.

6. See Christopher Duffy, *Fire and Stone: The Science of Fortress Warfare, 1660–1860* (Vancouver: David and Charles, 1975), 9. On the logical tendency of war to advance to the extreme, see Carl von Clausewitz, *On War* (Princeton: Princeton University Press, 1989), 77–78.

7. Michael Howard, "*Temperamenta Belli*: Can War Be Controlled?" in *Restraints on War*, ed. Michael Howard (Oxford: Oxford University Press, 1979), 3.

8. Aineias the Tactician, *How to Survive under Siege*, trans. David Whitehead (Oxford: Clarendon Press, 1990), 1–2. Hereafter cited as Aineias Tacticus.

9. Hans Volkmann, *Die Massenversklavungen der Einwohner eroberter Städte in der Hellenistisch-Römischen Zeit* (Stuttgart: Franz Steiner Verlag, 1990), 3.

10. Polybius 2.56.7.

11. Polybius 2.56.8–13.

12. Howard, "Can War Be Controlled?" 4.

1. FORTIFICATIONS AND SIEGE MACHINERY

1. Arther Ferrill, *The Origins of War from the Stone Age to Alexander the Great* (London: Thames and Hudson, 1985), 28; Yigael Yadin, *The Art of Warfare in Biblical Lands in the Light of Archeological Discovery* (London: McGraw-Hill, 1963), 33.

2. Yadin, *Warfare in Biblical Lands*, 33–34.

3. O'Connell, *Arms and Men*, 26–27.

4. Ferrill, *Origins of War*, 27–28.

5. See J. Bronowski, *The Ascent of Man* (Boston: Little, Brown, 1973), 64–72, for his discussion of the origins of agriculture at Jericho. See also O'Connell, *Arms and Men*, 31.

6. William H. McNeill, *The Rise of the West* (Chicago: University of Chicago Press, 1963), 11–13.

7. Ferrill, *Origins of War*, 30–31.

8. See Horst de la Croix, *Military Considerations in City Planning: Fortifications* (New York: Braziller, 1972).

9. McNeill, *Rise of the West*, 43.

10. A. Billerbeck, "Der Festungsbau im alten Orient," *Der Alte Orient*, 1 (1903), 26–27.

11. Hans Waschow, *4000 Jahre Kampf um die Mauer* (Postberg, Bottropiw: Buch- und Kunstdruckerei Wilk, 1938), 18.

12. Billerbeck, "Festungsbau," 16–19.

13. De la Croix, "Fortifications," 15–16.

14. Yadin, *Warfare in Biblical Lands*, 50–53; Ferrill, *Origins of War*, 34–37.

15. Yadin, *Warfare in Biblical Lands*, 53.

16. Alexander Badawy, *Architecture in Ancient Egypt and the Near East* (Cambridge: MIT Press, 1966), 57.

17. See Yadin, *Warfare in Biblical Lands*, 56–57, for an excellent discussion of gate designs.

18. Waschow, *Kampf um die Mauer*, 32–34. See Polybius 9.19.

19. See Polybius 9.19.6–7 for instruction on the proper length and placement of scaling ladders.

20. Yadin, *Warfare in Biblical Lands*, 54–55 and 147.

21. Waschow, *Kampf um die Mauer*, 45.

22. Ibid., 21–22.

23. C. J. Gadd, in *The Cambridge Ancient History*, 3rd ed. (Cambridge: Cambridge University Press), 1, pt. 2, 124. Hereafter cited as *CAH*.

24. Yadin, *Warfare in Biblical Lands*, 65.

25. See Alexander Badawy, *A History of Egyptian Architecture* (Berkeley: University of California Press, 1966), chap. 5, for a good description of the Nubian fortresses.

26. Yadin, *Warfare in Biblical Lands*, 65–66.

27. De la Croix, *Military Considerations*, 18–19; Badawy, *Egyptian Architecture*, 219–222; John Keegan, *A History of Warfare* (New York: Vintage, 1994), 142–143.

28. Yadin, *Warfare in Biblical Lands*, 68–69.

29. J. G. Macqueen, *The Hittites and Their Contemporaries in Asia Minor* (Boulder: Westview Press, 1975), 104.

30. See Yadin, *Warfare in Biblical Lands*, 91, for a detailed description of Hattussas. There is also a good description of Hattussas's walls in Macqueen, *Hittites*, 104–109.

31. Yadin, *Warfare in Biblical Lands*, 70 and 158–160.

32. Ferrill, *Origins of War*, 34–35; Alan Gardiner, *Egypt of the Pharaohs* (New York: Oxford University Press, 1961), 393.

33. *Ancient Near Eastern Texts*, 79.

34. Richard Humble, *Warfare in the Ancient World* (London: Cassell, 1980), 13 and 21.

35. Yadin, *Warfare in Biblical Lands*, 69–70.

36. Quoted in Yadin, *Warfare in Biblical Lands*, 70.

37. Quoted in Albert Ernest Glock, "Warfare in Mari and Early Israel" (Ph.D. diss., University of Michigan, 1968), 172.

38. Glock, "Warfare in Mari," 171–172; Billerbeck, "Festungsbau," 6.

39. Yadin, *Warfare in Biblical Lands*, 70.

40. Glock, "Warfare in Mari," 179.

41. Ibid., 143–144.

42. Quoted in ibid., 173.

43. Ibid., 173–175.

44. *CAH*, 124.

45. Glock, "Warfare in Mari," 92 and 185–186.

46. *Ancient Records of Assyria and Babylonia*, 2 vols., ed. Daniel David Luckenbill (New York: Greenwood Press, 1968), 1:38–48.

47. Waschow, *Kampf um die Mauer*, 49.

48. Quoted in Yadin, *Warfare in Biblical Lands*, 70–71.

49. Yadin, *Warfare in Biblical Lands*, 289.

50. O. R. Gurney, *The Hittites* (Baltimore: Penguin, 1969), 109.

51. Yadin, *Warfare in Biblical Lands*, 71.

52. Ibid., 228–229.

53. Ibid., 97.

54. *Ancient Near Eastern Texts*, 237. See also Ferrill, *Origins of War*, 57; Waschow, *Kampf um die Mauer*, 24.

55. John Keegan, *The Mask of Command* (New York: Penguin, 1987), 71–72.

56. Philo H. J. Houwink Ten Cate, "The History of Warfare according to Hittite Sources: The Annals of Hattusilis I," *Anatolica* 11 (1984), 67–68.

57. Quoted in Yadin, *Warfare in Biblical Lands*, 97.

58. Glock, "Warfare in Mari," 76.

59. Houwink Ten Cate, "Annals of Hattusilis I," 73.

60. Martin van Creveld, *Technology and War* (New York: Free Press, 1989), 34.

61. Houwink Ten Cate, "Annals of Hattusilis I," 69.

62. Yadin, *Warfare in Biblical Lands*, 98.

63. *Ancient Near Eastern Texts*, 22–23.

64. De la Croix, *Military Considerations*, 8; Creveld, *Technology and War*, 34–35; Waschow, *Kampf um die Mauer*, 13.

2. TREATMENT OF CAPTURED CITIES

1. Quoted in Glock, "Warfare in Mari," 77.

2. See ibid., 77–79.

3. I. J. Gelb, "Prisoners of War in Early Mesopotamia," *Journal of Near Eastern Studies*, 32 (1972), 95–96.

4. Quoted in Glock, "Warfare in Mari," 178.

5. Gelb, "Prisoners," 80–81.

6. Ibid., 74.

7. Ibid., 72.

8. Ibid., 90.

9. Ibid., 86–87.

10. Ibid., 92–93.

11. Ibid., 91. See also Glock, "Warfare in Mari," 77.

12. Gelb, "Prisoners," 95–96.

13. Ibid., 81–82.

14. *Ancient Records of Assyria*, 1:39–40.

15. Ibid., 49.

16. Ibid., 76.

17. Ibid., 85.

18. *Ancient Near Eastern Texts*, 237.

19. Ibid., 247.

20. Gurney, *The Hittites*, 113–116.

21. Houwink Ten Cate, "Annals of Hattusilis I," 66.

22. H. Craig Melchert, "The Acts of Hattusilis I," *Journal of Near Eastern Studies*, 37 (January 1978), 12.

3. ISRAEL, MESOPOTAMIA, AND THE PERSIANS

1. Abraham Malamat, "How Inferior Israelite Forces Conquered Fortified Canaanite Cities," *Biblical Archeological Review*, 8 (March–April 1982), 32.

2. Chaim Herzog and Mordechai Gichon, *Battles of the Bible* (New York: Random House, 1978), 25.

3. Exodus 13:17–18.

4. Numbers 13:1–20, 13:28, 14:1–4.
5. Albrecht Alt, *Die Landnahme der Israeliten Palästina* (Leipzig: Druckerei der Werkgemeinschaft, 1925), 8–10.
6. Judges 1:19.
7. Numbers 21:21–25; see Herzog and Gichon, *Battles of the Bible*, 24.
8. Herzog and Gichon, *Battles of the Bible*, 25.
9. Malamat, "Inferior Israelite Forces," 29.
10. Joshua 2:1–15.
11. Joshua 6:25.
12. Joshua 6:1–21.
13. Herzog and Gichon, *Battles of the Bible*, 28–29.
14. Malamat, "Inferior Israelite Forces," 25; see Alt, *Die Landnahme*, 33–34.
15. Gerhard von Rad, *Der heilige Krieg im alten Israel* (Göttingen: Vandenhoeck & Ruprecht, 1952), 43–44.
16. Ernest Ludwig Ehrlich, *A Concise History of Israel* (New York: Harper and Row, 1965), 22, n. 2.
17. Herzog and Gichon, *Battles of the Bible*, 29.
18. Joshua 7:2–4.
19. Malamat, "Inferior Israelite Forces," 29.
20. Joshua 7:5.
21. Joshua 8:3–8.
22. Joshua 8:10–24; Herzog and Gichon, *Battles of the Bible*, 31–34. See also Malamat, "Inferior Israelite Forces," 33–34.
23. Friedrich Schwally, *Semitische Kriegsaltertumer* (Leipzig: Dieterich'sche, 1901), 21–22.
24. Ziony Zevit, "The Problem of Ai," *Biblical Archeology Review*, 11 (March–April 1985), 58.
25. F. W. Albright, "The Israelite Conquest of Canaan," *Bulletin of the American School of Oriental Research*, 74 (1939).
26. Zevit, "The Problem of Ai," 61–62.
27. Herzog and Gichon, *Battles of the Bible*, 30–31.
28. Malamat, "Inferior Israelite Forces," 25–26.
29. Joshua 11:10–13.
30. Yadin, *Warfare in Biblical Lands*, 95.
31. Herzog and Gichon, *Battles of the Bible*, 48.
32. Judges 1:19.
33. Judges 1:21–33. Judges 1:8 reports the fall of Jerusalem, but this is contradicted by Judges 1:21.
34. Judges 1:22–26.
35. Yadin, *Warfare in Biblical Lands*, 254.
36. Judges 20:24–48.
37. Judges 9:42–45.
38. Yadin, *Warfare in Biblical Lands*, 261.
39. Ibid.
40. Judges 9:46–49.
41. Judges 9:50–54.
42. 2 Samuel 11:19–21.
43. Herzog and Gichon, *Battles of the Bible*, 77.
44. 2 Samuel 5:6.
45. *The Interpreter's One-Volume Commentary for the Bible*, ed. Charles M. Laymon (Nashville: Abingdon Press, 1971), 172; Yadin, *Warfare in Biblical Lands*, 269.
46. 1 Chronicles 11:6.
47. Yadin, *Warfare in Biblical Lands*, 268.

48. 1 Chronicles 11:8.

49. Joshua 10:31–32.

50. See Yadin, *Warfare in Biblical Lands*, 275–284, for a detailed discussion of the organization of David's army.

51. 1 Samuel 17:17–18.

52. Rad, *Der heilige Krieg*, 34–36 and 76.

53. This follows Yadin's interpretation of the Biblical text, which is not entirely clear. See Yadin, *Warfare in Biblical Lands*, 274–275.

54. 2 Samuel 12:26–29.

55. Yadin, *Warfare in Biblical Lands*, 24–25.

56. 2 Samuel 20:15. *The Interpreter's Dictionary of the Bible*, ed. George Arthur Buttrick, 4 vols. (New York: Abingdon Press, 1962), 1:366.

57. 1 Samuel 13:19–21.

58. Josephus, *Antiquities* 7.11.7.

59. 2 Samuel 20:16–22.

60. 1 Kings 9:15–16.

61. Yadin, *Warfare in Biblical Lands*, 287–290.

62. 1 Kings 9:17–19.

63. 1 Kings 9:20–22.

64. 1 Kings 5:13–18.

65. 1 Kings 16:24.

66. Herzog and Gichon, *Battles of the Bible*, 104.

67. 1 Kings 20.

68. 2 Chronicles 11:5–12.

69. 2 Chronicles 12:2–9.

70. 2 Chronicles 14:6–7.

71. 2 Chronicles 17:2, 12–13.

72. 2 Chronicles 26:9–10.

73. 2 Chronicles 26: 11–14.

74. 2 Chronicles 26:15.

75. Yadin, *Warfare in Biblical Lands*, 326–327.

76. E. W. Marsden, *Greek and Roman Artillery* (Oxford: Clarendon Press, 1969), 52–53. See also Mordechai Gichon's reservations in Herzog and Gichon, *Battles of the Bible*, 177.

77. 2 Chronicles 27:3–4.

78. 2 Chronicles 33:14.

79. See Yadin, *Warfare in Biblical Lands*, 322–324, for a detailed discussion of these changes in wall design.

80. This campaign is described in 2 Kings 3. See Herzog and Gichon, *Battles of the Bible*, 123–125, for an analysis of the Biblical account.

81. Yadin, *Warfare in Biblical Lands*, 317.

82. 2 Kings 3:27.

83. Herzog and Gichon, *Battles of the Bible*, 125.

84. 2 Kings 6:24–31.

85. 2 Kings 7:3–16.

86. 2 Kings 12:17–18.

87. 2 Chronicles 25:11–12.

88. 2 Kings 14:8–14; 2 Chronicles 25:17–24.

89. 2 Kings 16:5.

90. 2 Kings 16:7–9.

91. See William H. Hallo, "From Qarqar to Carchemish: Assyria and Israel in the Light of New Discoveries," *Biblical Archaeologist*, 23, no. 2 (1960), 34–61, for a careful review of Assyrian-Hebrew relations based on an analysis of both Assyrian and Biblical sources.

92. Ibid., 51.

93. 2 Kings 17:5–6.

94. 2 Kings 17:6.

95. *Ancient Assyrian Records*, 2:26.

96. 2 Kings 17:24.

97. 2 Chronicles 32:1–5.

98. 2 Kings 18:14–16.

99. Herzog and Gichon, *Battles of the Bible*, 142.

100. David Ussishkin, *The Conquest of Lachish by Sennacherib* (Tel Aviv: Tel Aviv University, Institute of Archeology, 1982), 11.

101. Ibid., 29–45.

102. Ibid., 49–57.

103. *Ancient Near Eastern Texts*, 288.

104. Ussishkin, *Conquest of Lachish*, 17.

105. 2 Kings 18:28–36.

106. 2 Kings 18:19–24.

107. 2 Kings 19:32–33.

108. 2 Chronicles 32:21.

109. 2 Kings 19:35; see A. T. Olmstead, *History of Assyria* (New York: Scribner, 1923), 308–309.

110. Herodotus 2.141.

111. Herzog and Gichon, *Battles of the Bible*, 142–143.

112. *Ancient Near Eastern Texts*, 288.

113. 2 Kings 24:10–17.

114. 2 Kings 25:27–30.

115. 2 Kings 25:1–21.

116. Jeremiah 52:28–30.

117. Sidney Smith, in *CAH*, 3:99.

118. Yadin, *Warfare in Biblical Lands*, 314.

119. Erkki Salonen, "Die Waffen der alten Mesopotämier," *Studia Orientalia*, 33 (1966), 181.

120. Yadin, *Warfare in Biblical Lands*, 315.

121. Johannes Hunger, "Heerwesen und Kriegführung des Assyrer auf der Hohe ihrer Macht," *Der Alte Orient*, 12 (1911), 27.

122. Yadin, *Warfare in Biblical Lands*, 314–315.

123. See especially the relief from the time of Ashurbanipal reproduced in Yadin, *Warfare in Biblical Lands*, 448.

124. Waschow, *Kampf um die Mauer*, 45.

125. Joel 2:7–9.

126. Salonen, "Die Waffen der alten Mesopotämier," 175.

127. Hunger, "Heerwesen und Kriegführung," 25.

128. Salonen, "Die Waffen der alten Mesopotämier," 197–198.

129. Hunger, "Heerwesen und Kriegführung," 29.

130. Yadin, *Warfare in Biblical Lands*, 317.

131. Florence Malbran-Labat, *L'Armée et l'Organisation Militaire de l'Assyrie* (Geneva: Libraire Droz, 1982), 61.

132. Ussishkin, *Siege of Lachish*, 52.

133. Ibid., 125.

134. Waschow, *Kampf um die Mauer*, 66–67.

135. *Ancient Records of Assyria*, 2:223.

136. Waschow, *Kampf um die Mauer*, 72.

137. Hunger, "Heerwesen und Kriegführung," 26–27.

138. This relief is reproduced in Yadin, *Warfare in Biblical Lands*, 392–393.

139. Yadin, *Warfare in Biblical Lands*, 316.
140. Salonen, "Die Waffen der alten Mesopotämier," 128.
141. *Ancient Records of Assyria*, 2:99 and 236.
142. Hunger, "Heerwesen und Kriegführung," 31.
143. H. W. F. Saggs, *The Might That Was Assyria* (London: Sidgwick and Jackson, 1984), 97.
144. Hunger, "Heerwesen und Kriegführung," 23.
145. *Ancient Near Eastern Texts*, 293.
146. Malbran-Labat, *L'Armée*, 8.
147. *Ancient Records of Assyria*, 1:112-113.
148. *Ancient Near Eastern Texts*, 655.
149. Ibid., 282 and 288.
150. 2 Kings 25:5.
151. *Ancient Near Eastern Texts*, 292.
152. Ibid., 298.
153. 2 Kings 6:24-30.
154. 2 King 18:27.
155. 2 Kings 25:3-4.
156. Lamentations 4:9-10.
157. Ezekiel 5:12.
158. See A. L. Oppenheim, "Siege-Documents from Nippur," *Iraq*, 17 (1955) for a careful analysis of these documents.
159. Ibid., 71.
160. Ibid., 80.
161. Herodotus 3.150.
162. *Ancient Near Eastern Texts*, 280.
163. Israel Eph'al, "On Warfare and Military Control in the Ancient Near Eastern Empires: A Research Outline," in *History, Historiography and Interpretations*, ed. H. Tadmor and M. Weinfeld (Jerusalem: Magnes Press, 1983), 101.
164. *Ancient Records of Assyria*, 2:87.
165. Olmstead, *History of Assyria*, 81.
166. Ibid., 268. Walther Manitius dated the establishment of a standing army to the time of Tiglath-pileser III in "Das stehende Heer der Assyrerkönige und seine Organisation," *Zeitschrift für Assyriologie*, 24 (1910), 117.
167. H. W. F. Saggs, "Assyrian Warfare in the Sargonid Period," *Iraq*, 25 (1963), 146.
168. Manitius, "Das stehende Heer," 118. See also Malbran-Labat, *L'Armée*, 59-60.
169. Saggs, "Assyrian Warfare," 148.
170. *Ancient Near Eastern Texts*, 286. See Manitius, "Das stehende Heer," 112-113, for a discussion of the significance of this campaign.
171. J. E. Reade, "The Neo-Assyrian Court and Army: Evidence from the Sculptures," *Iraq*, 34 (1972), 104; Hunger, "Heerwesen und Kriegführung," 7.
172. *Ancient Records of Assyria*, 1:153.
173. Malbran-Labat, *L'Armée*, 60, 82-84, 184-185.
174. Ibid., 90.
175. Salonen, "Die Waffen der alten Mesopotämier," 175, 196-197.
176. Hunger, "Heerwesen und Kriegführung," 29.
177. Manitius, "Das stehende Heer," 221-222.
178. Malbran-Labat, *L'Armée*, 11.
179. Ibid., 174.
180. Ibid., 172.
181. *Ancient Near Eastern Texts*, 299.

182. Herodotus 1.106.
183. Diodorus 2.27; Nahum 1.8; Saggs, *Assyria*, 120.
184. Herodotus 1.84.
185. G. B. Grundy, *The Great Persian War and Its Preliminaries: A Study of the Evidence, Literary and Topographical* (New York: AMS Press, 1969), 28. A little over three hundred years later, the Syrian emperor Antiochus III took Sardis in a similar way. A Greek in Antiochus's service, named Lagoras, discovered a place left unguarded because it was so inaccessible and managed to cross the wall there. Very likely it was the same place Cyrus's man had found. In this case, Lagoras did indeed use ladders. See Polybius 7.15–18.
186. Herodotus 1.190–191.
187. Herodotus 3.151–158.
188. Xenophon, *Cyropaedia* 7.2.2.
189. Xenophon, *Cyropaedia* 7.4.1.
190. Herodotus 1.162.
191. Aineias Tacticus 37.7.
192. Herodotus 4.200–201.
193. Yvon Garlan, *Recherches de Poliorcétique Grecque* (Paris: Boccard, 1974), 143.
194. A. R. Burn, *Persia and the Greeks: The Defense of the West* (London: Arnold, 1962), 203–205.
195. Herodotus 5.115.
196. Herodotus 7.22–23.

4 · TREATMENT OF CAPTURED CITIES

1. Joshua 5.
2. Joshua 8:24–29.
3. Deuteronomy 20:16–18.
4. Humble, *Warfare in the Ancient World*, 64.
5. A good example is Peter C. Craigie, *The Problem of War in the Old Testament* (Grand Rapids: Eerdmans, 1978).
6. Deuteronomy 20:10–15.
7. Rad, *Der heilige Krieg*, 70.
8. K. Lawson Younger, Jr., *Ancient Conquest Accounts* (Sheffield: Sheffield Academic Press, 1990), 228.
9. Joshua 10:29–39.
10. Deuteronomy 20:19–20.
11. Schwally argues that there was a host of beliefs concerning demons and spirits involved in planting, harvesting, and clearing forests. However, he does not discuss the prohibition of cutting fruit trees during sieges. See Schwally, *Semitische Kriegsaltertumer*, 81–91.
12. See Victor Davis Hanson, *Warfare and Agriculture in Ancient Greece* (Ann Arbor, 1980).
13. Joshua 7:1 and 7:10–26.
14. Josephus, *Jewish War* 6.439.
15. 2 Samuel 24:18–25.
16. 1 Kings 9:20.
17. 2 Samuel 12:30–31.
18. 1 Chronicles 22:7–10.
19. 1 Samuel 11:1–2.
20. 1 Samuel 11:11.
21. 1 Samuel 22:17–19.
22. For a thoughtful discussion of this problem from the perspective of

a moralist, see Michael Walzer, *Just and Unjust Wars* (New York: Basic Books, 1977), 160–170.

23. Deuteronomy 20:13–14.

24. Judges 9:56.

25. Paul D. Hanson, "War, Peace, and Justice in Early Israel," *Bible Review*, 3 (Fall 1987), 41.

26. Deuteronomy 28:52–57.

27. *Ancient Records of Assyria and Babylonia*, 1:144–145. Also quoted in Ferrill, *Origins of War*, 69.

28. Ferrill, *Origins of War*, 69.

29. Olmstead, *History of Assyria*, 625.

30. Saggs, "Assyrian Warfare," 149.

31. *Ancient Near Eastern Texts*, 278.

32. See *Ancient Records of Assyria* for the Assyrian annals on which this discussion is based.

33. *Ancient Records of Assyria*, 1:146.

34. See illustration in Erika Bleibtreu, "Grisly Assyrian Record of Torture and Murder," *Biblical Archaeology Review* (January–February 1991), 56.

35. *Ancient Records of Assyria*, 1:155.

36. Bleibtreu, "Grisly Assyrian Record of Torture and Murder," 57.

37. Bustenay Oded, *Mass Deportations and Deportees in the Neo-Assyrian Empire* (Wiesbaden: Ludwig Reichert, 1979), 2 and 19.

38. Ibid., 74.

39. Ibid., 25.

40. Saggs, *The Might That Was Assyria*, 263–264.

41. Oded, *Mass Deportations*, 27–28.

42. Ibid., 35–36.

43. Ibid., 38–39.

44. Ibid., 98.

45. Ibid., 90–91.

46. See illustration in Bleibtreu, "Grisly Assyrian Record of Torture and Murder," 61.

47. See illustration in Bleibtreu, "Grisly Assyrian Record of Torture and Murder," 55.

48. *Ancient Records of Assyria*, 2:119.

49. Ibid., 120.

50. Olmstead, *History of Assyria*, 129–130.

51. Yadin, *Warfare in Biblical Lands*, 320 and 425.

52. 2 Kings 18:13 to 19:19.

53. *Ancient Records of Assyria*, 2:152.

54. Ibid., 132.

55. Ibid., 268.

56. See the illustrations of these reliefs in Bleibtreu, "Grisly Assyrian Record of Torture and Murder," 53 and 56.

57. Judges 1:6–7.

58. *Ancient Near Eastern Texts*, 320.

59. See "Middle Assyrian Laws" in *Ancient Near Eastern Texts*, 180–188.

60. Herodotus 3.15.

61. Isaiah 44:28–45:3.

62. *CAH*, 3:129.

63. Herodotus 3.159.

64. Herodotus 1.86–90.

65. Herodotus 1.88–90.

66. Xenophon, *Cyropaedia* 7.2.11–14. Quoted in W. Kendrick Pritchett, *The Greek State at War* (Berkeley: University of California Press, 1971–1991), 5:305.

67. Herodotus 1.161.

68. Herodotus 1.163–164.

69. Herodotus 6.9.

70. Herodotus 3.145–147.

71. Herodotus 8.127.

72. Herodotus 6.19–20.

73. Burn, *Persia and the Greeks*, 215.

74. Herodotus 6.21.

75. Herodotus 9.102–104. See Raoul Lonis, *Les Usages de la Guerre entre Grecs et Barbares* (Paris: Les Belles Lettres, 1969), 33, n. 10.

76. Herodotus 6.101.

77. Herodotus 8.32–33.

78. Herodotus 8.50.

79. Herodotus 8.51–54. See Lonis, *Les Usages de la Guerre*, 76–77, for a discussion of Xerxes' motives in this incident.

80. Lonis, *Les Usages de la Guerre*, 72.

81. Herodotus 6.32.

82. Herodotus 4.202.

83. Herodotus 4.205.

84. Herodotus 5.25.

85. Curtius 5.5.6.

86. Aeschylus, *Persians* 93–99.

87. The relief is reproduced in Yadin, *Warfare in Biblical Lands*, 392–393.

88. Ussishkin, *Siege of Lachish*, 56.

89. Ezekiel 23.

90. Nahum 2:6–7.

91. Nahum 3:5.

92. Schwally, *Semitische Kriegsaltertumer*, 65.

93. Deuteronomy 21:10–14.

94. Schwally, *Semitische Kriegsaltertumer*, 107–108.

95. Olmstead, *History of Assyria*, 646–647. Olmstead misses the parallel with Deuteronomy and mistakenly implies that Hebrew treatment of women was worse than Assyrian.

96. Isaiah 13:16.

97. Zechariah 14:2.

98. Lamentations 5:11.

99. Quoted in Mordechai Cogan, " 'Ripping Open Pregnant Women' in Light of an Assyrian Analogue," *Journal of the American Oriental Society*, 103 (1983), 756.

100. Ibid., 755–756.

101. Ibid., 756.

102. 2 Kings 8:12.

103. Hosea 13:16.

104. 2 Kings 17:5–6.

105. *Ancient Near Eastern Texts*, 184–185.

106. Exodus 21:22.

107. Amos 1:13.

108. 2 Kings 15:16.

109. Cogan, " 'Ripping Open Pregnant Women,' " 757.

110. Isaiah 13:16.

111. Hosea 11:14.

112. Psalms 137:9.

113. Nahum 3:10.

5. EARLY SIEGES THROUGH THE PELOPONNESIAN WAR

1. On the early vase, see Pritchett, *The Greek State at War*, 5:57. On the siege of Camicus, see Herodotus 7.170.

2. Thucydides 1.11.

3. Thucydides 1.2.

4. Thucydides 1.7.

5. Herodotus 1.15.

6. Herodotus 1.18–19.

7. See Garlan, *Poliorcétique Grecque*, 142–143, for the siege ramp at Smyrna. On Croesus, see Herodotus 1.26–27.

8. Garlan, *Poliorcétique Grecque*, 116.

9. J. K. Anderson, *Military Theory and Practice in the Age of Xenophon* (Berkeley: University of California Press, 1970), 47–48.

10. On hoplite warfare in Greece, see Victor Davis Hanson, *The Western Way of War, Infantry Battle in Classical Greece* (New York: Knopf, 1989), and Robin Osborne, *Classical Landscape with Figures: The Ancient Greek City and Its Countryside* (Dobbs Ferry, N.Y.: Sheridan House, 1987), chap. 7. See also Josiah Ober, *Fortress Attica: Defense of the Athenian Land Frontier, 404–322 B.C.* (Leiden: E. J. Brill, 1985), 32–35.

11. F. E. Winter, *Greek Fortifications* (Toronto: University of Toronto Press, 1971), 15–18.

12. Ibid., 102–104. See also 152–153 on the difficult problem of dating towers.

13. Pausanias 8.8.7–8.

14. Winter, *Greek Fortifications*, 130–134.

15. Ibid., 54.

16. Pausanias 10.37.7; cited by Winter, *Greek Fortifications*, 50–51.

17. Winter, *Greek Fortifications*, 15–16.

18. Ibid., 293.

19. Ibid., 55.

20. Garlan, *Poliorcétique Grecque*, 148.

21. Winter, *Greek Fortifications*, 153–154.

22. Ibid., 294.

23. Herodotus 3.60.

24. Winter, *Greek Fortifications*, 297.

25. Herodotus 3.54–56.

26. Herodotus 7.154.

27. T. J. Dunbabin, *The Western Greeks* (Oxford: Clarendon Press, 1948), 401.

28. Winter, *Greek Fortifications*, 107.

29. Herodotus 6.132–136.

30. Herodotus 9.114–118; Thucydides 1.89.

31. Thucydides 1.98.

32. Pausanias 8.8.8–9.

33. Thucydides 1.98.

34. Thucydides 1.101.

35. Thucydides 1.116–117.

36. Plutarch, *Pericles* 28.

37. Thucydides 3.17.

38. G. B. Grundy, *Thucydides and the History of His Age* (Oxford: Basil Blackwell, 1948–1961), 1:257–258.

39. Diodorus 12.28 and Plutarch *Pericles* 27.

40. On Ephorus, see Felix Jacoby, *Die Fragmente der Griechischen Historiker* (Berlin: Weidmannsche, 1926), 2, c: 22-35.

41. Polybius 12.25.

42. Garlan, *Poliorcétique Grecque*, 131-134. See also A. W. Gomme, *A Historical Commentary on Thucydides* (Oxford: Clarendon Press, 1956), 1:354 (hereafter cited as *HCT*); Georg Busolt, *Griechische Geschichte* (Gotha: F. A. Perthes, 1893-1904), 3, pt. 1, 549-551, n. 2; and H. Droysen, *Heerwesen und Kriegführung der Griechen*, in K. F. Hermann's *Lehrbuch der Griechischen Antiquitäten* (Freiburg: Akademische Verlagsbuchhandlung von J. C. B. Mohr [Paul Siebeck], 1889), 208, n. 1. At least two modern scholars believe Ephorus. See Winter, *Greek Fortifications*, 85-86, and Grundy, *Thucydides and the History of His Age*, 284.

43. Pausanias 4.25.2.

44. Garlan, *Poliorcétique Grecque*, 131-134.

45. Thucydides 1.102.

46. Diodorus 12.3-4.

47. Thucydides 1.109. See also Diodorus 11.75.

48. Winter, *Greek Fortifications*, 57-58.

49. Ibid., 110.

50. Thucydides 1.93.

51. Thucydides 5.82.

52. Thucydides 1.107.

53. Thucydides 2.13.

54. Osborne, *Classical Landscape*, 153.

55. Winter, *Greek Fortifications*, 213.

56. Ibid., 305.

57. Ibid., 57.

58. Ibid, 111.

59. Ibid., 269-270.

60. Gomme, *HCT*, 2:16-19.

61. This account of the siege of Plataea draws heavily on my article "Military Technology and Ethical Values in Ancient Siege Warfare: The Siege of Plataea," *War and Society*, 6 (September 1988), and is used with the kind permission of the University of New South Wales.

62. Donald Kagan, *The Archidamian War* (Ithaca: Cornell University Press, 1974), 44.

63. John Buckler, *The Theban Hegemony, 371-362 B.C.* (Cambridge: Harvard University Press, 1980), 11-12.

64. Kagan, *Archidamian War*, 341-342.

65. Hanson, *Warfare and Agriculture*, 34-36.

66. On the Theban oligarchy and the Boeotian League, see the *Hellenica Oxyrhynchia*, XI. See also I. A. F. Bruce, *An Historical Commentary on the "Hellenica Oxyrhynchus"* (London: Cambridge University Press, 1967), 157-164, and Robert J. Buck, *A History of Boeotia* (Edmonton: University of Alberta Press, 1979), 124-125.

67. Kagan, *Archidamian War*, 44-45.

68. Herodotus 7.233. Plutarch, *De Hdt. mal.* 33, asserts that Herodotus was wrong and that Leontiades was not in command of the Thebans at Thermopylae, but the evidence supports Herodotus. See Robert J. Buck, "Boeotarchs at Thermopylae," *Classical Philology*, 69 (1974), 47-48.

69. Kagan, *Archidamian War*, 46.

70. Nancy H. Demand, *Thebes in the Fifth Century: Heracles Resurgent* (London: Routledge and Kegan Paul, 1982), 26-27.

71. Xenophon, *Hellenica* 5.2.25. On Xenophon's anti-Theban bias, see George Cawkwell, Introduction, in Xenophon, *A History of My Times*, trans. Rex Warner (New York: Penguin, 1979), 15–16.

72. Thucydides 3.62.

73. Thucydides 3.82.

74. Herodotus (7.233) puts the size of the advance force at 400 and says it was led by Eurymachus. Thucydides (2.2) estimates the guard at "rather over 300" and places it under the command of the Boeotarchs. Probably Eurymachus was the real leader with the Boeotarchs in official, but nominal, command. The presence of the Theban Boeotarchs may have provided a facade of Boeotian patriotism to the treacherous attack on Plataea.

75. Thucydides 3.56 and 3.65.

76. Thucydides 2.2–3.

77. Thucydides (3.68) dates the alliance ninety-three years before the fall of Plataea in 427 B.C. E. Kirsten believes his dating is accurate: "Plataia," in *Paulys Realencyclopaedie der classischen Altertumswissenschaften* (1950), 20:2286.

78. Herodotus 6.108, 8.1, and 9.30.

79. Kirsten, "Plataia," 2303–2304.

80. Thucydides 2.4.

81. Thucydides 2.5.

82. Thucydides 2.5.

83. Thucydides 2.5.

84. Gomme, *HCT*, 2:6.

85. *CAH*, 5:192.

86. Thucydides 2.6.

87. Modern scholars are unanimous in their verdict on the stupidity of the Plataean atrocity. See Busolt, *Griechische Geschichte*, 3, pt. 2:916; George Grote, *Greece* (New York: Peter Fenelon Collier, 1894), 6:119; and Kagan, *Archidamian War*, 47–48.

88. Thucydides 2.6 and 2.78.

89. Thucydides 2.71.

90. Thucydides 1.79–85.

91. Thucydides 2.72.

92. Thucydides 5.105.

93. Thucydides 3.68. Gomme believes that Thucydides took little stock in the sincerity of Archidamus's offer; *HCT*, 2:206.

94. Thucydides 2.73–74. On the suspicion that the Plataean women and children in Athens functioned as hostages, see Lonis, *Les Usages de la Guerre*, 134.

95. F. E. Adcock in *CAH*, 5:212.

96. Alfred Zimmern, *The Greek Commonwealth: Politics and Economics in Fifth-Century Athens* (New York: Oxford University Press, 1961), 434. E. Kirsten agrees with Zimmern's judgment; "Plataia," 2306–2307.

97. I follow Gomme's chronology outlined in *HCT*, 3: 716–718. Some accounts, such as Zimmern's, assume Pericles was already dead at the time of this decision.

98. Kagan, *Archidamian War*, 104–105. See also Busolt, *Griechische Geschichte*, 3, pt. 2:967.

99. G. B. Grundy, *The Topography of the Battle of Plataea* (London: John Murray, 1894), 61. The archeological evidence of ancient Plataea allows no definitive conclusion about the extent and location of the walls in the fifth century. In addition to Grundy, see Kirsten, "Plataia, 2272–2280.

100. On the fecundity of Boeotia, see Demand, *Thebes in the Fifth Century*, 10.

101. Thucydides 2.74.

102. Donald Lateiner, "Heralds and Corpses in Thucydides," *Classical World*, 71 (October 1977), 97–106.

103. Thucydides 1.102. G. B. Grundy, in "A Suggested Characteristic in Thukydides' Work," *Journal of Hellenic Studies*, 18 (1898), 220–221, suggests the Spartans, who were masters of a Peloponnese bristling with fortified towns, had never had to develop skills in siege warfare because the Peloponnesians could not afford to remain in their fortified towns while the Spartans destroyed their crops.

104. H. D. Westlake, *Individuals in Thucydides* (London: Cambridge University Press, 1968), 134.

105. This is the consensus of modern scholars. See Edmund F. Bloedow, "The Intelligent Archidamus," *Klio*, 65 (1983); Busolt, *Griechische Geschichte*, 3, pt. 2:967–968; and Grundy, "Suggested Characteristics in Thukydides," 227. Westlake dissents, believing Archidamus represented the typical Spartan: "slow, cautious, conventional, lacking in inspiration and imagination"; *Individuals in Thucydides*, 122.

106. Thucydides 1.80–81. Westlake is not alone in doubting the authenticity of this part of the speech, finding it wholly unconvincing that a Spartan king could express the Thucydidean and Periclean theory that wealth and naval power are the keys to military success. See Westlake, *Individuals in Thucydides*, 124–125. Thomas Kelly, while not rejecting the speech out of hand, is extremely cautious in his discussion of it in "Thucydides and Spartan Strategy in the Archidamian War," *American Historical Review*, 87 (February 1982), 36–37. Kagan, who accepts the authenticity of all the speeches in Thucydides' *History*, believes the speech accurately reflects Archidamus's thinking; *Archidamian War*, 300–304. So does Grundy, "Suggested Characteristics in Thukydides," 221.

107. Kelly, "Thucydides and Spartan Strategy in the Archidamian War," 25–54, argues vigorously that the Spartans had a far-reaching and aggressive naval strategy.

108. Thucydides 2.18–19.

109. Grundy, "Suggested Characteristics in Thukydides," 221–222.

110. Garlan, *Poliorcétique Grecque*, 147.

111. Thucydides 2.75. Grundy sees this conventional beginning as a concession by Archidamus to the ruling conservatism in the Spartan army; "Suggested Characteristics in Thukydides," 227.

112. See Hanson, *Warfare and Agriculture*, 55–65, for a detailed discussion of the problems the olive tree's size, toughness, extensive root system and amazing regenerative powers presented to the would-be ravager. He convincingly argues that the extent of the devastation caused by this tactic has been widely exaggerated.

113. Bloedow argues that Archidamus contemplated a long siege from the beginning and only went through routine motions before settling down to circumvallation; "The Intelligent Archidamus," 43. Bloedow is defending Archidamus against modern critics such as Westlake and is anxious to mitigate Archidamus's failure to avoid circumvallation. But the detail Thucydides lavished on Archidamus's methods shows that Thucydides found those methods far from routine and Bloedow's denial of Archidamus's originality actually works against his general argument that Archidamus was a far-sighted commander. Busolt believes that Archidamus did hope to avoid a passive siege; *Griechische Geschichte*, 3, pt. 2:967.

114. Thucydides 2.75. The Persians had used this method against Greek cities in Asia Minor in the second half of the sixth century. See Herodotus 1.162.

115. F. E. Winter finds it "difficult to believe that the Spartans were respon-

sible for such an innovation" and concludes that the Athenians had "most probably" already used ramps; *Greek Fortifications*, 307.

116. Grundy, *Topography*, 65.

117. Thucydides 2.75.

118. Gomme, *HCT*, 2:207–208. Kelly seems more willing to consider that the seventy days may be correct, seeing the tremendous effort as another piece of evidence that the Spartans did not confine their strategy to destroying the Attic countryside; "Thucydides and Spartan Strategy," 53, n. 103.

119. Thucydides 2.75–76.

120. Garlan, *Poliorcétique Grecque*, 146.

121. Thucydides 2.76.

122. This is the opinion of Busolt, who suggests that the word *mechane* that Thucydides used in his description of earlier sieges at Potidaea and Oenoe is so vague it may refer to devices less sophisticated than rams. Droysen believes the *mechane* the Spartans used at Oenoe were scaling ladders. However, Winter finds the idea that Thucydides would call a scaling ladder a machine to be implausible. See Busolt, *Griechische Geschichte*, 3, pt. 2:968–969; Droysen, *Heerwesen und Kriegführung*, 209, n. 1; Winter, *Greek Fortifications*, 85–86, n. 44.

123. Grundy, "Suggested Characteristics in Thukydides," 222.

124. Ibid. In another place Grundy speculates that Asia Minor was the source of the Spartan knowledge of siegecraft; *Thucydides and the History of His Age*, 1:289.

125. Winter, *Greek Fortifications*, 155–156.

126. See Droysen, *Heerwesen und Kriegführung*, 217–232, for a detailed description of the siege machinery and tactics necessary for a successful breach.

127. Thucydides 2.76.

128. Thucydides 2.77.

129. Thucydides 2.13; Gomme, *HCT*, 2:41–42.

130. W. W. Tarn, *Hellenistic Military and Naval Development* (New York: Biblo and Tannen, 1966), 85; O'Connell, *Arms and Men*, 48; F. E. Adcock, *The Greek and Macedonian Art of War* (Berkeley: University of California Press, 1957), 21.

131. Thucydides 3.2.

132. Thucydides 4.55.

133. Aeschylus, *Seven against Thebes* 158.

134. Thucydides 2.81.

135. Gomme, Andrewes, Dover, *HCT* 4:310; Thucydides 6.43.

136. Thucydides 4.100.

137. Grundy, "Suggested Characteristics in Thukydides," 227.

138. Grundy, *Thucydides and the History of his Age*, 1:261–262. See also R. T. Ridley, "The Hoplite as Citizen: Athenian Military Institutions in Their Social Context," *L'Antiquité Classique*, 48 (1979), 512, and Adcock, *Art of War*, 58.

139. Grundy, *Thucydides and the History of His Age*, 1:290.

140. Herodotus 3.55.

141. Herodotus 9.102.

142. Garlan, *Poliorcétique Grecque*, 20.

143. Plutarch, *Comp. Lys. et Sulla* 4.2–3.

144. Herodotus 9.70.

145. Diodorus 11.31.3.

146. Garlan, *Poliorcétique Grecque*, 130.

147. Thucydides 4.116.

148. Gomme, *HCT*, 3:591–592.

149. Diodorus 14.53.4.

150. Garlan, *Poliorcétique Grecque*, 106. See Thucydides 1.93 and 2.13.

151. Gomme, *HCT*, 2:211–212; Busolt, *Griechische Geschichte*, 3, pt. 2:969.

152. Thucydides 2.78. See Busolt, *Griechische Geschichte*, 3, pt. 2:969.

153. For a discussion of the problems of interpreting Thucydides' description of the wall, see Garlan, *Poliorcétique Grecque*, 115–116.

154. Thucydides 3.21.

155. Winter, *Greek Fortifications*, 161.

156. Grundy, *Topography*, 68. Busolt estimates the circumference of the wall was 2,500 meters, or a little more than one-and-a-half miles; *Griechische Geschichte*, 3, pt. 2:969. There are no archeological remains of this temporary wall. Thucydides tells us that the Plataean defenders could see the individual bricks in the wall of circumvallation. Grundy tested this fact with his own sight, which must have been keen, and concluded the distance from the city wall to the wall of circumvallation was between eighty and one hundred yards.

157. Thucydides 3.20–24.

158. Thucydides 3.52.

159. Thucydides 2.58.

160. Thucydides 7.43.

161. Thucydides 8.100.

162. Thucydides 3.51.

163. Gomme, *HCT*, 2:334.

164. Thucydides 4.100.

165. Thucydides 4.115.

166. For a convenient list of sieges and the methods used during the Peloponnesian War, see Grundy, *Thucydides and the History of His Age*, 1:286.

167. Garlan, *Poliorcétique Grecque*, 125–126.

168. Thucydides 8.28.

169. Thucydides 4.135 and 7.43.

170. Thucydides 4.110–113.

171. Garlan, *Poliorcétique Grecque*, 128–129.

172. Xenophon, *Hellenica* 1.6.13; Thucydides 3.18 and 8.100.

173. Diodorus 13.76.5.

174. Diodorus 12.56.4.

175. Garlan, *Poliorcétique Grecque*, 106.

176. Thucydides 5.75.

177. Thucydides 4.69.

178. Thucydides 6.44.

179. Thucydides 4.69.

180. Thucydides 7.43.

181. Garlan, *Poliorcétique Grecque*, 112–113.

182. Thucydides 4.4–5.

183. Thucydides 3.17; 7.13.

184. Rachel L. Sargent, "The Use of Slaves by the Athenians in Warfare," *Classical Philology*, 22 (1927), 203.

185. Thucydides 3.18.

186. Thucydides 6.97.

187. Thucydides 6.99.

188. Thucydides 4.69.

189. Thucydides 4.90.

190. Thucydides 4.5.

191. Thucydides 6.98.

192. Garlan, *Poliorcétique Grecque*, 113; see also Donald Kagan, *The Peace of Nicias and the Sicilian Expedition* (Ithaca: Cornell University Press, 1981), 267–268.

193. Busolt, *Griechische Geschichte*, 3, pt 2:969.

194. Thucydides 1.93.

195. Thucydides 4.4.

196. Thucydides 4.90.

197. Thucydides 4.69.

198. Xenophon, *Hellenica* 1.3.4.

199. Thucydides 2.70.

200. Thucydides 3.52.

201. Xenophon, *Hellenica* 1.3.19.

202. Thucydides 8.56.

203. Xenophon, *Hellenica* 2.2.11.

204. Garlan, *Poliorcétique Grecque*, 118.

205. Thucydides 4.26.

206. Thucydides 7.24.

207. Thucydides 2.58.

208. Thucydides 2.70.

209. Diodorus 13.12.1.

210. Aeschylus, *Agamemnon* 555–566.

211. Garlan, *Poliorcétique Grecque*, 122–125.

212. Thucydides 3.17.

213. Thucydides 2.70.

214. Thucydides 2.13.

215. Thucydides 2.24.

216. Thucydides 3.19.

217. Thucydides 3.46 and 7.47.

218. Xenophon, *Hellenica* 1.1.14. See Garlan, *Poliorcétique Grecque*, 128.

219. Garlan, *Poliorcétique Grecque*, 123.

220. Thucydides 4.86.

221. Thucydides 4.67–68.

222. Thucydides 4.110–112.

223. Thucydides 3.34.

224. See Andreas Panagopoulos, *Captives and Hostages in the Peloponnesian War* (Athens: Grigoris, 1978), 53–54, for a disapproving discussion of this atrocity. Paches' main defender is Gomme. See *HCT*, 2:297.

225. A. W. H. Adkins, *Moral Values and Political Behaviour in Ancient Greece, from Homer to the End of the Fifth Century* (London: Chatto & Windus, 1972), 55–56.

226. Ibid., 131.

227. Ibid., 146.

228. O'Connell, *Arms and Men*, 51.

229. Garlan, *Poliorcétique Grecque*, 40.

230. Thucydides 4.3. and 4.41.

231. Garlan, *Poliorcétique Grecque*, 41–44.

232. Ibid., 20; 98–99.

233. Thucydides 7.27.

234. David Whitehead, Introduction, Aineias the Tactician, *How to Survive under Siege* (Oxford: Clarendon Press, 1990), 9.

235. See Hermann Bennigston, "Die Griechische Polis bei Aeneas Tacticus," *Historia*, 11 (1962), for an analysis of Aeneas Tacticus's discussion of treachery as a revealing mirror of the Greek polis.

236. Whitehead, Introduction, 23–24.

237. Ibid., 9–12.

238. Aineias Tacticus 1.6–7.

239. Aineias Tacticus 22.4–10.

240. Aineias Tacticus 22.15.

241. Aineias Tacticus 22.19.

242. Aineias Tacticus 1.9–2.1.

243. Aineias Tacticus 18.3–22.
244. Aineias Tacticus 5.1.
245. Aineias Tacticus 7.1–2.
246. Aineias Tacticus 8.1–4.
247. Aineias Tacticus 10.1–2.
248. Aineias Tacticus 10.3.
249. Aineias Tacticus 15.2 and 16.2.
250. Aineias Tacticus 16.5–7.
251. Aineias Tacticus 16.11–12.
252. Aineias Tacticus 16.16–18.
253. Aineias Tacticus 10.6–17.
254. Aineias Tacticus 10.20.
255. Aineias Tacticus 10.23 24.
256. Aineias Tacticus 14.1.
257. Aineias Tacticus 26.8.
258. Aineias Tacticus 38.4.
259. Aineias Tacticus 16.15.
260. Garlan, *Poliorcétique Grecque*, 121. See 120–122 for a good discussion of Greek defensive tactics against siege during the Peloponnesian War.
261. Winter, *Greek Fortifications*, 235.
262. Thucydides 6.34. See Kagan, *Peace of Nicias*, 221; Busolt, *Griechische Geschichte*, 3, pt. 2:1300–1301.
263. Thucydides 6.37.
264. Thucydides 6.41.
265. Thucydides 6.50–52; Kagan, *Peace of Nicias*, 218.
266. Thucydides 6.49.
267. Thucydides 6.63–72.
268. Thucydides 6.43.
269. See Kagan, *Peace of Nicias*, 239–242, for an excellent discussion of this controversial decision.
270. Thucydides 6.88.
271. Thucydides 6.75.
272. Thucydides 6.73.
273. Thucydides 6.88.
274. Thucydides 6.94.
275. Thucydides 6.96.
276. Thucydides 4.131.
277. Thucydides 6.97.
278. Kagan, *Peace of Nicias*, 261.
279. Busolt, *Griechische Geschichte*, 3, pt. 2:1331.
280. Thucydides 6.97–98.
281. Thucydides 6.98.
282. Garlan, *Poliorcétique Grecque*, 121 and 146.
283. Busolt, *Griechische Geschichte*, 3, pt. 2:1333.
284. Thucydides 6.99–100; Kagan, *Peace of Nicias*, 265.
285. Thucydides 6.100.
286. Thucydides 6.101.
287. Thucydides 6.102.
288. Thucydides 6.103.
289. Thucydides 7.2.
290. Kagan, *Peace of Nicias*, 267–268.
291. Busolt, *Griechische Geschichte*, 3, pt. 2:1333.
292. Ibid., 1336–1337.
293. Ibid., 1330.

294. Grote, *Greece*, 6:259–260.

295. Busolt, *Griechische Geschichte*, 3, pt. 2:1340.

296. Ibid., 1342.

297. Grote, *Greece*, 7:260.

298. See Paul B. Kern, "The Turning Point in the Sicilian Expedition," *Classical Bulletin*, 65 (1989), 75–82.

299. Thucydides 7.48.

300. Thucydides 6.103.

301. Busolt, *Griechische Geschichte*, 3, pt. 2:1337.

302. Peter Green, *Armada from Athens* (Garden City, N.Y.: Doubleday, 1970), 205; Plutarch, *Nicias* 18.

303. Kagan, *Peace of Nicias*, 273–274.

304. Thucydides 7.3–4.

305. Thucydides 7.5–6.

306. Thucydides 7.11.

307. Green, *Armada from Athens*, 254–255; Thucydides 7.49.

308. Kagan, *Peace of Nicias*, 317.

309. Thucydides 7.1 and 7.43.

310. Thucydides 7.37.

311. Thucydides 7.42.

312. Thucydides 7.43.

313. E. A. Freeman, *History of Sicily* (Oxford: Clarendon Press, 1891–94), 3:312.

314. Thucydides 7.43–44.

315. Thucydides 7.48–49.

316. Thucydides 7.87.

317. Kagan, *Peace of Nicias*, chap. 11; Green, *Armada from Athens*, 207; Grote, *Greece*, 7:266.

6. TREATMENT OF CAPTURED CITIES

1. *Iliad* 9.590–594.

2. Pierre Ducrey, *Le Traitement de Prisonniers de Guerre dans la Grèce Antique* (Paris: Editions de Boccard, 1968), 314–315.

3. *Iliad* 6.447–465.

4. Heinrich Kuch, *Kriegsgefangenschaft und Sklaverei bei Euripides* (Berlin: Akademie Verlag, 1974), 27.

5. J. A. O. Larson, "Federation for Peace in Ancient Greece," *Classical Philology*, 39 (July 1944), 146–147.

6. Dunbabin, *The Western Greeks*, 357.

7. Athenaeus 521. d.

8. Diodorus 12.10.1.

9. Strabo 6.1.13.

10. Herodotus 6.21.

11. Dunbabin, *Western Greeks*, 365–366.

12. Ibid., 404.

13. Herodotus 7.156.

14. Diodorus 11.67.2.

15. Herodotus 7.33 and 9.120.

16. Plutarch, *Cimon* 9.

17. Herodotus 7.107.

18. Thucydides 1.98.

19. Thucydides 1.113.

20. Diodorus 11.65.5.

21. Thucydides 1.103.

22. Pausanias 4.25.2.
23. Thucydides 1.98; See Gomme, *HCT*, 1:282, for the terms.
24. Thucydides 1.101.
25. Diodorus 12.3.3.
26. Thucydides 1.117.
27. Donald Kagan, *The Outbreak of the Peloponnesian War* (Ithaca: Cornell University Press, 1969), 176.
28. Plutarch, *Pericles* 28. See Ducrey, *Le Traitement de Prisonniers de Guerre*, 212.
29. Thucydides 1.76.
30. Thucydides 2.70.
31. Thucydides 3.25–26.
32. Thucydides 3.19.
33. Thucydides 3.27–28.
34. Thucydides 3.36.
35. Euripides, *Hecuba* 106, trans. William Arrowsmith (Chicago: University of Chicago Press, 1958).
36. Euripides, *Hecuba* 132–133.
37. Thucydides 3.36.
38. Thucydides 3.39–40.
39. Thucydides 3.44.
40. Thucydides 3.45.
41. Thucydides 3.47.
42. Thucydides 3.49.
43. See Gomme, *HCT*, 2:325–326, for a discussion of these problems. Gomme does not find the circumstantial evidence strong enough to warrant changing the text.
44. Thucydides 3.52.
45. See Gomme, *HCT*, 2:289; Busolt, *Griechische Geschichte*, 3, pt. 2:1036; Grote, *Greece*, 6:260.
46. Thucydides 3.52.
47. Gomme, *HCT*, 2:354.
48. Thucydides 1.30. See Gomme, *HCT*, 1:164–65, for a discussion of these distinctions.
49. Thucydides 3.56. See Ducrey, *Le Traitement des Prisonniers de Guerre*, 61–64, for a discussion of the legalities and conventions governing this incident.
50. Thucydides 3.59.
51. Kagan, *Archidamian War*, 173; Busolt, *Griechische Geschichte*, 3, pt. 2:1037, n. 4.
52. Thucydides 3.68.
53. Thucydides 3.68.
54. Lonis, *Les Usages de Guerre entre Grecs et Barbares*, 35.
55. Thucydides 3.68.
56. See David Cohen, "Justice, Interest, and Political Deliberation in Thucydides," *Rechtshistorisches Journal*, 2 (1983), 258–259.
57. Ibid., 245.
58. Thucydides 3.82.
59. Thucydides 4.122–123.
60. Thucydides 4.130.
61. Thucydides 5.3.
62. Thucydides 4.123.
63. Thucydides 5.32.
64. Ducrey, *Le Traitement de Prisonniers de Guerre*, 121–122.

65. Thucydides 4.116.
66. Thucydides 5.83.
67. Thucydides 5.116.
68. Thucydides 5.84.
69. Thucydides 5.111.
70. Thucydides 5.89.
71. Thucydides 5.91.
72. Aristophanes, *The Birds* 186.
73. Xenophon, *Hellenica* 2.2.9; Plutarch, *Lysander* 14.4.
74. Ducrey, *Le Traitement de Prisonniers de Guerre*, 123-124.
75. Thucydides 4.57.
76. Diodorus 12.65.9. See Kagan, *Archidamian War*, 264-265.
77. Xenophon, *Hellenica* 2.2.3.
78. Xenophon, *Hellenica* 2.2.19.
79. Xenophon, *Hellenica* 1.6.14-15.
80. Diodorus 13.76.2.
81. Xenophon, *Hellenica* 2.1.15.
82. Diodorus 13.104.7.
83. Panagopoulos, *Captives and Hostages*, 169-170.
84. Xenophon, *Hellenica* 2.1.19; Diodorus 13.104.8.
85. Ducrey discusses the treatment of the population of Cedreae in the context of Greek treatment of barbarians and concludes that although the city's barbarian origins cannot be a complete explanation of Lysander's severity, it undoubtedly played a role. See Ducrey, *Traitement de Prisonniers de Guerre*, 274-275.
86. Thucydides 4.105-106.
87. Plutarch, *Alcibiades* 30.4-5.
88. Panagopoulos, *Captives and Hostages*, 215.
89. Diodorus 13.67.5-7.
90. Xenophon, *Hellenica* 2.2.1-2.
91. Xenophon, *Hellenica* 2.3.6-7.
92. Thucydides 6.62.
93. Thucydides 7.13.
94. Plutarch, *Nicias* 15.4; Athenaeus 13.588a-e.
95. Ducrey, *Le Traitement des Prisonniers de Guerre*, 111-112 and 140.
96. Ibid., 334-335.
97. Franz Kiechle, "Zur Humanität in der Kriegführung der Griechischen Staaten," *Historia*, 7 (1958), 143 and 156.
98. Panagopoulos, *Captives and Hostages*, 219.
99. Josiah Ober, "Classical Greek Times," in *The Laws of War: Constraints on Warfare in the Western World*, ed. Michael Howard, George J. Andreopoulos, and Mark R. Shulman (New Haven: Yale University Press, 1994), 13-14 and 18.
100. Ducrey, *Le Traitement des Prisonniers de Guerre*, 125-126 and 336.
101. Pritchett, *Greek State at War*, 5:203-205.
102. Lonis, *Les Usages de la Guerre entre Grecs et Barbares*, 36.
103. Pritchett, *Greek State at War*, 5:218-219.
104. Yvon Garlan, *War in the Ancient World: A Social History* (New York: Norton, 1975), 127-128; 164. See also François Chamoux, *The Civilization of Greece* (New York: Simon and Schuster, 1965), 198-204, for a good discussion of the decline of the citizen hoplite. Ober relates the decline of hoplite standards to the rise of the navy and democracy in Athens. But he also recognizes the significance of siege warfare in this decline. Ober, "Classical Greek Times," 18-21 and 24-25.

105. Ridley, "The Hoplite as Citizen," 512.

106. *Iliad* 6.492.

107. *Iliad* 18.285–287.

108. *Iliad* 7.76–91.

109. Aeschylus, *Seven against Thebes*, 181–203, trans. H. Weir Smyth (Cambridge: Harvard University Press, 1922).

110. Compare Thucydides 2.4 with Thucydides 3.74.

111. Aineias Tacticus 3.6.

112. Aineias Tacticus 40.4–5.

113. *Odyssey* 8.523–530, trans. Richmond Lattimore (New York: Harper & Row, 1968).

114. *Iliad* 6.55–60.

115. *Iliad* 6.61–62.

116. *Iliad* 24.725–736.

117. Euripides, *Andromache* 10–11, trans. John Frederick Nims (Chicago: University of Chicago Press, 1958).

118. Euripides, *The Trojan Women* 725–733, trans. Richmond Lattimore (Chicago: University of Chicago Press, 1958).

119. Euripides, *Trojan Women* 737–739.

120. Euripides, *Trojan Women* 787–789.

121. Euripides, *Trojan Women* 1167–1191.

122. Kuch, *Kriegsgefangenschaft und Sklaverei bei Euripides*, 73.

123. Aeschylus, *Seven against Thebes* 345–356.

124. *Iliad* 2.354–356.

125. *Iliad* 21.586–588.

126. *Iliad* 18.265.

127. *Iliad* 22.56–57.

128. *Iliad* 9.139–140.

129. *Iliad* 4.237–239.

130. On the concept of the antifuneral, see James M. Redfield, *Nature and Culture in the Iliad* (Chicago: University of Chicago Press, 1975), 183–186.

131. Thucydides 7.75.

132. Aeschylus, *Seven against Thebes* 321–339.

133. Aeschylus, *Seven against Thebes* 452–456.

134. Aeschylus, *Seven against Thebes* 333–344.

135. Euripides, *Hecuba* 488–492.

136. Aeschylus, *Seven against Thebes* 217–218.

137. Euripides, *Trojan Women* 26–27.

138. Euripides, *Trojan Women* 95–97.

139. Ducrey, *Le Traitement de Prisonniers de Guerre*, 319.

140. *Iliad* 23.703–705.

141. *Iliad* 1.113–114.

142. *Iliad* 9.342–343.

143. *Iliad* 19.297–300.

144. *Iliad* 19.291–294

145. Euripides, *Hecuba* 918–941.

146. Euripides, *Hecuba* 351–366.

147. See Kuch, *Kriegsgefangenschaft and Sklaverei bei Euripides*, 55–56, for an analysis of the social implications of Polyxena's lament. Kuch makes some interesting observations, but his rigid Marxist framework, in which he has to cast his discussion in terms of a class conflict between slaves and masters, limits its usefulness.

148. Euripides, *Andromache* 36–38.

149. Euripides, *Trojan Women* 202–204.

7. DIONYSIUS I

1. Diodorus 15.23.5; see Freeman, *History of Sicily*, 4:2.
2. Freeman, *History of Sicily*, 4:3-4.
3. Karl Friedrich Stroheker, *Dionysios I* (Wiesbaden: Franz Steiner, 1958), 178-179.
4. Ferrill, *Origin of War*, 149-150.
5. Brian Caven, *Dionysius I* (New Haven: Yale University Press, 1990), 31-32. See also Martin Märker, *Die Kämpfe der Karthager auf Sizilien in den Jahren 409-405 v. Chr.* (Weida i. Thür.: Druck von Thomas & Hubert, 1930), 9-16, for a detailed discussion of the relative strengths of the Carthaginian and Greek forces. He estimates that the Carthaginian army numbered 30-40,000.
6. Diodorus 13.54.1-5.
7. Thucydides 6.43-44.
8. See Ferrill, *Origin of War*, 146.
9. Joseph I. S. Whitaker, *Motya* (London: G. Bell, 1921), 17-18.
10. Märker, *Die Kämpfe der Karthager auf Sizilien*, 17.
11. Ibid., 28 and 31.
12. Ibid., 30.
13. Ibid., 21-23.
14. Diodorus 13.54-55.
15. Diodorus 13.55.4-5.
16. Diodorus 13.55.5. See R. K. Sinclair, "Diodorus Siculus and Fighting in Relays," *Classical Quarterly*, 60 (1966), 253.
17. Caven, *Dionysius*, 30.
18. Diodorus 13.44.1-2; Freeman, *History of Sicily*, 3:452-453.
19. Diodorus 14.9.1.
20. Diodorus 13.62.5.
21. Diodorus 13.55.
22. Diodorus 13.56.1-2.
23. Diodorus 13.56.3-8.
24. Märker, *Die Kämpfe der Karthager auf Sizilien*, 34.
25. Ibid., 38.
26. Diodorus 13.59.4-9.
27. Diodorus 13.60.
28. Garlan, *Poliorcétique Grecque*, 159.
29. Diodorus 13.61.
30. Diodorus 13.62.1-4.
31. Diodorus 13.62.6.
32. Diodorus 13.80.1-2.
33. Diodorus 13.80.5.
34. Caven, *Dionysius*, 45-46.
35. Diodorus 13.81.4-13.84.
36. Diodorus 13.84.5-6.
37. Diodorus 13.81.3; Caven, *Dionysius*, 46.
38. Freeman, *History of Sicily*, 3:520; Märker, *Die Kämpfe der Karthager auf Sizilien*, 60-61.
39. Diodorus 13.85.1-4.
40. Garlan, *Poliorcétique Grecque*, 160.
41. Diodorus 13.85.4. Märker believes the mercenaries were stationed on the acropolis because the hill he has identified as the Hill of Athena is not large enough to contain eight hundred mercenaries. But Diodorus is quite explicit on the point, and the layout of Acragas and its fortifications is uncertain. Märker, *Die Kämpfe der Karthager auf Sizilien*, 51-52.
42. Diodorus 13.85.5-86.1.

43. Diodorus 13.86.1–3.

44. Diodorus 13.86.3.

45. Märker, *Die Kämpfe der Karthager auf Sizilien*, 65–66.

46. Ibid., 13–14.

47. Caven, *Dionysius*, 47.

48. See Märker, *Die Kämpfe der Karthager auf Sizilien*, 67–68, for speculations about the exact location of this battle.

49. Diodorus 13.86.4–87.5.

50. Garlan, *Poliorcétique Grecque*, 156.

51. Diodorus 13.88.1–2.

52. Diodorus 13.88.3–5.

53. Märker, *Die Kämpfe der Karthager auf Sizilien*, 70.

54. Aineias Tacticus 10.19.

55. Aineias Tacticus 12.2.

56. Aineias Tacticus 13.1.

57. Aineias Tacticus 13.2–3.

58. Diodorus 13.88.3–8.

59. Diodorus 13.96.3.

60. Diodorus 13.89.1–91.1.

61. Stroheker, *Dionysios*, 44. Märker places the arrival at the end of August or the beginning of September. See Märker, *Die Kämpfe der Karthager auf Sizilien*, 71.

62. Diodorus 13.108 & 110.1.

63. Caven, *Dionysius*, 62.

64. Diodorus 13.109.

65. Märker, *Die Kämpfe der Karthager auf Sizilien*, 84.

66. Caven, *Dionysius*, 63.

67. The most convincing interpretation of this battle is that of Caven, *Dionysius*, 63–72. See also Stroheker, *Dionysios*, 45–46, for a good summary of the significance of Dionysius's plans.

68. Diodorus 13.110.

69. Stroheker, *Dionysios*, 46–47.

70. Diodorus 13.111.1–2. See Lonis, *Les Usages de la Guerre entre Grecs et Barbares*, 59.

71. Diodorus 13.111.

72. Diodorus 13.112–113.

73. Freeman, *History of Sicily*, 3:580.

74. Diodorus 13.114.

75. Diodorus 14.8.

76. Diodorus 14.9.

77. Lars Karlsson, *Fortification Towers and Masonry Techniques in the Hegemony of Syracuse* (Stockholm: Paul Astroms, 1992), 106.

78. Diodorus 14.7.2–3.

79. Stroheker, *Dionysios*, 64.

80. Caven, *Dionysius*, 13.

81. Diodorus 14.18.

82. Karlsson, *Fortification Towers and Masonry Techniques*, 12 and 106–107.

83. Diodorus 13.15.5.

84. Karlsson, *Fortification Towers and Masonry Techniques*, 14.

85. Stroheker, *Dionysios*, 63. Caven and Garlan agree. See Caven, *Dionysius*, 88–89. Garlan, *Poliorcétique Grecque*, 185.

86. See Garlan, *Poliorcétique Grecque*, 185–189, for a discussion of the chronology of these remains.

87. Freeman, 4:52–57; *CAH*, 6:119.

88. Diodorus 15.13.5.

89. Aineias Tacticus 40.1. See Garlan, *Poliorcétique Grecque*, 189–190.

90. Diodorus 14.41–42.1. Diodorus's claim that the catapult was invented at Syracuse has not gone unchallenged, especially by Orientalists. Their argument that the catapult had, at the very least, been anticipated by the Assyrians and then transmitted to Sicily by the Carthaginians carries a certain plausibility, but the evidence is too scarce and weak to refute Diodorus's convincing account. See Garlan, *Poliorcétique Grecque*, 164–166.

91. The most detailed and authoritative description of the first catapult is Marsden, *Greek and Roman Artillery*, 5–12. For a briefer description, see Ferrill, *Origins of War*, 170–171. The idea that Dionysius's catapults may have used torsion power is Garlan's. Garlan, *Poliorcétique Grecque*, 167–168.

92. Diodorus 14.50.4. Marsden, *Greek and Roman Artillery*, 12.

93. Diodorus gives a minimum of eighty thousand infantry, but Caven revises the figure downward. Caven, *Dionysius*, 100.

94. Garlan, *Poliorcétique Grecque*, 168.

95. Caven, *Dionysius*, 84; Stroheker, *Dionysios*, 67.

96. Stroheker, *Dionysios*, 154–156.

97. Diodorus 14.47.7.

98. Anderson, *Military Theory and Practice in the Age of Xenophon*, 47–48.

99. Garlan, *Poliorcétique Grecque*, 168.

100. Caven, *Dionysius*, 163–166; Stroheker, *Dionysios*, 161–162.

101. Diodorus 14.48.1–2.

102. The description of Motya's fortifications relies on Whitaker, *Motya*, 140–192.

103. Ibid., 87 and 121.

104. Diodorus 14.48.4.

105. Diodorus 14.51.1.

106. Whitaker, *Motya*, 132.

107. Freeman, *History of Sicily*, 4:71.

108. Garlan, *Poliorcétique Grecque*, 168.

109. Diodorus 14.50.4.

110. Marsden, *Greek and Roman Artillery*, 56.

111. Whitaker, *Motya*, 82–83.

112. Diodorus 14.50.

113. Garlan, *Poliorcétique Grecque*, 163.

114. Xenophon, *Cyropaedia* 6.1.52–54. See A. Burford, "Heavy Transport in Classical Antiquity," *Economic History Review*, Second Series, 13 (1960), 13, for a discussion and drawing of Cyrus's hitching method.

115. Whitaker, *Motya*, 85.

116. Diodorus 14.51.1–3.

117. Aineias Tacticus 33.3; 34.1.

118. Aineias Tacticus 33.4; 33.1–2; 35.1.

119. Aineias Tacticus 32.1; 37.3.

120. Garlan, *Poliorcétique Grecque*, 177–178.

121. Diodorus 14.51.4–52.4.

122. Diodorus 14.52.5–7.

123. Diodorus 14.53.4.

124. Stroheker, *Dionysios*, 157.

125. Diodorus 14.54.4–6. Again I follow Caven's downward revision of Diodorus's numbers. See Caven, *Dionysius*, 107.

126. Diodorus 14.55.

127. I follow Caven's interpretation of these events, including his opinion that Leptines pursued the Carthaginians with a squadron of quinqueremes, even though Diodorus says they were triremes. Caven, *Dionysius*, 108-109.

128. Diodorus 14.56.

129. Diodorus 14.57.

130. Diodorus 14.58.3.

131. Diodorus 14.58.1.

132. Polyainos 5.2.9; see Freeman, *History of Sicily*, 4:113.

133. Caven, *Dionysius*, 111.

134. Diodorus 14.61.1-3.

135. There is a lacuna here in Diodorus's text that causes confusion about the number, even if we could accept it as reliable. Diodorus 14.62.2.

136. Diodorus 14.62.

137. Stroheker, *Dionysios*, 77.

138. Diodorus 14.63.

139. Diodorus 14.64.

140. Diodorus 14.70.1-3.

141. Caven, *Dionysius*, 115-116.

142. Diodorus 14.70.

143. Diodorus 14.71.

144. Diodorus 14.72-73.

145. Diodorus 14.75.

146. Diodorus 14.76.4.

147. Diodorus 14.78.5-6. See Freeman, *History of Sicily*, 4:154, for a description of the fortifications of Tyndaris, whose ruins still stand.

148. Freeman, *History of Sicily*, 4:108-112.

149. Diodorus 14.87.

150. Diodorus 14.88.

151. Diodorus 14.96.4.

152. Diodorus 14.90.4-7.

153. Diodorus 14.100.

154. Diodorus 14.103.3.

155. Caven, *Dionysius*, 137.

156. Diodorus 14.106.

157. Diodorus 14.108.1-3.

158. Diodorus 14.108.3-6.

159. Diodorus 14.111.

160. Strabo 6.261.

161. Caven, *Dionysius*, 196.

162. Diodorus 15.73.2.

163. Garlan, *Poliorcétique Grecque*, 171-172.

164. Aineias Tacticus 32-37.

165. Garlan, *Poliorcétique Grecque*, 178-179.

166. Caven, *Dionysius*, 244.

167. See Raoul Lonis's expression of horror at the "unimaginable cruelty" in the war between the Greeks and Carthaginians in Sicily. Lonis, *Les Usages de Guerre entre Grecs et Barbares*, 37. See also 151-152, where Lonis singles out Sicily as a notable exception to his general conclusion that the usages of war between Greeks and barbarians did not differ significantly from those between Greeks and Greeks.

168. Diodorus 13.111.4. Freeman, *History of Sicily*, 3:572.

169. Diodorus 13.57.2.

170. Diodorus 13.57.3-6 and 13.58.3.

171. Diodorus 13.58.1-2.

172. Diodorus 13.58.2.
173. Diodorus 13.59.1–3.
174. Caven, *Dionysius*, 34.
175. Freeman, *History of Sicily*, 3:474–475.
176. Diodorus 13.62.1–4.
177. Diodorus 13.90.1–4; 13.108.2.
178. Diodorus 22.10.4.
179. Diodorus 14.53.1–4.
180. Diodorus 14.112.
181. Diodorus 14.15.2–3.
182. Stroheker, *Dionysios*, 59.
183. Caven, *Dionysius*, 86–87.
184. Diodorus 14.106.
185. Diodorus 14.107.2.
186. Diodorus 14.105.
187. Diodorus 14.105.4.

8. PHILIP II AND ALEXANDER THE GREAT

1. Arrian 7.9.2.
2. George Cawkwell, *Philip of Macedon* (London: Faber & Faber, 1978), 47–48.
3. Donald W. Engels, *Alexander the Great and the Logistics of the Macedonian Army* (Berkeley: University of California Press, 1978), 119–120.
4. Garlan, *Poliorcétique Grecque*, 201.
5. Cawkwell, *Philip*, 162.
6. Diodorus 16.74–76.
7. Marsden, *Greek and Roman Artillery*, 56–61.
8. Polyaenus 2.38.2.
9. Garlan, *Poliorcétique Grecque*, 208.
10. Cawkwell, *Philip*, 162.
11. Diodorus 16.74.4–5. See Marsden, *Greek and Roman Artillery*, 100–101 and 116.
12. Cawkwell, *Philip*, 139–140.
13. Diodorus 16.8.2.
14. Diodorus 16.31.6.
15. Cawkwell, *Philip*, 162.
16. Garlan, *Poliorcétique Grecque*, 212.
17. Diodorus 16.54.3.
18. Garlan, *Poliorcétique Grecque*, 202.
19. Robin Lane Fox, *Alexander the Great* (New York: Dial Press, 1974), 137.
20. John Keegan conveniently lists them in *Mask of Command*, 72.
21. Engels, *Logistics of the Macedonian Army*, 15.
22. N. G. L. Hammond, *Alexander the Great* (Park Ridge, N.J.: Noyes Press, 1980), 32–33.
23. Garlan, *Poliorcétique Grecque*, 205.
24. Ibid., 203–204.
25. Arrian 1.5.5.
26. Arrian 1.5.7.
27. J. F. C. Fuller, *The Generalship of Alexander the Great* (New York: Da Capo Press, 1960), 224.
28. Arrian 1.5.8.
29. Arrian 1.6.8.
30. Arrian 1.6.11.
31. Arrian 1.7.7–9.

32. Diodorus 17.11.1.

33. Arrian 7.8.1.

34. Diodorus 17.9.4. Grote argues that Diodorus is more credible here; *Greece*, 12:38–40. Hammond says that it is obvious that Diodorus's account is worthless; *Alexander the Great*, 60.

35. Diodorus 17.11.1–14.1. Arrian 1.7.7–8.8. Sinclair, "Diodorus Siculus and Fighting in Relays," 249–250.

36. Engels, *Logistics of the Macedonian Army*, 34.

37. Arrian 1.18.2–4.

38. Arrian 1.18.6.

39. Arrian 1.20.1–2.

40. Diodorus 17.24.1.

41. Diodorus 17.7.2.

42. Arrian 1.20.3; Diodorus 17.24.6.

43. J. F. C. Fuller places the camp to the southeast, near the sea. Lane Fox locates it to the northeast, nearer the place of the final assault. Fuller, *The Generalship of Alexander the Great*, 201. Lane Fox, *Alexander the Great*, 138. Donald Engels shows the logistical necessity of the location near the sea; *Logistics of the Macedonian Army*, 34–35.

44. Robin Lane Fox, *The Search for Alexander* (Boston: Little, Brown, 1980), 138; Peter Green, *Alexander of Macedon* (Harmondsworth: Penguin, 1974), 196.

45. Arrian 1.20.4; Diodorus 17.24.6.

46. Green, *Alexander of Macedon*, 196.

47. Arrian 1.20.5–7.

48. Green, *Alexander of Macedon*, 196.

49. Arrian 1.20.9.

50. Diodorus 17.24.4; Marsden, *Greek and Roman Artillery*, 101.

51. Arrian 1.20.9–10.

52. Arrian 1.21.4; Diodorus 17.24.4.

53. Diodorus 17.25.2.

54. Arrian 1.21.1–3.

55. Diodorus 17.25.5–6.

56. C. Bradford Welles makes this suggestion. *Diodorus of Sicily*, trans. C. Bradford Welles (Cambridge: Harvard University Press and London: Heinemann, 1963), 8:189.

57. See Victor Davis Hanson, *The Western Way of War*, chap. 11, for an imaginative argument based on admittedly precious little evidence that hoplites regularly drank before battles.

58. Diodorus 17.26.1–2.

59. Arrian 1.21.5.

60. Arrian 1.21.6; Diodorus 17.26.6. On the height of Alexander's towers, see Garlan, *Poliorcétique Grecque*, 225.

61. Arrian 1.22.1–2.

62. Marsden, *Greek and Roman Artillery*, 62. Garlan, *Poliorcétique Grecque*, 214–215.

63. Diodorus 17.27.1–3.

64. Arrian 1.22.

65. Fuller, *Generalship of Alexander*, 205–206.

66. Arrian 1.23.1–5.

67. Lane Fox, *Alexander*, 137; Engels, *Logistics of the Macedonian Army*, 46.

68. Arrian 1.23.5–6.

69. Arrian 2.5.7.

70. John Keegan, *The Mask of Command*, 71–72.

71. Quintus Curtius 4.2.9.

72. Arrian 2.17–18.2.

73. Engels, *Logistics of the Macedonian Army*, 56–57.

74. Garlan, *Poliorcétique Grecque*, 208–209.

75. Garlan, *Poliorcétique Grecque*, 211.

76. Quintus Curtius 4.2.20.

77. Diodorus 17.40.5.

78. Arrian 2.18.4. See John Keegan's shrewd remarks about the psychology of siege labor in *Mask of Command*, 74–75.

79. Quintus Curtius 4.2.8.

80. Fritz Schachermeyr, *Alexander der Grosse* (Graz: Verlag Anton Pustet, 1949), 181.

81. Diodorus 17.41.4. Marsden doubts the Tyrian catapults were torsion powered; *Greek and Roman Artillery*, 102.

82. Arrian 2.18.3–5.

83. Diodorus 17.42.3–4.

84. Diodorus 17.41.5.

85. Fuller, *Generalship of Alexander*, 210. See Diodorus 17.43.7.

86. Arrian 2.18.6.

87. Arrian 2.19.1–5.

88. Quintus Curtius 4.3.10.

89. Diodorus 17.42.5–6.

90. Arrian 2.19.6–2.20.4.

91. Arrian 2.20.6–10.

92. Arrian 2.21.1–7. The suggestion that Alexander used catapults to throw the stones into deep water comes from Marsden, *Greek and Roman Artillery*, 103.

93. Arrian 2.21.8–2.22.2.

94. Arrian 2.22.3–5.

95. Diodorus 17.42.7; 17.45.2.

96. Garlan, *Poliorcétique Grecque*, 223–224.

97. Quintus Curtius 4.3.23.

98. Diodorus 17.43.1.

99. Marsden, *Greek and Roman Artillery*, 61–62.

100. Diodorus 17.43.3–17.44.5.

101. Diodorus 17.45.7.

102. Fuller, *The Generalship of Alexander*, 212.

103. Arrian 2.23.1–2.

104. Sinclair, "Diodorus Siculus and Fighting in Relays," 252–253; Garlan, *Poliorcétique Grecque*, 160–162.

105. Arrian 2.24.4.

106. Arrian 2.23.4.

107. Arrian 2.23.4–6.

108. Arrian 2.24.1–2.

109. Marsden, *Greek and Roman Artillery*, 104.

110. Engels, *Logistics of the Macedonian Army*, 58–59.

111. Arrian 2.25.4–2.26.3.

112. Quintus Curtius 4.6.11.

113. Quintus Curtius 4.6.9.

114. Arrian 2.26.3–27.3.

115. Hammond, *Alexander the Great*, 116–117.

116. Grote, *Greece*, 12:143.

117. Fuller, *Generalship of Alexander*, 217.

118. Arrian 2.27.3–4.

119. Quintus Curtius 4.6.8.

120. Arrian 2.27.3–7.

121. Diodorus 17.48.7.
122. Arrian 2.27.5–7.
123. Hammond, *Alexander the Great*, 117.
124. Garlan, *Poliorcétique Grecque*, 211.
125. Engels, *Logistics of the Macedonian Army*, 17.
126. Arrian 4.2.1–6.
127. Fuller, *Generalship of Alexander*, 236.
128. Arrian 4.3.1–4.
129. Arrian 4.3.5.
130. Garlan, *Poliorcétique Grecque*, 205.
131. Arrian 4.18.4–4.19.4.
132. This is P. A. Brunt's suggestion in his translation of Arrian. See Arrian, *Anabasis Alexandri*, trans. P. A. Brunt (Cambridge: Harvard University Press and London: Heinemann, 1976–1983), 1:407.
133. Arrian 4.21.1–5.
134. Arrian 4.21.6–10.
135. Arrian 4.28.1–3. In 1926 Sir Aurel Stein identified the Rock of Aornos and made a close topographical survey in light of Arrian's account. See Fuller, *Generalship of Alexander*, 248–254, for a convenient summary of Stein's findings.
136. Arrian 4.28.8.
137. Diodorus 17.85.4–6; Quintus Curtius 8.11.3–4.
138. Arrian 4.28.8–4.29.6.
139. Arrian 4.29.7–4.30.4.
140. Arrian 4.27.5–9.
141. Arrian 5.23.6–5.24.4.
142. Arrian 4.27.9.
143. Fuller, *Generalship of Alexander*, 261.
144. Arrian 6.7.2–3.
145. Arrian 6.7.4–6.
146. Arrian 6.8.7–8.
147. Arrian 6.9.1–4.
148. Arrian 6.9.5–6.11.1.
149. Arrian 6.11.1–2.
150. Arrian 6.13.1–4.
151. Keegan, *Mask of Command*, 65.

9. TREATMENT OF CAPTURED CITIES

1. Volkmann, *Die Massenversklavungen*, 15–16.
2. Arrian 1.8.8.
3. Ulrich Wilcken, *Alexander the Great* (New York: Norton, 1967), 73.
4. Diodorus 17.14.2–3.
5. Arrian 1.9.9.
6. Diodorus 17.14.1–4. W. W. Tarn revised the number of Thebans sold into slavery downward to 7,500–8,000, but Hans Volkmann defends the traditional figure of 30,000. See Volkmann, *Massenversklavungen*, 110–112.
7. Polybius 38.2.13–14.
8. Volkmann, *Massenversklavungen*, 16–17.
9. A. H. Jackson, "Some Recent Works on the Treatment of Prisoners of War in Ancient Greece," *Talanta*, 2 (1970), 42.
10. Schachermeyr, *Alexander der Grosse*, 97.
11. Arrian 1.19.6; Diodorus 17.22.4.
12. Pausanias 6.18.2–4. Cited in Ducrey, *Traitement de Prisonniers de Guerre*, 280.
13. Arrian 1.23.4.

14. Arrian 1.23.6; Diodorus 17.27.6.

15. Diodorus 17.7.9.

16. This was the view of E. Bickermann in his article "Alexandre le Grande et les Villes d'Asie," *REG*, 47 (1934), 346–374. Bickermann was strenuously opposed by that indefatigable champion of Alexander, W. W. Tarn, in *Alexander the Great* (Cambridge: Cambridge University Press, 1948–1950), 2:199–227. See Ducrey, *Le Traitement des Prisonniers de Guerre*, 169–70. Fritz Schachermeyr agrees with Bickermann; *Alexander der Grosse*, 148–149.

17. Quintus Curtius 4.2.15.

18. Quintus Curtius 4.4.17; Diodorus 17.46.4.

19. Quintus Curtius 7.11.28.

20. Quintus Curtius 4.4.17; Diodorus 17.46.4; Justin 18.3.18.

21. Arrian 2.24.5. Because Arrian is the preferred source on Alexander, many classical scholars have accepted his figure of 30,000. See Volkmann, *Massenversklavungen*, 112.

22. See Diodorus 17.46.4 and Quintus Curtius 3.20 and 4.18 for conflicting accounts of this.

23. Quintus Curtius 4.15.

24. Schachermeyr, *Alexander der Grosse*, 183.

25. Fuller, *Generalship of Alexander*, 218.

26. Arrian 2.27.7.

27. Quintus Curtius 4.6.29.

28. Fuller, *Generalship of Alexander*, 104.

29. Lane Fox, *Search for Alexander*, 193.

30. Lonis, *Les Usages de la Guerre entre Grecs et Barbares*, 22–23.

31. Arrian 6.9.5–11.1.

32. Arrian 4.23.2–5.

33. Quintus Curtius 8.10.5.

34. Arrian 4.26.4–4.27.4.

35. Diodorus 17.84.

36. Plutarch, *Alexander* 59.3–4.

37. Plutarch, *Alexander* 59.4.

38. Arrian 5.24.5–7.

39. Arrian 6.6.3.

40. Arrian 6.6.2–5.

41. Arrian 6.16.5; Diodorus 17.103.8.

42. Arrian 6.17.1.

43. Quintus Curtius 9.1.19–23.

44. Quintus Curtius 6.6.34.

45. Arrian 6.16.2; Diodorus 17.102.5.

46. Diodorus 17.102.6.

47. Arrian 7.10.3.

48. Diodorus 17.69.1; Quintus Curtius 5.6.11.

49. Arrian 3.18.11–12.

50. Diodorus 17.70; Quintus Curtius 5.6.1–8.

51. Diodorus 17.72.

52. Quintus Curtius 7.5.28–35.

53. W. W. Tarn, *Alexander the Great* (Boston: Beacon Press, 1956), 67; Wilcken, *Alexander the Great*, 178; Grote, *Greece*, 12:203–204.

54. Arrian 4.2.1–6.

55. Quintus Curtius 5.3.11–15; Arrian 3.17.6.

56. Diodorus 17.96.3–5; Quintus Curtius 9.4.6–7.

57. Quintus Curtius 6.6.23–32. Engels, *Logistics of the Macedonian Army*, 87–89.

58. Diodorus 17.28.
59. Engels, *Logistics of the Macedonian Army*, 12–13.
60. Arrian 6.25.5.
61. Engels, *Logistics of the Macedonian Army*, 13.
62. Wilcken, *Alexander the Great*, 161 and 105.
63. Quintus Curtius 8.4.29–30.
64. Ducrey, *Traitement des Prisonniers de Guerre*, 170.
65. Arrian 4.19.5–6.
66. Quintus Curtius 3.12.21–23.
67. Arrian 4.20.1–3.
68. Quintus Curtius 4.10.29–34.
69. Plutarch, *Alexander* 21.3.
70. Plutarch, *Alexander* 21.3.
71. Plutarch, *Alexander* 30.1.
72. Quintus Curtius 8.10.34.
73. Quintus Curtius 6.2.1–2.
74. Quintus Curtius 6.2.6–9.

10. DEMETRIUS THE BESIEGER

1. Benedictus Niese, *Geschichte der griechischen und makedonischen Staaten seit der Schlacht bei Chaeronea* (Gotha: Friedrich Andreas Perthes, 1893), 1:324–325. See Richard M. Berthold, *Rhodes in the Hellenistic Age* (Ithaca: Cornell University Press, 1984), chap. 1, for a more detailed account of the early history of Rhodes.
2. Berthold, *Rhodes in the Hellenistic Age*, 34–35.
3. Ibid., 61–67.
4. Diodorus 20.92.1–5; Plutarch *Demetrius* 19.4.
5. Diodorus 20.92.2.
6. Diodorus 20.82.3.
7. Diodorus 20.82.4–5.
8. Diodorus 20.83.1–2.
9. Diodorus 20.83.3–4.
10. Diodorus 20.84.1–3.
11. Diodorus 20.84.3.
12. Berthold, *Rhodes in the Hellenistic Age*, 68.
13. Diodorus 20.84.4–6.
14. Diodorus 20.85.1–3.
15. Diodorus 20.85.4.
16. Diodorus 20.86.1–4.
17. Diodorus 20.86.4–88.1.
18. Berthold, *Rhodes in the Hellenistic Age*, 43.
19. Diodorus 20.88.1–6.
20. Diodorus 20.88.7–9.
21. Vitruvius 10.16.4.
22. Diodorus may have referred to a shorter Macedonian cubit, in which case the base would have been fifty feet square. See *Diodorus of Sicily*, trans. Russel M. Geer (Cambridge: Harvard University Press, 1962), 10:382–383.
23. Plutarch, *Demetrius* 21.
24. Diodorus 20.48.3.
25. Diodorus 20.91.2–7.
26. Diodorus 20.91.2–3.
27. Diodorus 20.91.7–8.
28. Plutarch, *Demetrius* 40.1.
29. Diodorus 20.91.8.

30. Diodorus 20.93.1.
31. Berthold, *Rhodes in the Hellenistic Age*, 42–43.
32. Marsden, *Greek and Roman Artillery*, 106.
33. Diodorus 20.93.2–5.
34. Berthold, *Rhodes in the Hellenistic Age*, 43.
35. Niese, *Geschichte der griechischen und makedonischen Staaten*, 33.
36. Berthold, *Rhodes in the Hellenistic Age*, 73. Berthold says twice, but he apparently does not count the Rhodian raid just as the siege was beginning.
37. Diodorus 20.93.6–7.
38. Diodorus 20.94.1–5.
39. *Diodorus of Sicily*, translated by Russel M. Geer, 10:395.
40. Diodorus 20.95.1.
41. Diodorus 20.95.4–5.
42. Diodorus 20.96.1–3.
43. Diodorus 20.96.3–7.
44. Diodorus 20.97.1–2.
45. Diodorus 20.97.3–4.
46. Diodorus 20.97.5–6 & 98.1.
47. Diodorus 20.97.7–98.3.
48. Plutarch, *Demetrius* 21.4.
49. Diodorus 20.98.4–9.
50. Diodorus 20.99.1–3.
51. Plutarch, *Demetrius* 20.1.

11. EARLY SIEGES AND THE PUNIC WARS

1. Livy 1.11.5–1.12.10.
2. Livy 1.15.3–5.
3. Livy 2.63.6.
4. Livy 2.65.7.
5. Livy 3.23.4.
6. Livy 2.17.1.
7. See Livy 2.16.8–9 and 2.22.5.
8. Livy 1.43.3.
9. Livy 4.22.3–6.
10. Livy 4.47.5.
11. Livy 4.59.4–6.
12. Livy 4.61.6–8.
13. Livy 5.22.8. See the discussion of the chronology of the siege of Veii by Hugh Last in *CAH*, 7:511–512.
14. Plutarch, *Camillus* 2.5.
15. Livy 5.2.1–4.
16. Livy 5.6.1–2.
17. Livy 5.10.4–6; 5.12.3–7.
18. Livy 5.13.6.
19. Livy 5.20.5–6.
20. Livy 5.6.7–10.
21. Theodor Mommsen, *Römische Geschichte* (Vienna: Phaidon, 1932), 100; *CAH*, 7:514.
22. Livy 5.1.8–9.
23. Livy 5.5.6.
24. Livy 5.7.2–3.
25. Mommsen, *Römische Geschichte*, 101.
26. Livy 5.16.8–5.17.4.
27. Livy 5.19.2.

28. Livy 5.19.9–11.
29. Livy 9.19.9.
30. Livy 5.21.4–13.
31. Livy 6.9.2.
32. Marsden doubts Livy's story but does not rule out the possibility that the Romans may have had nontorsion artillery at this time; *Greek and Roman Artillery*, 83.
33. Diodorus 23.2.1.
34. Livy 9.13.9–12; 15.3.
35. Polybius 28.11.1–2.
36. Livy 10.41.12–42.4.
37. Livy 44.9.3–9.
38. Mommsen comments that the Romans were "incapable of storming the strong city"; *Römische Geschichte*, 204. The estimate of 40,000 men is from Nigel Bagnall, *The Punic Wars* (London: Hutchinson, 1990), 55.
39. Polybius 1.17.
40. The best description of Agrigentum is in J. F. Lazenby, *The First Punic War: A Military History* (London: UCL Press, 1996), 55–57.
41. Polybius 9.27.
42. Diodorus 23.8.1.
43. Polybius 1.18.
44. Polybius 1.19.
45. Diodorus 23.9.1; Brian Caven, *The Punic Wars* (New York: St. Martin's Press, 1980), 23–26; Bagnall, *Punic Wars*, 55–58.
46. Diodorus 23.9.3–4; Polybius 1.24.11.
47. Polybius 1.24.12.
48. Diodorus 23.9.5.
49. Polybius 1.38.8–9.
50. Polybius 1.42.7; Diodorus 23.1.2. Vergil, *Aeneid* 3.706, trans. Lazenby, in *First Punic War*, 126.
51. Polybius 1.41.3–4; Diodorus 24.1.1. On the sizes of the forces engaged, see Bagnall, *Punic Wars*, 83–85, and Lazenby, *First Punic War*, 123–126.
52. Mommsen, *Römische Geschichte*, 217.
53. Polybius 1.42.8–13; Diodorus 24.1.1.
54. Polybius 1.43; Diodorus 24.1.1.
55. Polybius 1.44. Diodorus 24.1.2–3.
56. Polybius 1.45.
57. Polybius 1.48.
58. Polybius 1.46–47.
59. Polybius 1.48.10–11; Diodorus 24.1.4.
60. Diodorus 24.1.1–5.
61. Polybius 1.49.1.
62. Diodorus 24.1.4.
63. Diodorus 24.1.2.
64. Polybius 8.3.1–4.
65. Polybius 8.3.2.
66. Livy 24.34.7.
67. Polybius 8.4.
68. Polybius 8.5.1–3.
69. Polybius 8.5.4–6.6.
70. F. W. Walbank, *A Historical Commentary on Polybius* (London: Oxford: Clarendon Press, 1967), 2:71.
71. Polybius 8.3.2–3.
72. Polybius 8.7.1–5.

73. Polybius 8.7.5–10.
74. Polybius 8.37.2.
75. Livy 25.23.3.
76. Caven, *Punic Wars*, 164–165.
77. Livy 25.23.1–7.
78. Polybius 8.37.2–11; Livy 25.23.1–24.7.
79. See the appendix, "The Topography of Syracuse," in Livy, trans. Frank Gardner Moore (Cambridge: Harvard University Press and London: Heinemann, 1966), 6:505–510, for an excellent discussion of the layout of Syracuse at this time.
80. Livy 25.24.14–26.2.
81. Livy 25.25.11–26.6.
82. Livy 25.26.7–15.
83. Livy 25.27.
84. I have summarized what Livy described in considerable detail. See Livy 25.28–31.
85. Livy 25.31.5. Nigel Bagnall estimates two-and-a-half years; *Punic Wars*, 225.
86. Livy 25.31.12–15.
87. Livy 26.21.7–8.
88. Polybius 10.7–10. Polybius misunderstood the directions of the features around New Carthage. See H. H. Scullard, *Scipio Africanus: Soldier and Politician* (Ithaca: Cornell University Press, 1970), 48–52, for an excellent discussion, with maps, of the layout of ancient New Carthage.
89. Polybius 10.9.1.
90. Scullard, *Scipio Africanus*, 254, n. 34.
91. Polybius 10.9.7 and 10.11.1–3.
92. Livy 26.48.5–13.
93. Polybius 10.11.5–8.
94. Polybius stated emphatically that the wall was twenty stades long, but there is uncertainty over the length of a stade. Caven has twenty stades equal 2,300 yards, Scullard says 3,700 meters. Caven, *Punic Wars*, 198; Scullard, *Scipio Africanus*, 49.
95. Polybius 10.12.2–3.
96. Neither Polybius nor Livy mentions sambucae, but Livy tells us that marines did cross the walls, so it seems almost certain that the ships were equipped with sambucae. Livy 26.48.6.
97. Polybius 10.12.8. See Scullard, *Scipio Africanus*, 61, and Walbank, *Polybius*, 2:213–214.
98. Polybius 10.12
99. Polybius 10.12.10–11; 10.13.6–11.
100. Polybius attributed the receding waters to a low tide, but there are no tides in Cartagena Bay. I follow the argument of Scullard, *Scipio Africanus*, 53–59.
101. Livy 26.48.6.
102. Polybius 10.14–15.
103. Appian said that the Romans used battering rams at New Carthage, but his fanciful account of the siege is worthless. Appian 6.4.20.
104. Livy 28.3.
105. Livy 28.19.16–18; Appian 6.6.32.
106. Livy 21.7.10; Polybius 3.17.8.
107. Livy 39.21.3.
108. Polybius 10.13.1–5.
109. Livy 29.35.6–13.
110. Appian 8.5.30.

111. Livy 30.9–10.
112. Marsden, *Greek and Roman Artillery*, 174–176.
113. Livy 31.46.10–16.
114. Livy 36.25.
115. Livy 32.17.4–18.3.
116. Livy 32.23.7–13.
117. Livy 36.22–24.
118. Livy 38.29.1–8.
119. Polybius 21.26.1–6; Livy 38.4.1–4.
120. Polybius 21.27.1–6; Livy 38.5–6.
121. Polybius 21.28.1–2.
122. Polybius 21.28.
123. Polybius 21.29–32; Livy 38.9.9–14.
124. Livy 44.11–12.
125. Livy 37.53.16.
126. Livy 23.14.7.
127. Livy 24.2.8.
128. Livy 23.17.1–2.
129. Livy 23.17.5–6.
130. Livy 23.19–20.2.
131. Polybius 7.1.3; Livy 23.30.2–4.
132. Appian 7.5.29.
133. Livy 39.1.6.
134. Livy 37.33.2.
135. Livy 36.10.6–12.
136. Livy 21.61.10–11.
137. Appian 8.18.126.
138. Livy 24.46.2–7.
139. Livy 28.20.4.
140. Livy 29.1.12–13.
141. Polybius 6.39.5; Livy 26.48.5.
142. Livy 4.49.9–50.6.
143. Livy 4.53.9–11.
144. Livy 4.55.4–8.
145. Livy 4.59.8–10.
146. Livy 6.4.11.
147. Livy 7.17.8.
148. Livy 43.1.1–3.
149. Livy 5.19.4.
150. Livy 5.20.1–10.
151. Livy 5.23.8–12 and 5.32.7–9.
152. Livy 7.16.4.
153. Diodorus 23.19. See Lazenby, *The First Punic War*, 119.
154. Livy 8.29.13–14.
155. Livy 9.31.4–5.
156. Livy 10.17.1–10.
157. Polybius 14.7.2–3.
158. Polybius 10.16.
159. Polybius 10.17.1–5.
160. Tacitus, *History* 3.19.
161. Appian 7.5.29.
162. Livy 43.10.
163. Caesar, *Gallic War* 7.47–51.
164. Josephus, *Jewish War* 3.263.

12. THE AGE OF IMPERIALISM

1. Appian 8.12.80.
2. Bagnall, *Punic Wars*, 7.
3. Appian 8.14.95. Appian's description of the walls is sloppy. Serge Lancel, with the aid of modern archeology, sorts it out; *Carthage, a History* (Oxford: Blackwell, 1995), 415–417.
4. Appian 8.13.93.
5. Appian 8.13.93–94.
6. Appian 8.14.97.
7. Appian 8.14.97–98.
8. Appian 8.14.98.
9. Appian 8.14.99.
10. Appian 8.14.99–100.
11. Appian 8.15.101–103.
12. Appian 8.16.108–109.
13. Appian 8.16.110; Polybius 32.18; Zonaras 9.29.
14. Appian 8.16.111.
15. Appian 8.17.113–114. According to Zonaras, Scipio did not evacuate Mancinus's force. Rather his arrival enabled Mancinus to maintain his foothold in Megara until Scipio later mounted his own attack on Megara. Zonaras 9.29. I prefer Appian's account because it fits into the general narrative more comfortably.
16. Appian 8.17.114.
17. Appian 8.17.115.
18. Appian 8.17.116.
19. Appian 8.17.116–8.18.117.
20. Appian 8.18.117.
21. Appian 8.18.117 & 119.
22. Appian 8.18.119.
23. Appian 8.18.118.
24. Appian 8.18.120.
25. Appian 8.18.121.
26. Appian 8.18.121–122. The speculation on the Carthaginian reason for avoiding battle is from Caven, *Punic Wars*, 287.
27. Appian 8.18.122–123.
28. Appian 8.18.124.
29. Appian 8.18.125.
30. Appian 8.18.126; Zonaras 9.30.
31. Appian 8.19.127.
32. Bagnall, *Punic Wars*, 319.
33. Lancel, *Carthage*, 426.
34. Appian 8.19.128–130.
35. Appian 8.19.130–131.
36. Appian 8.20.133–135.
37. Sallust, *The War with Jugurtha* 75–76.
38. Sallust, *The War with Jugurtha* 89.4–90.1.
39. Sallust, *The War with Jugurtha* 91.1–5.
40. Sallust, *The War with Jugurtha* 92.2.
41. Marsden, *Greek and Roman Artillery*, 177.
42. Caesar, *Gallic War* 2.12.
43. Caesar, *Gallic War* 2.29–31.
44. Caesar, *Gallic War* 7.15 & 23.
45. Caesar, *Gallic War* 7.32.
46. Caesar, *Gallic War* 7.17.

47. Caesar, *Gallic War* 7.22.

48. J. F. C. Fuller doubts that the ramp could have been so large; *Julius Caesar: Man, Soldier, and Tyrant* (New Brunswick, N.J.: Rutgers University Press, 1965), 136.

49. Caesar, *Gallic War* 7.24.

50. Caesar, *Gallic War* 7.24–25.

51. Caesar, *Gallic War* 7.26.

52. Caesar, *Gallic War* 7.27–28.

53. Caesar, *Gallic War* 7.28.

54. Caesar, *Gallic War* 7.69.

55. Caesar, *Gallic War* 7.70–71. Fuller calls Vercingetorix's decision a "fatal blunder"; *Julius Caesar*, 150.

56. Fuller, *Julius Caesar*, 150.

57. Caesar, *Gallic War* 7.72–74.

58. Caesar, *Gallic War* 7.77–78.

59. Caesar, *Gallic War* 7.79–80.

60. Caesar, *Gallic War* 7.81–82.

61. Caesar, *Gallic War* 7.83–88.

62. Fuller, *Julius Caesar*, 157.

63. Caesar, *Gallic War* 8.32–33.

64. Caesar, *Gallic War* 8.34–37.

65. Caesar, *Gallic War* 8.40–43.

66. Caesar, *Civil Wars* 2.1–2.

67. Caesar, *Civil Wars* 2.9–10.

68. Caesar, *Civil Wars* 2.10.

69. Caesar, *Civil Wars* 2.11.

70. Caesar, *Civil Wars* 2.14.

71. Caesar, *Civil Wars* 2.15–16.

72. Caesar, *Civil Wars* 2.22.

73. Josephus, *Jewish War* 3.79.

74. Josephus, *Jewish War* 3.118.

75. Marsden, *Greek and Roman Artillery*, 183–185.

76. Marsden, *Greek and Roman Artillery*, 181.

77. Josephus, *Jewish War* 3.121.

78. Rupert Furneaux, *The Roman Siege of Jerusalem* (London: Hart-Davis MacGibbon, 1973), 101.

79. Josephus, *Jewish War* 2.546 and 2.553.

80. Josephus, *Jewish War* 3.162–168; 245–247.

81. Vergil, *Aeneid*, trans. Frank O. Copley (Indianapolis: Bobbs-Merrill, 1965), 12.921–922.

82. Jonathan J. Price, *Jerusalem under Siege* (Leiden: E. J. Brill, 1992), 120–121.

83. Josephus, *Jewish War* 3.169–178.

84. Josephus, *Jewish War* 3.186–192.

85. Josephus, *Jewish War* 3.213–228.

86. Josephus, *Jewish Wall* 3.234–252.

87. Josephus, *Jewish War* 3.254–257.

88. Josephus, *Jewish War* 3.266–282.

89. Josephus, *Jewish War* 3.283–288.

90. Josephus, *Jewish War* 3.316–331.

91. See Josephus, *Jewish War* 5.136–175, for a detailed description of Jerusalem.

92. For an exhaustive account of the factions that divided the Jews, see Price, *Jerusalem under Siege*.

93. See ibid., 125–127 and 243–247, for a discussion of Jerusalem's food supplies.

94. Ibid., 248–254.

95. Josephus provided exact dates, but their exact equivalents in our calendar are uncertain. See ibid., 210–213.

96. Brian W. Jones, *The Emperor Titus* (London: Croom Helm and New York: St. Martin's Press, 1984), 22–23, 53–54.

97. Josephus, *Jewish War* 5.54–66.

98. Josephus, *Jewish War* 5.69–97.

99. For a discussion of the archeological evidence of Jerusalem's walls, see Price, *Jerusalem under Siege*, 290–292.

100. Josephus, *Jewish War* 5.258–273.

101. Price, *Jerusalem under Siege*, 130–131.

102. Josephus, *Jewish War* 5.275–302.

103. E. Mary Smallwood, *The Jews under Roman Rule* (Leiden: E. J. Brill, 1976), 320.

104. Josephus, *Jewish War* 5.303–347.

105. Dio 61.5.3.

106. Price, *Jerusalem under Siege*, 207–208.

107. Josephus, *Jewish War* 5.356–375.

108. See Price, *Jerusalem under Siege*, 271–280, for a detailed analysis of Josephus's description of the famine in Jerusalem.

109. Josephus, *Jewish War* 5.420–452.

110. Josephus gave seventeen days as the time necessary to build the ramps (Josephus, *Jewish War* 5.466), but that figure raises chronological problems. Seven days seems to be a sensible emendation of the text. Price, *Jerusalem under Siege*, 226–228.

111. Josephus, *Jewish War* 5.466–490.

112. Josephus, *Jewish War* 5.491–501.

113. Josephus, *Jewish War* 5.502 510.

114. Josephus, *Jewish War* 5.571.

115. Josephus, *Jewish War* 5.511–519.

116. Josephus, *Jewish War* 5.522–523. The estimate of travel time is from Emil Schürer, *The History of the Jewish People in the Age of Jesus Christ* (Edinburgh: T.&T. Clark, 1973–1987), 1:505.

117. Josephus, *Jewish War* 6.5–22.

118. Josephus, *Jewish War* 6.23–28.

119. Josephus, *Jewish War* 6.29–33.

120. Josephus, *Jewish War* 6.54–67.

121. Josephus, *Jewish War* 6.68–80.

122. Josephus, *Jewish War* 6.93–94. The text of Josephus says the sacrifices ceased because of a lack of men, but this makes little sense. An emendation of the text to read *arnon* (sheep) instead of *andron* (men) is convincing. See Price, *Jerusalem under Siege*, 229–230, and Smallwood, *The Jews under Roman Rule*, 322.

123. Josephus, *Jewish War* 6.130–147.

124. Schürer, *History of the Jewish People*, 1:506.

125. Josephus, *Jewish War* 6.149–168.

126. Josephus, *Jewish War* 6.177–185.

127. Josephus, *Jewish War* 6.220–222.

128. Josephus, *Jewish War* 6.193–213.

129. Josephus, *Jewish War* 6.223–227.

130. Price, *Jerusalem under Siege*, 170.

131. Josephus, *Jewish War* 6.228–266.
132. Josephus, *Jewish War* 6.392–408.

13. TREATMENT OF CAPTURED CITIES

1. See Ober, "Classical Greek Times," for an extensive discussion of the laws of war in ancient Greece.
2. Xenophon, *Cyropaedia* 7.5.73.
3. Polybius 2.58.10.
4. Livy 31.30.2–3.
5. Dionysius of Halicarnassus 6.19.4.
6. Vergil, *Aeneid* 6.852–853.
7. Livy 37.32.8–14.
8. Polybius 9.10. Livy agreed with Polybius's judgment; 25.40.2–4.
9. Livy 26.31–32.
10. Livy 5.27.7.
11. Livy 5.21.13; 22.1.
12. Livy 26.33.7–8.
13. Livy 26.14.9–15.10.
14. Livy 26.14.8.
15. Livy 26.16.7.
16. Livy 26.34.2–13.
17. Livy 26.16.9–10.
18. Livy 26.16.11–12.
19. Livy 31.29.11.
20. Polybius 3.86.11.
21. Polybius 3.100.4.
22. Livy 21.13.9.
23. Livy 21.57.13–14.
24. Polybius 28.14.
25. Livy 43.1.1–3. See Volkmann, *Massenversklavungen*, 27.
26. Sallust, *The War with Jugurtha* 91.6–7.
27. Polybius 21.29–32; Livy 38.9.13–14.
28. Livy 38.43.4.
29. Livy 38.44.6.
30. Livy 39.4.8–13.
31. Livy 39.5.6–16.
32. Livy 45.27.1–4.
33. Livy 45.35.5–9.
34. Volkmann, *Massenversklavungen*, 92.
35. Livy 29.8–9.
36. Livy gives a detailed account of the controversy in the Senate, including a long Locrian speech to the senators; 29.16–22.
37. Livy 43.4.13.
38. Livy 43.4.8–13; Zonaras 9.22.
39. Caesar, *Gallic War* 2.13 & 7.11.
40. Caesar, *Gallic War* 2.32–33.
41. Caesar, *Gallic War* 7.28.
42. Caesar, *Gallic War* 7.89.
43. Volkmann, *Massenversklavungen*, 52; 116. The total number of people captured by Caesar in Gaul may have reached a million.
44. Caesar, *Gallic War* 8.44.
45. Fuller, *Julius Caesar*, 165.
46. Caesar, *Civil Wars* 2.22.

47. Tacitus, *Annals* 12.17.
48. Volkmann, *Massenversklavungen*, 37–38.
49. Dionysius of Halicarnassus 4.50.4.
50. Livy 1.53.2–3.
51. Livy 2.17.6.
52. Livy 2.16.9.
53. Livy 2.22.2.
54. Livy 2.25.5.
55. Livy 2.30.14–15.
56. Livy 2.31.4.
57. Livy 3.23.5.
58. Livy 6.9.11.
59. Livy 6.10.1–5.
60. Plutarch, *Camillus* 22.6.
61. Livy 4.58.3.
62. Livy 4.59.6–8.
63. Livy 6.22.4.
64. Livy 9.12.5–8.
65. Livy 9.31.2–3.
66. Livy 9.24.14–15.
67. Livy 9.43.1.
68. Livy 7.15.10.
69. Livy 7.19.1–3.
70. Livy 9.16.9–10.
71. Diodorus 19.101.3.
72. Polybius 1.7.
73. Volkmann, *Massenversklavungen*, 49.
74. Livy 9.14.11–16.
75. Livy 9.26.1–5.
76. Livy 28.19.10–17; Appian 6.6.32.
77. Livy 28.20.6.
78. Appian 6.6.32.
79. Scullard, *Polybius*, 98.
80. Livy 28.20.8–12.
81. Livy 26.40.7–14.
82. Caesar, *Civil Wars* 3.80–81.
83. Livy 9.25.4–9.
84. Livy 24.19.6–11.
85. Diodorus 23.18.5.
86. Polybius 10.15.4–6.
87. Polybius 10.15.8.
88. Appian 8.3.15.
89. Caesar, *Civil War* 1.21.
90. Tacitus, *History* 3.32–34.
91. Livy 7.17.8–9.
92. Volkmann, *Massenversklavungen*, 39.
93. Livy 42.51.6.
94. Polybius 1.19.15; Diodorus 23.9.1; Zonaras 8.10.
95. Diodorus 23.9.5. Volkmann accepts Zonaras's interpretation; *Massenversklavungen*, 55.
96. Zonaras 8.11. See Volkmann, *Massenversklavungen*, 55–56.
97. Livy 23.37.12–13.
98. Volkmann, *Massenversklavungen*, 20.

99. Livy 31.27.3–4.

100. Volkmann supplies an exhaustive list of examples during the first and second Macedonian wars; *Massenversklavungen*, 20 ff.

101. Polybius 30.15. Volkmann says this number "is not to be doubted"; *Massenversklavungen*, 115.

102. Polybius 33.10.1–3.

103. Livy 42.63.10–11.

104. Sallust, *The War with Jugurtha* 54.6.

105. Sallust, *The War with Jugurtha* 26.

106. Polybius 10.17.6–15. Livy adds a few details to Polybius's account of the treatment of New Carthage; 26.47.1–3.

107. Polybius 15.4.1–2.

108. Livy 32.24.7.

109. Appian 8.19.130 and 135; Zonaras 9.31.

110. Josephus, *Jewish War* 1.351–352.

111. Josephus, *Jewish War* 3.133–134.

112. Josephus, *Jewish War* 3.303–306.

113. Josephus, *Jewish War* 3.329–331.

114. Josephus, *Jewish War* 3.336–339.

115. *Chicago Tribune*, August 20, 1997.

116. Josephus, *Jewish War* 4.82.

117. Josephus, *Jewish War* 4.78–83.

118. Josephus, *Jewish War* 4.92.

119. Josephus, *Jewish War* 4.93–96.

120. Josephus, *Jewish War* 4.97–111.

121. Josephus, *Jewish War* 4.112–120.

122. Josephus, *Jewish War* 5.289.

123. Tacitus, *Annals* 13.41.

124. Josephus, *Jewish War* 5.450–451.

125. Josephus, *Jewish War* 7.196–206.

126. Josephus, *Jewish War* 5.455–456.

127. Josephus, *Jewish War* 5.334.

128. Josephus, *Jewish War* 5.550–561.

129. Josephus, *Jewish War* 6.255–259.

130. Josephus, *Jewish War* 6.271–272.

131. Josephus, *Jewish War* 6.281–284.

132. Josephus, *Jewish War* 6.321–322.

133. Josephus, *Jewish War* 6.350.

134. Josephus, *Jewish War* 6.352–353.

135. Josephus, *Jewish War* 6.378–386.

136. Josephus, *Jewish War* 6.404–406.

137. Josephus, *Jewish War* 6.414–420; 6.429.

138. Josephus, *Jewish War* 7.1–2.

139. Livy 3.42.4.

140. Livy 3.60.8.

141. Livy 37.32.5.

142. Vergil, *Aeneid* 9.142–144.

143. Vergil, *Aeneid* 9.146–155.

144. Livy 1.53–54.

145. Livy 1.53.4.

146. Livy 5.27.

147. Livy 1.57.1–5. See Livy 1.57–60 for the whole story.

148. Sallust, *The War with Jugurtha* 67.1–2.

149. Livy 1.39.5–6.
150. Livy 3.57.3.
151. Livy 3.61.4.
152. Polybius 10.18.3–15.
153. Polybius 10.19.5.
154. Polybius 10.19.3–7.
155. Livy 26.50.
156. Appian 6.2.12; Diodorus 25.15.
157. Livy 21.14.3–4.
158. Polybius 16.30.2–3.
159. Polybius 16.32.4.
160. Polybius 16.33.4–5.
161. Polybius 16.34.8–12.
162. Diodorus 25.17.
163. Livy 31.18.6.
164. Livy 28.22–23.
165. Appian 6.6.33.
166. Sallust, *The War with Jugurtha* 76.6.
167. Livy 41.11.
168. Livy 26.13.15.
169. Livy 26.14.1–5.
170. Livy 26.15.11–15. There was more than one tradition about the letter and about Taurea Vibellius. See Livy 26.16.1–4.
171. Josephus, *Jewish War* 7.334.
172. Josephus, *Jewish War* 7.337–339.
173. Josephus, *Jewish War* 7.386.
174. See Arthur J. Droge and James D. Tabor, *A Noble Death: Suicide and Martyrdom among Christians and Jews in Antiquity* (New York: Harper, 1992), 53–60.
175. Josephus, *Jewish War* 7.399–401.
176. Josephus, *Jewish War* 7.405–406.
177. Josephus, *Jewish War* 3.369.
178. Josephus, *Jewish War* 3.376–377.
179. See Droge and Tabor, *Noble Death*, 86–96, for a different and somewhat tendentious discussion of Josephus's attitude toward suicide.

EPILOGUE

1. Margaret Atwood, "The Loneliness of the Military Historian." From *Morning in the Burned House* by Margaret Atwood, copyright 1995. Published by McClellend & Stewart in Canada, by Houghton Mifflin in the United States, and by Virago in the United Kingdom.
2. Edward N. Luttwak, *The Grand Strategy of the Roman Empire* (Baltimore: Johns Hopkins University Press, 1976), chap. 3.
3. Erich Sander argues most vigorously for a decline; "Der Verfall der Römischen Belagerungskraft," *Historische Zeitschrift*, 149 (1934), 457–476. Friedrich Lammert argues that there was no decline in Roman technical siege capabilities but acknowledges that offensive siege warfare became less important; "Die Antike Poliorketik und ihr Weiterwirken," *Klio*, 31, 389–411.
4. Marsden, *Greek and Roman Artillery*, 195–198.
5. Jim Bradbury, *The Medieval Siege* (Woodbridge, Suffolk: Boydell Press, 1992), 1.
6. Ibid., 324.
7. Ibid., 53–54.
8. Ibid., 318.

9. Ibid., 302.

10. *Henry V* 3.4.7–41.

11. Bradbury, *Medieval Siege*, 304.

12. Ibid., 309.

13. Charles Cheney Hyde, *International Law* (Boston: Little, Brown, 1947), 3:1802–1803, quoted in Walzer, *Just and Unjust Wars*, 166.

14. *Times Literary Supplement*, May 8, 1992, 11.

15. Josephus, *Jewish War* 5.367–369.

16. Lucien Poznanski, *La Chute du Temple Jerusalem* (Brussels: Editions Complexe, 1991), 88.

SELECTED
BIBLIOGRAPHY

ANCIENT SOURCES

Aeschylus. *Agamemnon*. Translated by Herbert Weir Smith. Loeb Classical Library. London: Heinemann, 1971.

———. *Seven against Thebes*. Translated by H. Weir Smyth. Loeb Classical Library. Cambridge: Harvard University Press, 1922.

Aineias the Tactician. *How to Survive under Siege*. Translated by David Whitehead. Oxford: Clarendon Press, 1990.

Ancient Near Eastern Texts Relating to the Old Testament. Edited by James B. Pritchard. Princeton: Princeton University Press, 1969.

Ancient Records of Assyria and Babylonia. 2 vols. Edited by Daniel David Luckenbill. New York: Greenwood Press, 1968.

Appian. *Roman History*. Translated by Horace White. 4 vols. Loeb Classical Library. London: Heinemann and New York: Macmillan, 1912–1913.

Arrian. *Anabasis Alexandri*. Translated by P. A. Brunt. Loeb Classical Library. Cambridge: Harvard University Press and London: Heinemann, 1976–1983.

Caesar. *The Gallic War*. Translated by H. J. Edwards. Loeb Classical Library. Cambridge: Harvard University Press and London: Heinemann, 1966.

———. *The Civil Wars*. Translated by A. B. Peskett. Loeb Classical Library. Cambridge: Harvard University Press and London: Heinemann, 1961.

Quintus Curtius Rufus. *History of Alexander*. Translated by John C. Rolfe. Loeb Classical Library. Cambridge: Harvard University Press, 1946.

Diodorus of Sicily. Translated by C. H. Oldfather, Charles L. Sherman, Bradford Welles, Russel M. Geer, and F. R. Walton. Loeb Classical Library. Cambridge: Harvard University Press and London: Heinemann, 1933–1967.

Dionysius of Halicarnassus. *Roman Antiquities*. 7 vols. Translated by Earnest Cary. Loeb Classical Library. Cambridge: Harvard University Press and London: Heinemann, 1960–1963.

Euripides. *Andromache*. Translated by John Frederick Nims. Chicago: University of Chicago Press, 1958.

———. *Hecuba*. Translated by William Arrowsmith. Chicago: University of Chicago Press, 1958.

———. *The Trojan Women*. Translated by Richmond Lattimore. Chicago: University of Chicago Press, 1958.

Hellenica Oxyrhynchia. Translated by P. R. McKechnie and S. J. Kern. Warminster, Wiltshire, England: Aris and Phillips, 1988.

Herodotus. Translated by A.D. Godley. Loeb Classical Library. Cambridge: Harvard University Press, 1960–1963.

Homer. *Iliad*. Translated by Richmond Lattimore. Chicago: University of Chicago Press, 1951.

————. *Odyssey*. Translated by Richmond Lattimore. New York: Harper & Row, 1968.

Josephus. *Jewish Antiquities*. Translated by H. St. J. Thackeray and Ralph Marcus. Loeb Classical Library. London: Heinemann and New York: Putnam, 1930–1963.

————. *The Jewish War*. Translated by H. St. J. Thackeray. Loeb Classical Library. Cambridge: Harvard University Press and London: Heinemann, 1961.

Livy. Translated by B. O. Foster, Frank Gardner Moore, Evan T. Sage, and Alfred C. Schlesinger. 14 vols. Loeb Classical Library. London: Heinemann and Cambridge: Harvard University Press, 1960–1967.

Pausanias. *Description of Greece*. Translated by W. H. S. Jones. Loeb Classical Library. Cambridge: Harvard University Press and London: Heinemann, 1917–1935.

Plutarch. *Plutarch's Lives*. Translated by Bernadotte Perrin. Loeb Classical Library. London: Heinemann and New York: Putnam, 1919–1920.

Polybius. *The Histories*. Translated by W. R. Paton. Loeb Classical Library. Cambridge: Harvard University Press and London: Heinemann, 1927–1954.

Sallust. *The War with Jugurtha*. Translated by J. C. Rolfe. Loeb Classical Library. Cambridge: Harvard University Press and London: Heinemann, 1965.

Tacitus. *The Complete Works*. Translated by Alfred John Church and William Jackson Brodribb. New York: Modern Library, 1942.

Thucydides. *History of the Peloponnesian War*. Translated by C. F. Smith. Loeb Classical Library. Cambridge: Harvard University Press and London: Heinemann, 1975–1976.

Vergil. *The Aeneid*. Translated by Frank O. Copley. Indianapolis: Bobbs-Merrill, 1965.

Xenophon. *Cyropaedia*. Translated by Walter Miller. Loeb Classical Library. Cambridge: Harvard University Press and London: Heinemann, 1914.

————. *Hellenica*. Translated by Carleton L. Brownson. Loeb Classical Library. Cambridge: Harvard University Press and London: Heinemann, 1918–1922.

MODERN SOURCES

Adcock, F. E. *The Greek and Macedonian Art of War*. Berkeley: University of California Press, 1957.

Adkins, A. W. H. *Moral Values and Political Behaviour in Ancient Greece, from Homer to the End of the Fifth Century*. London: Chatto and Windus, 1972.

Albright, F. W. "The Israelite Conquest of Canaan." *Bulletin of the American School of Oriental Research*, 74 (1939).

Alt, Albrecht. *Die Landnahme der Israeliten Palästina*. Leipzig: Druckerei der Werkgemeinschaft, 1925.

Anderson, J. K. *Military Theory and Practice in the Age of Xenophon*. Berkeley: University of California Press, 1970.

Badawy, Alexander. *Architecture in Ancient Egypt and the Near East*. Cambridge: MIT Press, 1966.

————. *A History of Egyptian Architecture*. Berkeley: University of California Press, 1966.

Bagnall, Nigel. *The Punic Wars*. London: Hutchinson, 1990.

Bennigston, Hermann. "Die Griechische Polis bei Aeneas Tacticus." *Historia*, 11 (1962).

Berthold, Richard M. *Rhodes in the Hellenistic Age*. Ithaca: Cornell University Press, 1984.

Bickermann, E. "Alexandre le Grande et les Villes d'Asie." *Revue des Etudes Grecques*, 47 (1934).

Billerbeck, A. "Der Festungsbau im alten Orient." *Der Alte Orient*, 1 (1903).

Bleibtreu, Erika. "Grisly Assyrian Record of Torture and Murder." *Biblical Archaeology Review* (January–February 1991).

Bloedow, Edmund F. "The Intelligent Archidamus." *Klio*, 65 (1983).

Bradbury, Jim. *The Medieval Siege*. Woodbridge, Suffolk: Boydell Press, 1992.

Bronowski, J. *The Ascent of Man*. Boston: Little, Brown, 1973.

Bruce, I. A. F. *An Historical Commentary on the "Hellenica Oxyrhynchia."* London: Cambridge University Press, 1967.

Buck, Robert J. *A History of Boeotia*. Edmonton: University of Alberta Press, 1979.

Buckler, John. *The Theban Hegemony, 371–362 B.C.* Cambridge: Harvard University Press, 1980.

Burford, A. "Heavy Transport in Classical Antiquity." *Economic History Review*. Second Series, 13 (1960).

Burn, A. R. *Persia and the Greeks: The Defense of the West*. London: Arnold, 1962.

Busolt, Georg. *Griechische Geschichte*. Gotha: F. A. Perthes, 1893–1904.

Bustenay, Oded. *Mass Deportations and Deportees in the Neo-Assyrian Empire*. Wiesbaden: Reichert, 1979.

Caven, Brian. *Dionysius I, Warlord of Sicily*. New Haven: Yale University Press, 1990.

——. *The Punic Wars*. New York: St. Martin's Press, 1980.

Cawkwell, George. *Philip of Macedon*. London: Faber & Faber, 1978.

Chamoux, François. *The Civilization of Greece*. New York: Simon and Schuster, 1965.

Clausewitz, Carl von. *On War*. Princeton: Princeton University Press, 1989.

Cogan, Mordechai. " 'Ripping Open Pregnant Women' in Light of an Assyrian Analogue." *Journal of the American Oriental Society*, 103 (1983).

Cohen, David. "Justice, Interest, and Political Deliberation in Thucydides." *Rechtshistorisches Journal*, 2 (1983).

Craigie, Peter C. *The Problem of War in the Old Testament*. Grand Rapids: Eerdmans, 1978.

Creveld, Martin van. *Technology and War*. New York: Free Press, 1989.

de la Croix, Horst. *Military Considerations in City Planning: Fortifications*. New York: Braziller, 1972.

Demand, Nancy H. *Thebes in the Fifth Century: Heracles Resurgent*. London: Routledge and Kegan Paul, 1982.

Droge, Arthur J., and James D. Tabor. *A Noble Death: Suicide and Martyrdom among Christians and Jews in Antiquity*. New York: Harper, 1992.

Droysen, H. *Heerwesen und Kriegführung der Griechen*. In *K. F. Hermann's Lehrbuch der Griechischen Antiquitäten*. Freiburg: Akademische Verlagsbuchhandlung von J. C. B. Mohr (Paul Siebeck), 1889.

Ducrey, Pierre. *Le Traitement de Prisonniers de Guerre dans la Grèce Antique*. Paris: Editions de Boccard, 1968.

Duffy, Christopher. *Fire and Stone: The Science of Fortress Warfare, 1660–1860*. Vancouver: David and Charles, 1975.

Dunbabin, T. J. *The Western Greeks*. Oxford: Clarendon Press, 1948.

Engels, Donald W. *Alexander the Great and the Logistics of the Macedonian Army*. Berkeley: University of California Press, 1978.

Eph'al, Israel. "On Warfare and Military Control in the Ancient Near Eastern Empires: A Research Outline." In *History, Historiography and Interpretations*. Edited by H. Tadmor and M. Weinfeld. Jerusalem: Magnes Press, 1983.

Ehrlich, Ernest Ludwig. *A Concise History of Israel*. New York: Harper and Row, 1965.

Ferrill, Arther. *The Origins of War from the Stone Age to Alexander the Great*. London: Thames and Hudson, 1985.

Freeman, E. A. *History of Sicily*. 4 vols. Oxford: Clarendon Press, 1891–1894.

Fuller, J. F. C. *The Generalship of Alexander the Great.* New York: Da Capo Press, 1960.

――――. *Julius Caesar: Man, Soldier, and Tyrant.* New Brunswick, N.J.: Rutgers University Press, 1965.

Furneaux, Rupert. *The Roman Siege of Jerusalem.* London: Hart-Davis MacGibbon, 1973.

Gardiner, Alan. *Egypt of the Pharaohs.* New York: Oxford University Press, 1961.

Garlan, Yvon. *Recherches de Poliorcétique Grecque.* Paris: Boccard, 1974.

――――. *War in the Ancient World: A Social History.* New York: Norton, 1975.

Gelb, I. J. "Prisoners of War in Early Mesopotamia." *Journal of Near Eastern Studies,* 32 (1972).

Glock, Albert Ernest. "Warfare in Mari and Early Israel." Ph.D. dissertation, University of Michigan, 1968.

Gomme, A. W., A. Andrewes, and K. J. Dover. *A Historical Commentary on Thucydides.* 5 vols. Oxford: Clarendon Press, 1945–1981.

Green, Peter. *Alexander of Macedon.* Harmondsworth: Penguin, 1974.

――――. *Armada from Athens.* Garden City, N.Y.: Doubleday, 1970.

Grote, George. *Greece.* 12 vols. New York: Peter Fenelon Collier, 1894–1900.

Grundy, G. B. *The Great Persian War and Its Preliminaries: A Study of the Evidence, Literary and Topographical.* New York: AMS Press, 1969.

――――. "A Suggested Characteristic in Thukydides' Work." *Journal of Hellenic Studies,* 18 (1898).

――――. *Thucydides and the History of His Age.* 2 vols. Oxford: Basil Blackwell, 1948–1961.

――――. *The Topography of the Battle of Plataea.* London: John Murray, 1894.

Gurny, O. R. *The Hittites.* Baltimore: Penguin, 1969.

Hallo, William H. "From Qarqar to Carchemish: Assyria and Israel in the Light of New Discoveries." *Biblical Archeologist,* 23, no. 2 (1960).

Hammond, N. G. L. *Alexander the Great.* Park Ridge, N.J.: Noyes Press, 1980.

Hanson, Paul D. "War, Peace, and Justice in Early Israel." *Bible Review,* 3 (Fall 1987).

Hanson, Victor Davis. *Warfare and Agriculture in Ancient Greece.* Ann Arbor: UMI, 1982.

――――. *The Western Way of War: Infantry Battle in Classical Greece.* New York: Knopf, 1989.

Herzog, Chaim, and Mordechai Gichon. *Battles of the Bible.* New York: Random House, 1978.

Houwink Ten Cate, Philo H. J. "The History of Warfare according to Hittite Sources: The Annals of Hattusilis I." *Anatolica,* 11 (1984).

Howard, Michael, ed. *Restraints on War.* Oxford: Oxford University Press, 1979.

Howard, Michael, George J. Andreopoulos, and Mark R. Shulman, editors. *The Laws of War: Constraints on Warfare in the Western World.* New Haven: Yale University Press, 1994.

Humble, Richard. *Warfare in the Ancient World.* London: Cassell, 1980.

Hunger, Johannes. "Heerwesen und Kriegführung des Assyrer auf der Hohe ihrer Macht." *Der Alte Orient,* 12 (1911).

Jackson, A. H. "Some Recent Works on the Treatment of Prisoners of War in Ancient Greece." *Talanta,* 2 (1970).

Jones, Brian W. *The Emperor Titus.* London: Croom Helm and New York: St. Martin's Press, 1984.

Kagan, Donald. *The Archidamian War.* Ithaca: Cornell University Press, 1974.

――――. *The Outbreak of the Peloponnesian War.* Ithaca: Cornell University Press, 1969.

――――. *The Peace of Nicias and the Sicilian Expedition.* Ithaca: Cornell University Press, 1981.

Karlsson, Lars. *Fortification Towers and Masonry Techniques in the Hegemony of Syracuse.* Stockholm: Paul Astroms, 1992.

Keegan, John. *A History of Warfare.* New York: Vintage, 1994.

———. *The Mask of Command.* New York: Penguin, 1987.

Kelly, Thomas. "Thucydides and Spartan Strategy in the Archidamian War." *American Historical Review,* 87 (February 1982).

Kern, Paul B. "Military Technology and Ethical Values in Ancient Greek Warfare: The Siege of Plataea." *War and Society,* 6 (September 1988).

———. "The Turning Point in the Sicilian Expedition." *Classical Bulletin,* 65 (1989).

Kiechle, Franz. "Zur Humanität in der Kriegführung der Griechischen Staaten." *Historia,* 7 (1958).

Kirsten, E. "Plataia." *Paulys Realencyclopaedie der classischen Altertumswissenschaften,* 20 (1950).

Kuch, Heinrich. *Kriegsgefangenschaft und Sklaverei bei Euripides.* Berlin: Akademie Verlag, 1974.

Lancel, Serge. *Carthage, a History.* Oxford: Basil Blackwell, 1995.

Lane Fox, Robin. *Alexander the Great.* New York: Dial Press, 1974.

———. *The Search for Alexander the Great.* Boston: Little, Brown, 1980.

Larson, J. A. O. "Federations for Peace in Ancient Greece." *Classical Philology,* 39 (July 1944).

Lateiner, Donald. "Heralds and Corpses in Thucydides." *Classical World,* 71 (October 1977).

Lazenby, J. F. *The First Punic War: A Military History.* London: UCL Press, 1996.

Lonis, Raoul. *Les Usages de la Guerre entre Grecs et Barbares.* Paris: Les Belles Lettres, 1969.

McNeill, William H. *The Rise of the West.* Chicago: University of Chicago Press, 1963.

Macqueen, J. G. *The Hittites and Their Contemporaries in Asia Minor.* Boulder: Westview Press, 1975.

Malamat, Abraham. "How Inferior Israelite Forces Conquered Fortified Canaanite Cities." *Biblical Archeological Review,* 8 (March–April 1982).

Malbran-Labat, Florence. *L'Armée et l'Organisation Militaire de l'Assyrie.* Geneva: Libraire Droz, 1982.

Manitius, Walther. "Das stehende Heer der Assyrerkönige und seine Organisation." *Zeitschrift für Assyriologie,* 24 (1910).

Märker, Martin. *Die Kämpfe der Karthager auf Sizilien in den Jahren 409–405 v. Chr.* Weida i. Thür.: Druck von Thomas & Hubert, 1930.

Marsden, E. W. *Greek and Roman Artillery.* Oxford: Clarendon Press, 1969.

Melchert, H. Craig. "The Acts of Hattusilis I." *Journal of Near Eastern Studies,* 37 (January 1978).

Mommsen, Theodor. *Römische Geschichte.* Vienna: Phaidon, 1932.

Niese, Benedictus. *Geschichte der griechischen und makedonischen Staaten seit der Schlacht bei Chaeronea.* Part 1. Gotha: Friedrich Andreas Perthes, 1893.

Ober, Josiah. *Fortress Attica: Defense of the Athenian Land Frontier, 404–322 B.C.* Leiden: E. J. Brill, 1985.

O'Connell, Robert. *Of Arms and Men: A History of War, Weapons, and Aggression.* New York: Oxford University Press, 1989.

Olmstead, A. T. *History of Assyria.* New York: Scribner, 1923.

Oppenheim, A. L. "Siege Documents from Nippur." *Iraq,* 17 (1955).

Osborne, Robin. *Classical Landscape with Figures: The Ancient Greek City and Its Countryside.* Dobbs Ferry, N.Y.: Sheridan House, 1987.

Panagopoulos, Andreas. *Captives and Hostages in the Peloponnesian War.* Athens: Grigoris Publications, 1978.

Poznanski, Lucien. *La Chute du Temple Jerusalem*. Brussels: Editions Complexe, 1991.

Price, Jonathan J. *Jerusalem under Siege*. Leiden: E. J. Brill, 1992.

Pritchett, W. Kendrick. *The Greek State at War*. 5 vols. Berkeley: University of California Press, 1971–1991.

Rad, Gerhard von. *Der heilige Krieg im alten Israel*. Göttingen: Vandenhoeck & Ruprecht, 1952.

Reade, J. E. "The Neo-Assyrian Court and Army: Evidence from the Sculptures." *Iraq*, 34 (1972).

Redfield, James M. *Nature and Culture in the Iliad: The Tragedy of Hector*. Chicago: University of Chicago Press, 1975.

Ridley, R. T. "The Hoplite as Citizen: Athenian Military Institutions in Their Social Context." *L'Antiquité Classique*, 48 (1979).

Saggs, H. W. F. *The Might That Was Assyria*. London: Sidgwick and Jackson, 1984.

Salonen, Erkki. "Die Waffen der alten Mesopotämier." *Studia Orientalia*, 33 (1966).

Sargent, Rachel L. "The Use of Slaves by the Athenians in Warfare." *Classical Philology*, 22 (1927).

Schachermeyr, Fritz. *Alexander der Grosse*. Graz: Anton Pustet, 1949.

Schürer, Emil. *The History of the Jewish People in the Age of Jesus Christ*. 3 vols. Edinburgh: T.&T. Clark, 1973–1987.

Schwally, Friedrich. *Semitische Kriegsaltertumer*. Leipzig: Dieterich'sche, 1901.

Scullard, H. H. *Scipio Africanus: Soldier and Politician*. Ithaca: Cornell University Press, 1970.

Sinclair, R. K. "Diodorus Siculus and Fighting in Relays." *Classical Quarterly*, 60 (1966).

Smallwood, Mary E. *The Jews under Roman Rule*. Leiden: E. J. Brill, 1976.

Stroheker, Karl Friedrich. *Dionysios I*. Wiesbaden: Franz Steiner, 1958.

Tarn, W. W. *Alexander the Great*. 2 vols. Cambridge: Cambridge University Press, 1948–1950.

———. *Hellenistic Military and Naval Development*. New York: Biblo and Tannen, 1966.

Ussishkin, David. *The Conquest of Lachish by Sennacherib*. Tel Aviv: Tel Aviv University, Institute of Archeology, 1982.

Volkmann, Hans. *Die Massenversklavungen der Einwohner eroberter Städte in der Hellenistisch-Römischen Zeit*. Stuttgart: Franz Steiner, 1990.

Walbank, F. W. *A Historical Commentary on Polybius*. 3 vols. London: Clarendon Press, 1957–1979.

Walzer, Michael. *Just and Unjust Wars*. New York: Basic Books, 1977.

Waschow, Hans. *4000 Jahre Kampf um die Mauer*. Postberg, Bottropiw: Buch- und Kunstdruckerei Wilk, 1938.

Westlake, H. D. *Individuals in Thucydides*. London: Cambridge University Press, 1968.

Whitaker, Joseph I. S. *Motya*. London: G. Bell, 1921.

Wilcken, Ulrich. *Alexander the Great*. New York: Norton, 1967.

Winter, F. E. *Greek Fortifications*. Toronto: University of Toronto Press, 1971.

Yadin, Yigael. *The Art of Warfare in Biblical Lands in the Light of Archeological Discovery*. London: McGraw-Hill, 1963.

Younger, K. Lawson, Jr. *Ancient Conquest Accounts*. Sheffield: Sheffield Academic Press, 1990.

Zevit, Ziony. "The Problem of Ai." *Biblical Archeology Review*, 11 (March–April 1985).

Zimmern, Alfred. *The Greek Commonwealth: Politics and Economics in Fifth-Century Athens*. New York: Oxford University Press, 1961.

INDEX

Paul Bentley Kern is Professor of History
at Indiana University Northwest.